Farming Systems
in the Tropics

Farming Systems in the Tropics

HANS RUTHENBERG

with contributions by J. D. MacArthur,
H. D. Zandstra, and M. P. Collinson

THIRD EDITION

CLARENDON PRESS · OXFORD
1980

Oxford University Press, Walton Street, Oxford OX2 6DP

OXFORD   LONDON   GLASGOW
NEW YORK   TORONTO   MELBOURNE   WELLINGTON
KUALA LUMPUR   SINGAPORE   JAKARTA   HONG KONG   TOKYO
DELHI   BOMBAY   CALCUTTA   MADRAS   KARACHI
NAIROBI   DAR ES SALAAM   CAPE TOWN

First Edition 1971
Second Edition 1976
Third Edition 1980

*Published in the United States by
Oxford University Press, New York*

British Library Cataloguing in Publication Data

Ruthenberg, Hans
   Farming systems in the tropics. – 3rd ed.
   1. Agriculture – Tropics
   I. Title
   630'.913   SB111   79–41135
   ISBN 0–19–859481–X
   ISBN 0–19–859482–8 PK

*Printed and bound in Great Britain by
Morrison & Gibb Ltd, London and Edinburgh*

Preface to the Third Edition

Seek complexity and order it.

Clifford Geertz (1969)

THIS book arose from an introductory course of lectures on the characteristics of tropical farming systems for undergraduates at Stuttgart-Hohenheim and Göttingen Universities and for graduates in general agriculture who were preparing for applied work in the tropics. Its purpose is first to provide factual information, secondly to introduce the reader to the system approach to farms, and thirdly—by the way tables are ordered—to submit a methodology in ordering farm-management data which is suitable for the analysis of smallholder farming. This third edition has been considerably changed from the two earlier ones: more recent examples have been used, more and better figures are included, and the description of the various farming systems has been supplemented by notes on the methodology of cropping and farming system research by H. G. Zandstra and M. P. Collinson.

Beyond its use as a teaching aid, the book is intended to provide the agricultural development worker with a framework into which he can fit the various types of farming that he will encounter, and at the same time to introduce him to the management problems that arise in particular farming systems. The subject is considered in the agro-economic setting. Emphasis is laid on the interactions between technical and economic aspects of farming. Information about soils, climate, and socio-institutional factors is deliberately kept short.

In recent years, much literature has become available about tropical agriculture, dealing with such crucial topics as soil fertility, irrigation, and practices in plant and animal husbandry. Land-tenure problems, farmers' behaviour, and agricultural-policy measures are increasingly dealt with in a growing number of publications. I have, therefore, preferred to restrict this book to the farm-management aspects. The primary economic characteristics of the more important tropical farming systems and their broad lines of development are described. The description is based on the structural approach to farms as systems, as developed by Brinkmann (1924) and Woermann (1959). My experience in teaching is that the structural approach is effective in promoting an initial understanding of a farm system and its relations. It should then be supplemented by the functional approach, which is required for normative work.

The subject is by no means dealt with exhaustively, and many more farming systems than those depicted deserve consideration. I have tried in the illustrations to cover the major tropical crops, but no emphasis has been laid on a geographical balance. Almost all examples have been taken from areas that I have visited and where I was in a position to confirm the information cited in the relevant literature.

To describe the economic situation of each farming system I have used much quantitative information from existing farms. The reader should be aware of the problems connected with aggregations, the use of averages, and weighting coefficients. Land, labour, and livestock are not homogeneous inputs, and such products as rice, cotton, or cattle are not homogeneous outputs. The use of data about labour input and energy output per hour is nothing other than an effort to provide an idea about orders of magnitude. The monetary figures require particular caution. They are difficult to compare because of wide differences in price levels and price relations between the various countries, price changes over time, and distorted exchange rates. It is felt nevertheless, that quantitative and monetary data are helpful in showing which relations are important in the various systems and for this reason they are used extensively.

In the complex task of writing this book I have been greatly assisted by several co-workers. The main debt is, of course, to the numerous smallholders, farmers, and estate managers who so willingly discussed their work and problems. They in fact provided the information that is presented in this book. Most valuable advice and criticism were given by my friend John D. MacArthur, who read the manuscript of all editions. To him I am indebted for a great number of suggestions and ideas, and for Chapter 2, which he wrote. Mr. and Mrs. P. C. Brown deserve my thanks for having translated the original German version of the first edition. I owe a special word of thanks to the Volkswagen Foundation, the Afrika-Studienzentrum, IFO-Institut, Munich, and the Fritz Thyssen Foundation. The book would not have been possible without the research funds that I received from them. The inclusion of the plates was made possible by a grant from Mr. Heinrich Herrmann, Stuttgart. Finally I am most grateful to Mrs. Inge Breitschwerdt for the typing of many drafts and much editorial work.

Stuttgart-Hohenheim   HANS RUTHENBERG
1979

*Dieses Buch widme ich Emil Woermann, Professor
für landwirtschaftliche Betriebslehre in Göttingen,
dem ich das Interesse für die Organisation und
die Genesis von Bodennutzungssystemen verdanke.*

Contents

List of Tables

List of Figures

List of Plates

1. Introduction

1.1. The tropics and tropical climates

GEOGRAPHICALLY this book deals with tropical farming systems only, although much of the information is valid also for subtropical farming. The tropical regions are generally defined as those lying between 23½° north and 23½° south of the equator. This, however, is too rigid a definition for the purpose of dealing with farming systems. I therefore follow Manshard (1968), who, using temperatures and types of vegetation as criteria, defined the tropics as given in Fig. 1.1.

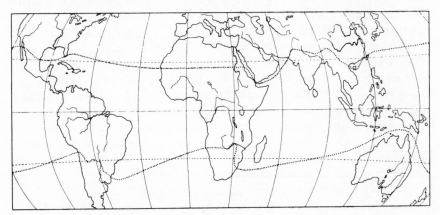

FIG. 1.1. The tropics, defined according to the principles of economic geography (from Manshard 1968, p. 18).

Tropical areas principally differ from each other in respect of climate. The most relevant criterion for classifying tropical lowland climates is rainfall. In a humid month rainfall exceeds evapotranspiration (usually more than 200 mm of rain). In classifying climates according to the number of humid months, I follow Troll (1966):†

1. Humid climates have 7 or more humid months (usually more than 1400 mm per annum)—a very humid climate would have 9 wet months.

† See IRRI (1974) for an example of a detailed agroclimatic classification for farm-system analysis.

2. Semi-humid (wet and dry) climates are those with $4\frac{1}{2}$–7 humid months—subclasses have to be formed according to rainfall distribution (months with 100–200 mm, climates with unimodal or bimodal patterns of rainfall).
3. Semi-arid climates are those with 2–$4\frac{1}{2}$ humid months—subclasses refer to rainfall distribution and whether the rains fall in the cooler or warmer season.

Tropical highlands are a distinct class in themselves and subclasses have to be formed according to rainfall, altitudes, and latitudes.

1.2. Farm systems and farming systems

1.2.1. *The relevance of the farm-system approach*

Choosing policies for agricultural development requires the use of information about the existing farming situation. The collection of information presupposes the ordering of the great number of phenomena which can be observed in a given rural area into entities which are meaningful in terms of development, and these entities are systems, i.e. sets of related elements. System theory is therefore employed as the guideline for farm-system description and analysis (see von Bertalanffy 1973; Dent and Anderson 1971; Emery and Frist 1971; Forrester 1972; Fuchs 1973; Kirsch 1974; Laszlo 1972)

There are many ways of defining goal-orientated systems in rural areas. Watersheds represent useful systems to the hydrologist. Geographers tend to consider regions as systems, while the sociologist will look for village or family systems. This book deals with farms, which are clearly systems because several activities are closely related to each other by the common use of the farm's labour, land, and capital, by risk distribution, and by the joint use of the farmer's management capacity. Farms are considered here as economic units, and the relations of the farm system with the environment are consequently expressed in economic terms.

The analysis of farms is quite important to the subject of development, because the farm is a major decision point in agricultural development. It is both an ecosystem (Walter 1973, p. 61), and an independent unit of economic activity. It is obvious, however, that farming systems are only one aspect of rural life. Ordering observable phenomena into ecological, social, and political systems may also be crucial. Clearly the farm and farming system approach is not the only relevant one. It is more useful to think in terms of a division of labour, and this book seeks to present information in a way which facilitates the linking of the farm approach with the approach of those concerned with other rural systems.

1.2.2. *A hierarchy of systems*

Any farm is part in a hierarchy of systems, belonging first to the larger system of the rural area and consisting, secondly, of various activities, which are systems themselves. Also such inputs as workers, soils, animals, or tractors are systems with rather distinct boundaries. The farm as a system thus consists of a set of related subsystems which form a hierarchy of systems: micro-organisms in the soil are a subsystem of the soil system. The soil system again is a subsystem of the crop-producing system (activity) and the crop system is again a subsystem of the farm system. Efforts to explain the working of the various subsystems are the task of natural and social scientists in the respective areas. This book deals with crop, livestock, and processing activities only. The systems lower down in the hierarchy are treated as 'black boxes' which require inputs and produce outputs but which need not be fully understood (Kirsch 1974, p. 22). It is, however, important to remain aware of the fact that some of the subsystems involved are 'machine' systems (tractors) and others are biological systems (plants, soils, animals), while the workers belong to social systems. The farm is therefore a 'man–biological–machine' system (i.e. a hybrid system) and is accordingly much more complex than plant or soil systems which, as machine or biological systems, are determined or stochastic in nature; social systems are of an essentially undefined nature (Fuchs 1973, p. 61).

Policy measures concern aggregates. It is useful therefore to group into classes farms which are similar in their structure and which can be expected to produce on similar production functions, and we consequently deal with the class of 'shifting systems' or 'grazing systems'. A distinction has to be made between any given farm unit which, as it stands, is a system, and classes of similarly structured farms which are classified as belonging to a certain 'farming system'. In both cases we speak of systems, because the variance of the whole is less than the sum of the variance of the parts (Laszlo 1972, p. 41). Subclasses of each major class of systems are frequently called 'types of farming'. Wheat–rice farms in northern India or groundnut–millet farms in Senegal are such types of farming (for a detailed typology see Kostrowicki (1974)).

1.2.3. Characteristics of farms

(a) *Objectives and decisions.* Following Woermann (1959), a farm is taken to be an organized decision-making unit in which crop and livestock production is carried out with the purpose of satisfying the farmer's goals. A farm is thus a goal-orientated system. Relevant to the organization of a given farm are the goals of the decision-makers, who are, in most cases, the farmers. The fact that society leaves the function of decision-making to the farmers indicates that, by pursuing their own objectives, they are also expected to promote the common interest.

A farmer usually has several goals, and the emphasis varies from case to case. In larger holdings market production and profits are the main objectives, whereas for the smallholder—and most of the tropical land is farmed by smallholders—the farm and the household are closely linked with regard to operations and objectives. As important as, or even more important than, saleable products on such a farm is the provision of food for the household, together with goods for personal consumption like tobacco, or raw materials like wool or materials for building huts. Other important aims are safe-guarding the future by the accumulation of capital in the form of animals or plantations and the increasing of social status by the accumulation of wealth or by special technical, social, and economic achievements. The farm is consequently a multi-objective system.

The hypothesis in this book is that farmers are intentionally rational in organizing their systems, given their objectives.† In describing farming systems and their possible avenues of development, three working hypotheses are consequently adopted in this book.

1. There is always a reason why farming is carried out in one way rather than another, and the reason is often but certainly not always an economic one.

2. Change in the environment of the system generally produces discre-pancies between the actual farming system employed and the optimum solution, given the farmer's possibilities and preferences.

3. Farmers tend to adapt their farming to changing circumstances provided the change is 'satisfying' in terms of the additional benefits involved.

(b) *Boundaries.* Any system has a 'boundary', which separates the system from the 'environment'. In the case of the farm unit, the system embraces all workers and resources (elements of the system) which are under the management control of one decision-maker, the farmer, whether he is the owner, a tenant, or an authorized manager. The farmer, the family members working on the farm, and the hired labourers are supplied to the production unit by rural households. In Africa south of the Sahara a lot of households exist whose individual members cultivate small plots independently. In these cases all fields, animals, and other resources are considered as belonging to the farm which are used by those who share a common kitchen.

All land used wholly or partly for agricultural purposes, including grazing land other than communal grazing, is considered as belonging to the farm unit.‡ In shifting cultivation the farm area includes the cropped parcels and the fallow land claimed by the farmers. Processing units (cocoa fermentation, copra kilns, etc.) are part of the farm provided they are under the management control of the farmer. Establishments engaged in the production of livestock, even if they have no land under their control (nomads), are also considered as farm units.

---

† Rational action implies the use of knowledge in attaining given ends. Other types of action may be called emotional or traditional.

‡ Farm area: productive land (cropped fields, pastures, plantations, fish-ponds, forests, etc.) + indirectly productive land (farm building site, ditches, roads, etc.) + fallow land + unproductive land (rocks, desert, water) within the farm boundaries.

The boundaries of the farm unit with its economic environment are defined by the purchase or procurement of inputs and the sale or disposal of outputs. The farm receives labour from the household and the labour market. It purchases production inputs and obtains capital from the markets for investments. It supplies outputs to the household and the markets, and the household may supply capital to the farm and/or to the capital market.

In smallholder farming, farm and household are very closely related. Most smallholders in the tropics aim at producing their subsistence food requirements on their holdings, and food supply from farm to household is organized according to the needs of the household, which itself depends on the phases of family development. The household, on the other hand, is a distinct labour-supplying unit, and the amount and kinds of labour made available vary with the family cycle. Closely related objects should be in one system. The argument that it is more appropriate to look at 'household–farm systems' instead of production systems is a strong one particularly in smallholder areas. The tables in this book try to take account of this. It is admitted, however, that the description of farm–household interactions remains rudimentary, simply because a fuller consideration would have made the subject too complex for its purpose. The shortcomings which are involved with the emphasis on production economics should be kept in mind by the reader.

(c) *Activities and their relations.* The activities of the farm serve to transform inputs into outputs. Several kinds of activity have to be distinguished: (1) the activities which produce crops; (2) the activities which turn crops into livestock products; (3) processing activities which transform crop or livestock products into a factory product; (4) procurement activities including investments and farm maintenance works (e.g. drainage); and (5) marketing activities.

In system theory the terms 'inputs' and 'outputs' mean all manner of things which enter or leave the farm system or the subsystem of an activity. For our purpose these are classified into economic and non-economic inputs and outputs:

(i) The economic inputs and outputs are those which are usually registered in a farm-management accounting sheet, and they comprise all items which are either bought or sold or which have a value in terms of opportunity costs. Labour, land, and means of production are the inputs, and goods sold and consumed in the farm household are the outputs.

(ii) The non-economic inputs and outputs are free goods from the point of view of the farmer. Inputs of this type are, for instance, solar energy and rainfall, and outputs may be salts leaving the land through drainage or oxygen.

The analysis made in later chapters does not deal directly with non-economic inputs and outputs. They are, however, very much behind the phenomena observed and measured: labour, land, yields, etc.

The various activities receive some or most of their inputs from the system's environment and deliver the output either to other activities or to the environment. The farm is therefore an open system. A significant proportion of the inputs of the household–farm system is, however, produced within the system (labour, seeds, fodder, manure). The proportion of system-produced inputs in relation to total inputs is an important criterion for classification. The proportion is usually high in subsistence systems and low in fully commercialized large farms in an industrialized environment.

The classification of activities has to be very specific. A classification based simply on a crude definition of the type of crops or animals kept is insufficient. Growing rice in winter is certainly not the same activity as growing rice in summer. Growing upland and lowland rice are different activities, and the same applies to rice growing using the transplanting technique or the seeding technique. Every production of a specific crop on a specific soil type, in a specific season of the year, with a specific husbandry technique, is a distinct activity. Even farms with only one crop (e.g. rice) usually are units with more than one activity, because rice of different kinds may be grown on different soil types and at different seasons.

The various activities of a farm are all related to one another by the joint use of the farmer's management capacity, but this is not the only relationship. Normally there are many more and, much stronger ones.

1. Different crops grown at the same time on a given field (intercropping) are related to each other by competition for light, water, land, labour, soil nutrients, etc. The relationship need not be only competitive in nature. Complementary relationships exist due to symbiotic effects (take nitrogen fixation and use), weed control, disease control, etc. Under certain conditions the relationships are such that intercropping is more profitable than sole stands.

2. Crops grown at different times on a given field are related through the residual effects on soil fertility, of the earlier crop, plant diseases, weed growth, water use, salt balance, etc. This is one of the reasons for crop rotations.

3. Crops grown at the same time on different fields are related to each other because they compete for land, labour, machinery, water, etc. Crops on different soil types of a catena, or irrigated and upland fields, show complementary relations to each other. The fields in the valley bottom, for instance, may supply fodder and employment at a time when this is not available from the upland. Similar relations hold between arable crops and permanent crops. This is another reason for farm diversification.

4. Livestock activities are related to each other and to crop production. The relationship with crops may be competitive with regard to labour and capital, but complementary through the use of manure, the utilization of crop residues, the reduction in risk, etc.

5. Processing activities (e.g. a sugar factory) are clearly related to the production of the crop that is processed, but there may be other complementary relations. A case in point is the supply of by-products as feed to livestock or of mulch to tree crops.

The various relationships extend over space and time. They are numerous, direct and indirect, rarely linear, never determined, often stochastic, and are to some degree undefined because of the human element involved. Measured by the number of relations between activities, the farm is a very complex system.

The relationships between activities are usually measurable. The marginal effect on one activity of a one-unit change in another indicates the strength of complementary relationships. The strength of competitive relationships is measured by marginal rates of substitution. The strength of the relationships between activities can differ widely. A poultry enterprise, relying on purchased feeds, hired labourers, and the sale of manure, is unlikely to be closely related to crop activities on the same farm. Dairying, however, processes farm-grown fodder and supplies manure to crops, and is therefore closely related to a great number of activities. Generally the strength of relationships between activities is the more pronounced the scarcer the land, the lower the producer prices, the higher the prices for inputs and the less favourable the natural conditions.

(d) *External relations.* The structure of any farm at any point in time is the result of interactions between the internal relations and the preceding state of the environment. The environment influences the farm system through the external relations which can be classified as follows (see also Duckham and Masefield 1970, p. 96).

1. The natural conditions (climate, soil, diseases, etc.) constrain the ecologically feasible activities.

2. The state of knowledge and information about agricultural techniques (innovations) determines the possible physical production functions of the various activities.

3. The farmer's choice of the ecologically feasible activities and possible techniques clearly depends on the institutional environment (land tenure, farm size, taxation systems, labour laws, credit and extension services, etc.).

4. The input combination, output mix, and input intensity in any activity on the farm depends on the economic environment which influences the system through the prices of inputs and outputs. Three relations are relevant in this context:

   (i) The combination of inputs going into the farm depends on the price relations between inputs (e.g. labour and machinery).

   (ii) The combination of outputs depends on the price relations between outputs (e.g. between crop or animal products).

(iii) The intensity of farming, i.e. the amount of input per unit of land (fertilizer per hectare), depends on the price relations between inputs and outputs.

5. The combination of activities in smallholder farming is to a high degree a function of the farm household demand for food, fibre, fuel, and other needs that the farm can meet, and this demand depends on the culture and the socio-political state of the society concerned.

The structure of a farm system at any time thus depends on all technical, economic, social, cultural, and political influences that impinge on the farmer, his household members, and his hired workers. Changes in the socio-political setting can be decisive for the organization of the farm system.

The importance of a change in external relations is measured by the change of the economic output in relation to one unit of change in the environment. The change may be a technical innovation, a change in prices, in land-tenure arrangements, in the rural-service structure, or in cultural values. The elasticity of farming systems to changes in their environment differs widely: some—e.g. shifting systems—are very inelastic; others—take irrigation farming—are usually quite elastic.

(e) *The farm structure as a function of internal and external relations.* The external relations differ from place to place. Each location with its particular climate, soils, and price relations is best suited to only one or a few crops. The various external relations listed before tend to differentiate farming into regions which *specialize* in certain crops.

These regional differences in cropping pattern are clearly recognizable, but within each region most farms remain *diversified*, each containing a variety of activities. This is for a number of reasons, which are behind the internal relations (Brinkmann 1924, p. 959; Woermann 1959, p. 476).

1. Given amounts of solar energy, land, and water are available on the farm, and the crops compete for them. Farmers face the task of attaining their objectives with given natural resources, and they normally make best use of these given resources by crop diversification.

2. The various activities of the farm make joint use of the available farm resources, labour, machinery, and capital. In smallholder farming the situation often is one where a given stock of labour can be utilized. The various crop and livestock activities differ in the time and nature of their requirements of labour, machines, buildings, irrigation facilities, etc. The return to these inputs is usually higher on a farm with several activities instead of only one or two.

3. Farms are diversified because human and stock-food needs are 'aseasonal', while most crops produce in certain seasons only (Duckham and Masefield 1970, p. 479). The household members and the livestock produce more effectively with an even supply all the year round, which is

more easily achieved by growing several crops that produce at different times.

4. Diversified farming reduces risks.

The existing structure of farms thus reflects the kind and the strength of the external relations in interaction with the kind and the strength of the internal relations. However, no full explanation of what is found can be expected from an analysis of the various relations involved. Farm systems are not fully determined by their environment and the nature of their internal relations.

1. A farmer's decision depends on his information about and perception of the factors which are relevant for him. His decisions may be taken to be intentionally rational, but are certainly not always fully rational.

2. The state of the environment is related to decisions on the farm. The impact of a single farm on the environment is certainly very weak, but the impact on the environment of a larger class of farms comprising hundreds of thousands or millions of units is definitely strong. The farm system is thus a function of its environment, but the environment of a farm unit also depends on the organization of farming.

1.2.4. *Dynamics of farms*

(a) *The mechanisms for change.* Farms exist in an environment with two powerful mechanisms for change.

1. Farmers use land which originally was part of a natural system, and most ecological systems untouched by man show a 'zonal vegetation' (Walter 1973, p. 31) which is typical for the given natural conditions and which is close to a 'steady state', i.e. the state of the system remains largely the same over time, maintained by a continuous flow of natural inputs entering the system (e.g. solar energy or rainfall). Outputs are to a minor degree economic (e.g. game) but mostly non-economic (e.g. water joining groundwater). Natural systems, however, are unproductive in terms of human objectives. The basic principle of farming is to change the natural system into one which produces more of the goods desired by man. The man-made system is an artificial construction which requires continuous economic inputs obtained from the environment to maintain its output level. Farming thus implies the abolition of an unproductive 'steady state' in favour of a man-created, more productive but unstable 'state', and much of the farm input (tillage, fertilizers, weeding, etc.) is nothing but an effort to prevent the new state from declining towards an unproductive low-level steady state.

Without sufficient human effort the system inevitably falls back to either the original state or, and this is more often the case, into another state which is at a lower level than the original one, which means that it is less complex and less productive in terms of meeting human needs. A case in point is a virgin forest which is cultivated by shifting cultivators and which after several cropping cycles degenerates into a derived savanna. All farm systems which are not sufficiently maintained tend to drift into a low-output steady state. Extended areas of the tropics show that this mechanism is a most powerful one, the more powerful the warmer and more humid the climate.

2. The other driving power behind change is the fact that time, given the existence of human beings, produces innovations. They allow a more effective use of inexhaustible flow resources (solar energy) and usually involve an increasing rate of mobilization of exhaustible resources (fossil energy, phosphates, etc.). These innovations, by interacting with population growth, capital formation, and economic development, change the institutional, economic, and socio-cultural environment of the farm. The natural conditions are also changed, because economic development influences the ecosystem in a negative or positive way. In times of rapid population growth and industrialization, changes in the environment of the farm system are usually very powerful.

(b) *Types of state*. The two mechanisms for change may be more or less powerful and they may or may not balance themselves. Consequently a distinction has to be made between farms which are in a 'steady state' and those which are in a 'moving state'. Farms which are in a steady state remain over time as they are. There is no change in inputs, activities, or outputs (von Bertalanffy 1973, p. 142). Two types of steady state can be distinguished.

1. Some farming systems remain in a steady state at a comparatively high level of soil fertility and output. A case in point is wet rice in the flood plains of central Thailand.

2. Many more farming systems have traditionally been, and still are, in a steady state at a low level of soil fertility and output. A case in point is permanent upland farming in large parts of India.

Any change in the environment can disturb a 'steady state'. In the case of a minor or temporary disturbance, the system may go back to its old situation. But in farming there usually is a sequence of important changes (disturbances) over time, and most productive farm systems are not very stable. Most farm systems are therefore in a moving state. The move may take the following directions.†

1. The inputs, activities, and outputs of the system change although a state of balance is retained during the change, so that there is no loss in the stock of elements which are relevant for the long-term maintenance of the system. Such a situation is called a 'quasi-equilibrium'. Several criteria may be applied in order to find out whether a given farm shows a quasi-equilibrium or not: soil fertility, yield levels, livestock numbers, farm capital, etc. Economic

† A given farm may be dissolved with the purpose of re-assembling the elements in new systems (new farms) with different sets of activities. This is done, for example, by land reform and land consolidation programmes. These types of system change are not dealt with in this book.

criteria are highly variable owing to changes in prices. I propose therefore to use a physical criterion for farm-system maintenance: soil fertility.†

In a balanced system (a conservation system) there is little erosion, and nutrient and humus levels in the soil are maintained at a level which allows high outputs for the foreseeable future. Most European and Japanese farming situations would fall into this class.

2. The system changes in a way which improves the stock of elements which are relevant for system maintenance, in particular soil fertility. This may occur if, for instance, more 'resource modifiers' (drainage, fertilizers, irrigation, etc.) are applied than are required for maintenance.

3. The system changes, but inputs are insufficient to maintain the stock of elements which are relevant for system maintenance. This applies, for instance, to dry farming on slopes. The farmer, in selecting his adaptations to changing circumstances, may prefer not to stop or reduce erosion, because this is too expensive for him. The system moves towards lower levels as stocks (in terms of soil fertility) are transformed into output or leave the system through leaching and erosion. The land is mined. There are two final states of the process: (i) production ceases (man-made deserts) or (ii) the system reaches a low-level steady state of the kind referred to previously.

Each type of state is found in tropical agriculture. Some farming systems in the tropics are balanced, a few are improving; many more are soil mining and in extended areas farming has reached a low-level steady state, and each type of option may be reasonable in terms of development.

Most professionals in agriculture have a preference for balanced systems, but soil mining may be good economics, and optimum solutions in economic terms may be soil-mining solutions. We mine coal and iron; we mine phosphates and potash which supply us with fertilizers. Why should we not mine soils? Soil mining in one system may well yield the net returns with which to improve soil fertility elsewhere, e.g. in turning deserts into fertile irrigation lands. Much of the wealth of the industrialized nations has been achieved through the exploitation of natural resources in terms of nutrients and humus

† Output is not a suitable yardstick for the maintenance of a farm system. Soil mining, i.e. the consumption of the stock of soil, nutrients, or humus, is an inherent feature of a great number of farming systems and may well produce constant or even increasing yields over a long time span. A case in point is dry farming on sloping land, which is usually accompanied by a significant amount of erosion and which may well yield constant returns over centuries. Indeed the returns may even increase owing to new varieties and mineral fertilizer. Nevertheless, the loss in soil occurs, and in the very long term only rocks will remain. Such a system, although producing high outputs for a long time, is certainly not balanced.

The same line of thought could be applied to the maintenance of a farm's stock in terms of machinery, buildings, livestock, etc. These stocks, however, are manufactured inputs and may be reproduced according to society's needs, while soil is available in given quantities only, and soil loss may be an irreversible process. However, this explanation does not go deep enough. There is clearly an emotional element (Mother Earth) in using maintenance of soil fertility as the major yardstick for measuring the maintenance of a farm system.

which were used to feed labourers cheaply and to facilitate industrial capital formation. Thereafter, induced by changing price relations, farmers in industrialized nations changed to 'balanced' and 'improving' systems because they became economic. We may hypothesize from past events that economic development apparently passes through the stage of soil mining in the early phases of industrialization to the stage of 'soil improvement' in a highly industrialized environment. This book is, nevertheless, biased in favour of balanced and improving systems. At the same time, it is recognized that soil mining may be inevitable in economic development, and to ask for soil-fertility maintenance as a *conditio sine qua non* in economic development is to ask for the impossible, because the requirements in terms of inputs, and the short-term outputs forgone, are simply too high. The extreme conservationist tends to waste funds which are badly needed for more productive forms of capital development, and he may be as dangerous in terms of a stable world as the land miner. The disquieting aspect, in my opinion, is not that soil mining occurs, but the world-wide extent of the phenomenon and the speed of the mining process in most of the tropics.

It must not be overlooked in this context, that maintaining or improving a farm system requires a flow of inputs from the environment. Some of these inputs are obtained from inexhaustible resources (for instance solar energy), some flows are obtained by recycling (nitrogen), although recycling requires energy which is usually obtained from exhaustible resources, and some flows are directly obtained from exhaustible resources, as for instance fossil energy, potash, or phosphate. Farm maintenance and improvement consequently takes place at the expense of stocks in the environment (Randall 1975, p. 804). Nutrients can only be accumulated on a given field at the expense of other areas or stores. The income-creating capacity of farming is thus a function of the effectiveness of the transfer of resources from places where they are useless or of low effect in terms of human objectives, to those where they serve man's goals.

(c) *Implications of change.* In time the farm as a system may improve, maintain itself, or decline; the same thing may occur to regions and countries. However, some tendencies can be considered as general characteristics of farm systems in changing environments.

1. *The tendency towards satisfying solutions.* Farms are organized by learning farmers and have therefore to be considered as learning systems.

(i) In a static environment a process of trial and error would lead farmers, who are assumed to be intentionally rational (over generations), to the most satisfying solution for their system, which would be their optimum solution, given the options open to them and their preferences. In the very long run this optimum would become a 'steady state' either at a high or a low level.

(ii) However, a static situation is very uncommon and it is likely that the coming decades will bring even more dynamics to the tropics than the past has shown. Smallholders are usually very skilful in adapting traditionally known techniques to a gradual change in the environment, but they often fail to move into something known but outside their normal habit. A case in point are the desperate efforts of the farmers in Eastern Nigeria (see 4.4.1) to maintain the productivity of their upland fallow system, while ignoring for decades the great potential of valley-bottom development along Asiatic lines. Farmers are limited in their capacity to obtain and absorb information. Their goals are numerous and their order of preferences in a situation of changing value systems is not always clearly structured. Changes required to bridge the gap between the existing situation and the economic optimum are risky and require information and experience. Farmers' analyses of the problems in their farms are usually only partial. They generally shy away from complete and sudden changes and prefer incremental adaptations to a new environment. This being so, few farms are likely to be found with optimal systems (see § 1.2.3(a)).

2. *The tendency towards more open systems.* In pursuing their objectives and adapting their farms to changes in the environment, farmers tend to rely increasingly on purchased industrial inputs and to specialize in outputs demanded by the market (von Bertalanffy 1973, p. 70). Most traditional systems are relatively closed. They include a great number of activities so as to cover subsistence needs or local demand, and the activities are closely related to each other. Time produces innovations. The innovations relevant to farming are embodied in new industrial inputs, and they tend to foster open systems and to reduce the strength of internal relations. This general tendency comes very much to the fore in periods of rapid economic development, when farmers commercialize and specialize (progressive differentiation). The idea of a 'steady state' is useful as a concept, but it is foreign to socio-economic systems which find themselves in an ever-moving environment. The tendency of the changes made in the environment is towards more open, more productive,† more dependent, and more vulnerable systems, because each system increasingly depends on others. This implies a strengthening of the hierarchical order in the economy. A shifting cultivator, producing for subsistence only, is largely independent of input supply and marketing systems, but a strongly commercialized holding is highly dependent on both.

---

† More open systems, supplied with a lot of industrial inputs, are more productive in economic terms than less open ones, as shown in many examples given in this book. They may be less productive in other terms. The ratio of food gained to human energy expended on a coral island is about 18:1, and there is no other energy involved in the food-production process. In modern British cereal production the ratio is about 5:1, and half of that if processing is included (Duckham 1974). This indicates that it depends on the social value system whether the change from more closed to more open farm systems brings a gain in productivity.

1.3. Classification of farming systems

The described properties of farms apply to each unit. In each case there is a system which transforms inputs into agricultural outputs and which undergoes changes over time. In the process of adapting cropping patterns and farming techniques to the natural, economic, and socio-political conditions of each location and the aims of the farmers, more-or-less distinct farm systems have developed. In fact, no farm is organized exactly like any other, but farms producing under similar natural, economic, and socio-institutional conditions tend to be similarly structured. For the purpose of agricultural development and to devise meaningful measures in agricultural policy it is advisable to group farms with similar structural properties into classes. It is important in this context that relevant criteria for the purpose of classification are used and no single criterion allows the formation of meaningful classes. In this book the great variety of farms that exist in the tropics are grouped into a few major classes, and a number of subclasses as given in the following classification scheme.

1.3.1. *Collecting*

Collecting is the most direct method of obtaining plant products. It may include either regular or irregular harvesting of uncultivated plants. Hunting and fishing usually go hand in hand with collecting. In prehistoric times, activities of this kind were a major source of food supply. In some regions these activities still provide rather important additions to the subsistence food gained from organized production in arable farming and animal husbandry. Only in a few cases (the wild oil-palms in some parts of West Africa, the gum arabic of the Sudan, the wild honey of Tanzania) is collecting a major cash-earning activity.

1.3.2. *Cultivation*†

Much more important than collecting are the numerous types of cultivation. The different cultivation systems may be classified according to a number of particular features. The most important classifications that have been adopted by recent authors are as follows.

(a) *Classification according to the type of rotation*‡ (after Aereboe 1919; Brinkmann 1924; Woermann 1959). Natural fallow system describes a

† The word *cultivation* is used in this book in the sense of the preparation and use of land for growing crops. This does not necessarily imply that the ground around the plants must be loosened.

‡ The word *rotation* has two meanings according to the time period involved. There is the long-term alternation between various types of land use, such as arable farming, tree farming, grassland use, fallow, etc. In this connection, rotation means the sequence of these basic types of land use on a given field. Within arable farming, there is also the term *crop rotation*, which means the short-term sequence of different arable crops on one field. In this section, the word is used in the first of these two senses.

situation where cultivation alternates with an uncultivated fallow. The natural fallow vegetation may take the following forms.

1. A forest fallow comprises woody vegetation with trunks and a closed canopy in which the trees are ecologically dominant.

2. A bush (thicket) fallow comprises dense wood vegetation without trunks.

3. A savanna fallow comprises a mixture of fire-resistant trees and grasses in which the grasses are ecologically dominant.

4. A grass fallow comprises grasses without woody vegetation. (FAO/ SIDA 1974, p. 17.)

The term *ley system* describes those cases where grass is planted or establishes itself on land that has carried crops for some years. The grass is allowed to remain for several years and is used for grazing. In the savannas there are extensive areas of *wild* or *unregulated* ley farming. In these areas, we find that, after a period of cultivation of several years, the field is covered with grass and shrubs for several more years and serves as rough pasture. *Regulated* lay farming, with established swards during the non-cropping period, is rare in the tropics, although it is found in some highlands (e.g. in Kenya) and in Latin America.

Field systems occur where one arable crop follows another, and where established fields are clearly separated from each other. The grassland associated with field systems is usually treated as permanent grassland, whether it is rough or well cared for, and it is separated from the arable land.

Systems with perennial crops, that is with crops that cover the land for many years, are in a separate category. Within this category we distinguish between perennial field-crops like sugar-cane and sisal, bush-crops like tea and coffee, and tree-crops proper like oil-palm and rubber. All kinds of rotation are found. In some cases tree-crops are alternated with fallow, in others with arable farming, grazing, or other perennial crops.

(b) *Classification according to the intensity of the rotation* (after Terra 1958; Nye and Greenland 1961; Faucher 1949). The fallow and ley systems display considerable variations and degrees of intensity. This is especially true for field–bush rotations and for unregulated ley systems in the African savannas. A relatively simple and appropriate criterion for classification is the relationship between crop cultivation and fallowing within the total length of one cycle of land utilization. Following the suggestion of Joosten (1962), we define the symbol R as the number of years of cultivation multiplied by 100 and divided by the length of the cycle of land utilization. The length of the cycle is the sum of the number of years of arable farming plus the number of fallow years. The characteristic R indicates the proportion of the area under cultivation in relation to the total area available for arable farming. If, for instance, 40 per cent of the available arable land in one holding is cultivated, then R is 40.

As long as fallow farming has an extensive character, in which many fallow years follow a short period of cultivation, R remains very small. If for example, 18 fallow years succeed 2 years of cultivation, as is frequently the case in the rainforest, R amounts to 10. This extensive type of fallow farming is generally designated *shifting cultivation*, because the shifting of fields within a broad area of wild vegetation usually results in the gradual relocation of the farming population. On the other hand, it should not be forgotten that there are a number of regions where stationary populations practise shifting cultivation.

The larger R becomes, the higher is the percentage of the area cultivated annually in relation to the total area available for arable farming, and the more stationary the character of the farming becomes. When cultivation is extended so far at the expense of fallowing that the characteristic R reaches or exceeds the value 33, then we can hardly speak of a shifting of the fields any more. A level of intensity of land utilization has been achieved that Terra and Nye and Greenland designate *semi-permanent cultivation*, and that Faucher designates *stationary cultivation with fallowing*. A characteristic R value of 50 is obtained, for example, if 7 fallow years succeed 7 years of cultivation. This book uses the term 'fallow systems' (see Fig. 1.2).† When

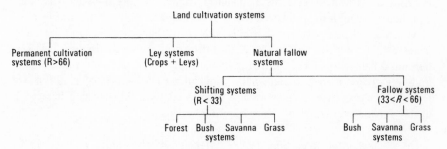

FIG. 1.2. Classes of cultivation systems (after FAO/SIDA 1974, p. 17).

the R value exceeds 66, and the soil is cultivated nearly every year or even more often, then permanent farming is being practised. Permanent farming may again be conveniently classified according to the degree of multiple cropping. An R value of 150 would indicate that 50 per cent of the area is carrying two crops a year, and a value of 300 would indicate that three crops a year are being grown.‡

† In the first edition the term 'semi-permanent' farming was used. This edition follows the Food and Agriculture Organization proposal (see Fig. 1.2).

‡ The R value is useful for the rough ordering of major rotation systems. It becomes difficult to apply in the case of crops with great differences in the length of vegetation cycle or where intercropping is practised.

(c) *Classification according to the water supply.* One of the first steps in classifying land utilization is to question whether farming is practised with or without irrigation. In *irrigation farming*, to ensure that the moisture level of the soil is higher than would occur naturally, a water supply is directed into the fields.

Farming without irrigation is widely referred to as *dry farming*. In the interest of precise definition, however, it is advisable to speak of dry farming only in semi-arid climates. A distinction between irrigated farming and rain-fed farming is therefore preferred. An important part of tropical agricultural production occurs on 'wet' land in valley bottoms, lowlands, or terraces with 'wet' rice, where water is impounded but without proper artificial irrigation. The term 'upland' farming is used to differentiate clearly between farming under hydromorphic conditions in lowlands or on wet-rice terraces and farming under dry field conditions.

(d) *Classification according to the cropping pattern and animal activities.* A most important aspect of the definition of farming systems is usually the classification according to the leading crops and the livestock activities of the holdings. Each activity has different requirements as to climate, soils, markets, and inputs. Therefore those farms can be grouped together whose gross returns (sales plus household consumption plus changes in stock) are similarly constituted, to give, for example, coffee–banana holdings or rice–jute holdings.

(e) *Classification according to the implements used for cultivation.* In addition, holdings are occasionally classified according to the main implements used. In various parts of the world, the land is cultivated by methods that require either no implements, or a few very simple tools. Millet is sown without fire-farming or soil preparation by a few nomads in the Sahara. Shifting cultivators frequently sow in ashes without touching the soil either beforehand or afterwards. Rice-growers in Madagascar, Sri Lanka, and Thailand make use of the treading of animals: a large number of cattle are driven across the moist field to trample down the soil until it becomes a mire ready for planting. In some parts of the world, planting-sticks and digging-sticks have not yet been replaced by hoes, spades, or ploughs. However, with the exception of these pre-technical methods, the main divisions are (1) hoe-farming or spade-farming; (2) farming with ploughs and animal traction; and (3) farming with ploughs and tractors.

(f) *Classification according to the degree of commercialization.* There are a number of distinct types according to the percentage of sales in relation to the gross return. The 1970 World Census of Agriculture classifies farms into three groups based on the destination of the agricultural output:

1. Subsistence farming—if there is virtually no sale of crop and animal products.

2. Partly commercialized farming—if more than 50 per cent of the value of the produce is for home consumption.

3. Commercialized farming—if more than 50 per cent of the produce is for sale.

1.3.3. *Grassland utilization*

The low yields of grassland areas in the arid and semi-arid tropics necessitate either nomadism or semi-nomadism, or the development of ranching systems. It seems advisable, therefore, to classify the different types of livestock farming according to the degree of stationariness of both the animals and those who tend them. According to Dittmar (1954), five main types can be distinguished.

1. *Total nomadism* covers systems in which the animal owners do not have a permanent place of residence. They do not practise regular cultivation and their families move with the herds.

2. *Semi-nomadism* is a related system, where the animal owners have a permanent place of residence near which supplementary cultivation is practised. However, for long periods of time they travel with their herds to distant grazing areas.

3. *Transhumance* is the situation in which farmers with a permanent place of residence send their herds, tended by herdsmen, for long periods of time to distant grazing areas.

4. *Partial nomadism* is characterized by farmers who live continuously in permanent settlements, and who have herds at their disposal which remain in the vicinity.

5. *Stationary animal husbandry* occurs where the animals remain on the holding or in the village throughout the entire year.

1.3.4. *Procedure in this book*

The definition, classifications, and criteria outlined above are employed in the following chapters. Collecting systems are omitted because of their economic insignificance. The book is restricted to cultivation systems and grazing systems, and its main chapters are organized according to the type and intensity of land use. As major cultivation systems, a distinction is thus made between shifting-cultivation systems, fallow systems, systems with regulated ley farming, systems with permanent upland cultivation, systems with arable irrigation farming, and systems with perennial crops. Under the heading of grazing systems, only nomadic grassland use and ranching are considered. The other forms of grassland use are described as part of the relevant cultivation system.

2. Some characteristics of farming in a tropical environment†

by J. D. MacArthur

THE tropics are characterized by enormous variability. Climates, soils, and altitude differ widely, and markedly differing conditions can occur, even over short distances. Large stretches or relatively uniform ecological conditions like the North American prairies, or the black-soil belt in the USSR, are not found in the tropics. It is therefore even more difficult to generalize about farming in the tropics than about farming in temperate climates. However, some generalizations can be made which help understanding of the possibilities for agriculture and of the difficulties facing farmers—small farmers especially—which influence and are reflected in the complex and specialized farming systems that have been developed.

2.1. A high potential for photosynthesis

The basic principle of agricultural production is the conversion of solar energy into food and other products useful to man. The application of 'support energies' (in the form of human and animal energy, fuel, machinery, fertilizer, etc.) by agriculturalists maintains and improves this conversion process. The more solar energy received per hectare, the greater the agricultural potential, provided sufficient water, nutrients, and labour are available. Temperate climates receive 80 to 120 kcal solar energy per cm^2 per year. In the subtropics the figures are 140 to 190, and in the tropics 130 to 220 kcal, depending on cloud cover and altitude (Trewartha 1968). The tropics and subtropics therefore receive almost twice as much solar energy as the temperate zones. Their potential for agricultural production is consequently higher.

Actual production in the tropics is not, however, as high as these figures suggest is possible. The other necessary conditions for high outputs are often

† Only a very cursory summary is possible in this chapter of a book about farming systems, not about the natural science background to farming. Fuller details can be found in Davies and Skidmore (1966), Webster and Wilson (1967), Williamson and Payne (1965), Wrigley (1971), Tempany and Grist (1958), and others.

lacking or do not coincide. Solar energy is ample in the dry season when moisture is lacking, and is less in the cloudy rainy season when moisture is ample, so a high level of incident radiation seldom coincides with the peak availability of other necessary conditions for high yields. The highest yields are obtained in latitudes some distance from the equator. Kassam and Kowal (1973), Chang (1968), and others have shown that incident radiation and potential crop production are highest in the tropics in the non-equatorial, non-humid areas. Temperate areas receive in summer roughly the same amount of solar energy as many humid tropical areas (Holliday 1976, p. 139). De Wit found a higher photosynthesis rate in summer in the Netherlands than in Uganda (1965, p. 48). The fact remains, however, that more solar energy is available.

In the tropics the thermal growing season extends throughout the year. Food-grain production in a temperate climate uses only about 50 per cent of the relatively short thermal growing season, while multiple cropping in Taiwan uses 93 per cent of a growing season extending throughout the year. The hydrological growing season, defined by the rainy seasons, is usually longer in the tropics than in the subtropics. With sufficient water the main-tenance of a fully assimilating leaf area is possible during the whole year. In the temperate zones solar energy is not used at all during winter; is only partially used in spring, because plants cannot develop their leaves until the winter is over; and in late summer and autumn solar energy can be utilized only poorly, because maturing plants show declining rates of assimilation. Furthermore, tropical conditions allow the growing of crops with C_4 photo-synthetic systems (maize, sugar-cane, sorghums, some millets, and most tropical forage grasses) which are able to recover more solar energy as chemical energy edible by man than can C_3 crops, such as wheat, soya-beans, rice, and most root crops (Roberts 1976, p. 96).

Holliday (1976, p. 140) indicates the potential dry-matter production of natural vegetation in different climates. In the humid tropics the potential is 146 t ha^{-1}; in bimodal rainfall areas of the semi-humid tropics it is 104 t ha^{-1}; in monomodal semi-arid tropics 37–72 t; and in the temperate climate of the UK roughly 50 t dry matter ha^{-1}. The yields of well-fertilized and irrigated fodder grasses which are photosynthetically fully active throughout the year reflect actually realized performances. In temperate zones, maximum yields are 20 to 25 t dry matter ha^{-1} per year, as against 40 to 80 t in the tropics (Cooper 1970, p. 8). The tropical potential for conversion of solar energy into plant dry matter can be two to three times as great as that of temperate regions. Cases such as sugar cane in Hawaii (10 to 15 t sugar ha^{-1} per year), oil palms in Malaysia (5 to 6 t oil ha^{-1} per year), and rice in southern India (three crops producing 15 t paddy ha^{-1} per year) show that economically rewarding high levels of productivity can be achieved in tropical commercial production.

The highest recorded conversion rates of annual photosynthetic solar energy were achieved with elephant grass in Puerto Rico (110·6 t ha^{-1} of dry matter and a 5·32 per cent conversion rate of usable solar energy). Two crops of maize in Uganda yielded 38·2 t ha^{-1} of dry matter (4·7 per cent conversion rate), which is more than the conversion rate of solar energy in a mass culture of algae in Japan which yielded 53·1 t ha^{-1} dry matter and a conversion rate of 4·3 per cent (Holliday 1976, p. 142).

2.2. The difficulties of the natural environment

The great potential of the tropics is not reflected in actual performance, which is generally poor. Intensive farming in temperate climates achieves conversion rates to dry matter of about 2 per cent of the photosynthetically active solar energy received, while in most tropical farming not more than 0.2 per cent is achieved (Holliday 1976, p. 141). Conversion to edible dry matter is even less favourable. In intensive temperate farming, and in the developed parts of the subtropics where improved varieties are used, the edible proportion amounts to 30 to 56 per cent, while in traditional tropical farming it is usually not more than 5 to 35 per cent (Holliday 1976, p. 134; Kassam 1977, p. 19).

Temperate-climate farming produces more dry matter, a much higher proportion of which is edible. Duckham, Jones, and Roberts (1976) estimate that the food actually produced per cropped hectare in the tropics (low-income countries) is about half of what is produced in temperate areas (high-income countries) (p. 470), and this in spite of the more ample supply of energy. The low efficiency of tropical farming is clearly related to environmental difficulties.

2.2.1. *Climatic difficulties*†

To utilize solar energy, plants require sufficient soil-moisture, and those areas in the tropics where rainfall or irrigation would allow crop production throughout the year are very limited. Most of the tropics experience a seasonal pattern of rainfall, either bimodal (close to the equator) or monomodal (at some distance from the equator). Without irrigation, arable crop production is limited to the rainy season and there is no store of soil moisture when the cropping season begins, as in temperate climates and in much of the subtropics after their cold and relatively wet winters. Moreover, the proportion of rainfall in the tropics available for crop production is often less

† A fuller account of the influence of environmental factors on crops is given in Spedding (1975), while tropical climatic conditions are discussed *inter alia* in Duckham and Masefield (1970, Chapters 1.2, 1.3, and 1.4); and Williams and Joseph (1973, Chapters 2–5). Grist (1970) gives a brief account of climatic conditions in each of the major tropical zones.

than in temperate climates, run-off and evapo-transpiration losses being much greater under tropical conditions, where many soils seal easily and show low rates of moisture absorption. Tropical rainfall patterns are also characterized by great variations between one year and another, and between adjacent places, especially where mean annual rainfall is low. Considerable uncertainty attaches to the level of rain that will fall in any season, and when it will fall. In temperate climates the coefficient of variation of rainfall is about 15 per cent, while 30 per cent is typical for the tropics (Duckham 1976, p. 68). The moisture-holding capacity of the soils is often low, and, given the high rates of evapo-transpiration on cloudless days in the tropics, short dry spells of only a few days may destroy a crop or cause heavy losses. Moreover, the rains often come in a series of heavy storms, much rain falling in a short time, so much is lost in run-off, while the intensity of rainfall may cause flood damage and erosion.

Wind is another main feature of the tropical climate. The impact on temperature and rainfall of some main rain-bearing winds, especially monsoons induced by large continental land masses, is well known, as are the more sensational hurricanes and typhoons. However, winds blowing off the hotter land masses—especially the larger deserts—can have an intense drying effect. In many tropical areas the dry season is a time of fairly high winds, which not only reduce opportunities for plant growth, but also bring a threat of wind-induced erosion.

Other climatic effects can be important. Within tropical latitudes day-length varies little between one part of the year and another; on the equator the length of all days is virtually the same. Within the day, the intensity of solar radiation and the air temperature may undergo fairly rapid change, although both effects are less pronounced in places where and at times when the relative humidity of the atmosphere is high, and when there is a high incidence of cloudiness. This means that, whilst the hours of daylight are limited to something like the length of a typical early spring or autumn day in temperate climates (i.e. 13–15 hours), the heat and intensity of radiation may at times exceed levels that some crops and livestock need or can tolerate. Animals and workers then seek shade, and can pursue productive activities for only a limited period. Altitude reduces the temperature effect, but can intensify the radiation problem beyond comfortable levels.

Climate influences farming not only through the control that it imposes on the timing and volume of plant growth, but also through its effect on livestock production. Excessive heat and humidity can upset animal physiology, causing distress, reducing appetites, and limiting productivity. High temperature and/or humidity can make the conservation of harvested produce very difficult, creating extreme problems of fodder supply in the non-productive dry season, and leading to high losses amongst crop reserves kept for use outside the productive months.

2.2.2. *Edaphic difficulties*†

The functions of land in the agricultural production process are defined by Duckham (1978) as follows:

Land is a factory floor or net which (a) catches (i) low entropy energy from solar radiation and (ii) rainfall, (b) provides rooting capacity for anchorage and nutrient foraging, (c) provides limited temporary storage capacities for nutrients and, on some soils, water and heat, and often moderates the impact of weather/ climate extremes, (d) provides a surface for leaves of man-sown crop plants (the crop canopy) to convert atmospheric carbon dioxide (CO_2, water, some atmospheric nitrogen (N), and unorganic mineral nutrients (Phosphates, potash and 18 others) plus any added fertilizer (N, P, K) plus nutrients in animal excreta, leaf fall, etc., from or on the soil into (e) low entropy living material, with the result (f) that a widely varying proportion of this low entropy photosynthate can be and is harvested by man and his livestock and, if properly stored, processed and cooked, becomes food for man or industrial raw material.

Tropical soils, being like all others the product of climate, parent material, and age, vary enormously in type and suitability for farming. Those that are geologically fairly recent, being volcanic or sedimentary in origin can be very suitable in terms of the functions described by Duckham, but others are derived from old igneous material and are low in pH, heavily leached, and their moisture and nutrient-holding capacity is usually very low. Despite their great variation, however, a few useful generalizations about tropical soils can be made.

Perhaps the most widespread feature is the extent to which precipitation in sudden and heavy rainstorms leads to extensive leaching, with most soluble plant nutrients carried below the root zone. Consequently, many soils, particularly in humid areas, have a low level of natural potential fertility (phosphates and nitrogen particularly are commonly deficient), a large proportion of the valuable nutrients in the crop zone being contained in the vegetative material (Williams and Joseph 1973, p. 13; cf. Spedding 1975, p. 91). Moreover, the downward movement of soil minerals can lead to the formation of subsurface hard-pans which commonly impede drainage and restrict root growth. Also detracting from fertility status is the typically low level of organic material. Under warm tropical conditions, most organic matter breaks down quickly whenever it is moist. In areas of alternating wet and dry seasons, this produces a rapid but short-term release of nutrients, often at the onset of the rains.

Another common characteristic detracting from their suitability for exploitation is the poor structure of many tropical soils. Because of this feature, they may not be resilient under intensive cultivation, which increases the risk of water and wind erosion, especially when the vegetative cover is

† See also Duckham and Masefield (1970; Chapter 3); Williams and Joseph (1973; Chapter 7).

removed. Erosion by surface water is a widespread reality and constant danger, especially because of the intensity of tropical storms and the impermeability of many soils, particularly after rain. For all of these reasons, soil management in the tropics can be relatively difficult.

2.2.3. *Biotic difficulties*

Partly within the sphere of human control is the biological environment, which contains natural or introduced species of all kinds. In tropical conditions both types are particularly numerous and varied. In the warmth, wherever moisture levels are high enough, crop production can be very high. However, many kinds of weed, fungi, and parasites also flourish, so domesticated plants and livestock are constantly subject to severe competition or parasitic infection and disease. This presents many farming problems. Not only is cropping beset by the need for clearing, weeding, and the removal of secondary vegetation, but all farm production processes are exposed to high risks from pests and diseases of many kinds. Some of the principal scourges, like the tsetse fly and ticks, are present constantly, as are some of the virus and bacterial agents that cause serious crop and livestock diseases. Others are not a constant threat, but many build up to plague proportions in a very short time. This is especially true of insect pests, but applies also to other species, including rodents and even small birds, which can create sudden and tremendous havoc amongst growing and harvested crops. In much of the tropics weed growth is so prolific and so difficult to control that it is the major yield-depressing factor.

2.3. The implications for tropical farming

Farmers who work under the climatic, edaphic, and biotic constraints of a tropical environment are faced with several operational problems which derive from the natural environment and which influence the systems of farming they follow.

2.3.1. *High costs of soil-fertility maintenance*

The importance of soil fertility and conservation is clearly recognized by most tropical farmers, and most farming systems involve specially adopted measures to preserve or increase the fertility of cropped land. Essentially, the problem is one of making available in the upper layers of the soil sufficient nutrients in the right condition to allow plants to take them up.

Under natural conditions an equilibrium develops in which losses through leaching, surface run-off, and gaseous escape are made up by the chemical breakdown of soil minerals, the supply of available nutrients from lower soil layers that deep-rooted plants maintain, the fixation of atmospheric elements, and, in some places, water-borne materials. In contrast to this natural situation, cropping or other forms of farming are exploitive, involving the removal of nutrient-bearing materials from the plants where they were produced. Often the residues remaining after human use are not returned to the land, but are carried away in drainage water. Natural replenishment may in some circumstances keep pace with this loss, whilst farmers may 'concentrate' naturally available nutrients by bringing crop residues, manure, or natural vegetation to places intended for cropping. However, all of these effects are wasteful in that, through the deliberate removal and translocation of produce and by disturbance of the natural order in ways that increase the rate of loss, nutrients are removed from the soil and are not replaced. Sooner or later such a system must lead to either temporary or permanent reductions in soil-nutrient availability.

In the tropics agricultural production is, of its very nature, more complex and demanding than in temperate zones. The tropical farmer replaces the usually 'balanced' natural system by others that give the production he wants, but either consumes soil fertility or requires much labour and high material inputs if both high output and soil fertility are to be maintained. The warmer and more humid it is, the speedier are biological processes, the quicker is organic matter decomposed and nutrients leached, the more quickly is the fertility of the soil consumed, and the greater must be the farmer's efforts to maintain fertility so that the tendency to increasing entropy is slowed down. Tropical farmers, therefore, face a fertility management choice (Ruthenberg 1977).

The soil may be 'mined'. Food can then be produced cheaply, with little support energy used (and the poverty of tropical countries makes that important), but yields decline. Under most conditions they decline in time to very low levels—food grain yields of about 500–1000 kg/ha^{-1} are typical—and utilization of solar energy is very inefficient. The result is stagnation at a low level. In temperate climates most farming leads to dry-matter production per hectare similar to that of natural vegetation (Snaydon and Elston 1976, p. 50). In traditional tropical production it is no more than 20–50 per cent of the dry-matter production of natural vegetation. The discrepancy between potential and realized output seems to be much greater in traditional tropical farming than in traditional temperate agriculture. This unproductive 'low level equilibrium' farming may, however, be very efficient in the use of support energy. Leach established that traditional low input farming produces 15 to 60 units of energy output per unit of support energy, while the ratio in modern intensive farming is between 1 and 4 at the farm gate level, and much less if the energy of processing and delivery is considered (p. 375).

Alternatively, the farmer maintains and increases the fertility of his land. This requires high support energies to counteract the 'unnaturalness' of cropping, but it yields high returns. This kind of approach allows efficient utilization of the natural energy potential, but is expensive in economic terms. Intensive systems are relatively inefficient in terms of edible output per unit of support energy (Snaydon and Elston 1976, p. 52). In view of the price conditions in countries in which people are poor and need cheap food, intensive and (solar) energy-efficient production is often too expensive to be economically profitable. Also it tends to consume rapidly the limited world stock resources of energy and nutrients. The fertility/energy problem of the tropics thus is that modes of farming which are efficient in the use of solar energy tend to be wasteful of support energies, and modes of farming which are efficient in the use of support energies tend to be wasteful of solar energy.

2.3.2. *Coping with risk*

Additional operational problems derive from the high levels of production uncertainty that surround natural processes in the tropics, especially farming at low levels of technology. In smallholder farming a main objective of every householder, whether commercialized or not, is to produce most of the basic food and other natural-product needs of the family. It is therefore essential that a minimum amount of food of different kinds should be available throughout the year. In meeting this objective, farmers have not only to consider the production risks that may arise from climatic variation and the effect of pests and diseases; they have to allow also for the effect of these same influences on stored produce where destruction and deterioration can be very rapid.

The particularly dramatic nature of variation of this kind associated with tropical production causes farmers to adopt a number of devices to spread risks and ensure a constant supply of produce, which, because of consumer preference and to avoid storage losses, must be as fresh as possible. Diversification of production to grow a range of crops is a typical risk-spreading device used the world over. The planting of a particular crop at different times over an extended period, which will ensure some production in either an early or a late season, whichever should occur; the combination of different species in crop mixtures; and the culture of small areas of especially reliable though non-preferred crops are all common practices with a strong element of risk-avoidance in their adoption.

From the point of view of farm operation and resource allocation, the need to provide for a minimal production in the very worst possible season can lead to an apparent misallocation of resources, in that the target levels of production for a particular commodity may provide a surplus in a normal or good year, which might not always find a market. At the same time, some crops may be grown that are not well suited to the natural environment. Thus, to overcome risk to domestic food supply, farmers may need to employ more resources for subsistence production than would normally be necessary, while they also grow some products for which their land or skills are not well adapted.

Similar factors apply to commercial enterprises, militating against mono-culture or specialization that could lead to the greatest production efficiency. Problems of uncertainty substantially reduce the flexibility of farm systems and their ability to evolve rapidly along apparently attractive development paths. These problems particularly apply where small farming units and low incomes make risk-avoidance especially important.

2.3.3. *Problems of seasonality*

Farm operation and labour productivity are further hindered by the acute seasonality of many climates, in which wide differences exist between the wet and dry seasons, or seasons with and without irrigation water. Where sufficient soil moisture is available in only a few months, all crop work needs to be carried out very quickly, often beginning only after first rains soften the soil, and continuing with cultivation, planting, and weeding for as long as crop growth is possible. The urgent need for high labour inputs at this time often coincides with the period when fresh high-quality food is most scarce, and low nutritional levels inhibit workers and draught animals from working to full capacity. Certainly this season contrasts strongly with the other months, in which little work to assist crop production can be carried out, harvesting and maintenance being the main activities. At such times, a lot of labour cannot find productive work, leading to low average returns to labour and low incomes.

Similar problems arise in livestock management. During the rainy season, fodder is abundant and of high quality. Animals can then recover condition and health, grow, produce milk and other products, and undertake sustained work. In the dry weather, the grazing deteriorates. Animals may lose con-dition, whilst their productivity declines, often to nil, as lactating animals dry off. Only under exceptional climatic and economic conditions can this problem be overcome by fodder conservation.

One consequence of the shortness of the cropping season, and the low productivity of manual and animal power at this time, is that the mechaniza-tion of cropping can bring high returns, if properly managed. Usually, where they can be kept, draught animals can bring an improvement over manual labour. Draught animals, however, require a lot of land for their food needs, and suffer in conditions of high humidity and heat and accompanying diseases. They also find it much more difficult than tractors to cope with rapidly hardening soils. The greater operating speed of tractors, and the fact that they can work at times when soil conditions prevent hand workers or draught animals from carrying out cultivation, crucially influence the volume of production where the best use must be made of a small amount of rain. This does not imply that the use of tractors in the tropics is always more practicable than hand work or draught animals, but in the context of farm-management characteristics it is important to note that the physical rate of substitution of animals by tractors is comparatively high.

2.3.4. *Low labour productivity*

Closely related to the problem of uncertainty is the low productivity of labour in tropical agriculture, whether measured in hours worked per available worker, or by the physical performance per hour of work, or by the return per hour of work or per man-year. In the majority of situations, even where land is limited in supply and advanced material inputs such as mineral fertilizer and pesticides are not used, the most critical farming input is the labour of the farmer and his family. Much of the work is both tedious and strenuous, so not surprisingly farmers show a marked preference for leisure; hired workers are often employed to replace family members once incomes exceed very modest levels. However, in some situations, despite the use of seasonal labour, farmers have limited scope for increasing output by making higher labour inputs, through the shortness of the growing season during which useful farm work can be done, and the low levels of technology at which they typically operate.

Where, as very often is the case, labour is the main variable input into farming, its availability can be the limiting factor to production, especially at certain critical times of the year and in view of the demands of both time and effort that domestic duties make upon certain household members. Consequently farmers seek—and usually quickly take up—new crops, production methods, and production combinations that offer higher returns to labour and are also acceptably free from risks. It is therefore those processes that offer higher returns to labour that many farmers actively seek, and the development of small-scale production must largely depend on the discovery and adoption of practices that will substantially improve returns to work. From the farmer's point of view, labour productivity and work rationalization are of great importance, not only because labour is a major input but because its use is the subject of acute personal experience.

2.4. The dynamic nature of tropical farming

Despite the difficulties of the natural environment, and the limitations that arise from unfavourable price relations, unstable markets, and the traditional socio-institutional conditions of agricultural production in low-income countries, tropical farming, which consists mainly of smallholder farming systems, is a dynamic business, in which small but numerous adjustments are continually being made to the production processes. In the following chapters of this book, seven basically different main types of farming systems are considered and smallholders are dominant in each. In every situation, the importance of the three major problems of soil fertility, uncertainty, and labour productivity is highlighted, together with a consideration of the ways in which technological and institutional changes can help overcome these basic problems. Discussions indicate how production can be transformed from its traditional pattern to a more productive level. Examples of dramatic change should not mask the fact that innovation and change, however slow, are normal features of traditional farming. Without a continuous series of small adjustments, the diverse and often well-adapted farming systems to be described could never have developed.

3. Shifting cultivation systems

3.1. Definition

SHIFTING cultivation is the name we use for agricultural systems that involve an alternation between cropping for a few years on selected and cleared plots and a lengthy period when the soil is rested. Cultivation consequently shifts within an area that is otherwise covered by natural vegetation. The intensity of shifting cultivation varies widely. A relatively simple and appropriate criterion of land-use intensity is the relation between the period of cultivation and the period of fallow (see § 1.3.2.).

The shorter the time between the cultivation periods, or the greater the annually cultivated area in relation to the total area, the more stationary does farming become. In the course of this development, short-fallow systems replace long-fallow systems. If the characteristic R is more than 33, i.e. if 33 per cent of the arable and temporarily used land is cultivated annually, we no longer speak of shifting systems but of fallow systems.

3.2. Types of shifting cultivation and their geographical distribution

Table 3.1 shows that shifting cultivation is practised principally in humid and semi-humid climates and in thinly populated savannas. It is found, however, in all climates and in very varied economic conditions. It was the elementary and pioneering cropping system used by the early agricultural occupants of many forested regions all over the world. Shifting cultivation is today practised by peoples in the interior of New Guinea just coming out of the Stone Age, and also by a large number of partly commercialized farmers in Asia, Africa, and Latin America. There are, however, few pure subsistence holdings, even among the shifting cultivators. In most cases supplementary cash cropping is practised. Shifting cultivation is often a technology of expediency, used because it will work better under a given set of conditions than will any other system available to the practitioner (Spencer 1966, p. 2). In other cases it is a land-use system which may belong to the past but is still carried on because the transition to more productive land-use systems demands more favourable price relations, knowledge, capital, and time.

The forms assumed by shifting cultivation are more varied than in practically any other land-use system. It is therefore useful to consider it in relation to several criteria; here a distinction is made between vegetation systems, migration systems, rotation systems, clearance systems, cropping systems, and tool systems.

3.2.1. *Vegetation systems*

We may distinguish on the basis of vegetation between the shifting cultivation of the forest, of bush (thicket) areas, of savannas, and of the grassland areas. Shifting cultivation in the forest is typical of rain-fed farming in humid areas of low population density (New Guinea, Borneo, Burma, Zaire, Amazon Basin, etc.). Rotations with bush fallowing are typical for humid and semi-humid areas with a somewhat higher population density (much of the West African lowlands). Rather distinct are the features of shifting cultivation in the derived savannas of Africa and Latin America, where forests have been replaced by grasses (frequently *Imperata cylindrica*). Alternation between cultivation and a natural grass vegetation is found in some tropical high-altitude areas (African Highlands and the Andes) and in semi-arid climates.

3.2.2. *Migration systems*

In most systems with shifting cultivation, the continual movement of cropping results in a slow migration of the population. The cultivated plots move slowly away from the previous clearing and the vicinity of the hut. At the same time the cost of transporting the harvest increases, especially where root crops are grown.† Beyond a certain distance, it becomes advantageous to build a new hut near the field instead of carrying the harvest such a long way.

Shifting of plots is made easier by the very nature of tropical housing; it is often more economic to build a new hut than to repair an old one. The frequency of hut changing varies greatly. In the rainforest, in some cases, a change occurs every 2 or 3 years; in the savanna less frequently, sometimes only once every one or two decades.

Allan (1967, p. 4) describes the process in central Africa as follows:

A village or other group within such a concentration will use up the accessible land in its neighbourhood before the area first cultivated has had time to regenerate and a move must then be made to regenerated land . . . A series of pictures or aerial photographs made at suitable intervals would show a gradual movement of the larger community across the countryside in a manner suggestive of the progress of an amoeba on a microscope slide.

---

† According to Jurion and Henry (1967, p. 320), a family of four, planting annually 0·5 ha of maize, rice, groundnuts, and bananas with an everage distance of 1·5 km between the hut and the plot, spends annually 120–140 man-days on transport if the carrying load is 35–45 kg and 4–5 journeys are made per day.

TABLE 3.1

The length of the crop and fallow periods under shifting cultivation

Place	Rain (mm per year)	Crops	Fallow	Normal Crop	Normal Fallow	Excessive Crop	Excessive Fallow	Typical value for R	Remarks
			Moist evergreen forest zone						
Sarawak	About 3800	Hill rice	Forest	1	>12	2	12	7	Early abandonment of land necessary to prevent invasion of *Imperata*
Guatemala	3400	Maize	Forest	1	>4			20	'Ando' type soil
Liberia	2000–4500	Rice, manioc	Forest	1–2	8–15			11	
Sierra Leone	2300–3300	Rice, manioc	Forest	1½	8	1½	5	12	Grasses (especially *Chasmopodium* spp.) invade with excessive cultivation
Assam	2500	Rice–millet, maize, rice	Forest	2	10–12	2	<7	15	
Sumatra	About 2300	Rice, root crops	Forest	2	10–16			13	*Imperata* invades, but may give place to forest
Philippines	2500	Rice, root crops, maize	Forest	2–4	8–10			25	
Nigeria (a) Umuahia	About 2300	Yams, maize, manioc	*Acioa barteri*	1½	4–7	1½	2–2½	21	Loam derived from tertiary sands and clays; stumps of fallow carefully preserved
(b) Alayi	About 2300	Yams, maize, manioc	*Anthonotha* spp.	1½	7			18	Very loose sandy soil
Central Zaire	1800	Rice, maize, manioc	Forest	2–3	10–15				

Table header (spanning the vegetation and numeric columns):

Moist semi-deciduous and dry forest zone (including humid zone of derived savanna)

Location	Date	Crops	Vegetation						Notes
West Africa	1500–2000	Maize, manioc	Moist semi-deciduous forest	2–4	6–12			25	
North Burma	1300–1800	Hill rice	Grassland and pine forest			5	10	33[a]	Kochin Hills area at about 2000 m
West Nile, Uganda	1400	Eleusine, sorghum, simsim, maize	Grass, mainly *Setaria* spp.	2–3	8–15	3	3	18	Refers to 'outside fields'
Abeokuta, Nogeria	About 1300		Thicket			2	4–5	30	Soil derived from tertiary sand; evidence of nitrogen deficiency
Ilesha, Nigeria	About 1300		Thicket	2	6–7	1	2	24	Soil derived from granite
Central Uganda	About 1300		Elephant grass	3	8		2	27	
Ivory Coast	About 1300		Elephant grass	3	3	9	6	50[a]	
Zambia	About 1300		Thicket	6–12	6–12			50[a]	'Chipya' forest soil
Savanna zone									
Ivory Coast	1200		*Imperata*	2–3	6–10	2–3	4–6	24	
Uganda	About 1100		Andropogoneae	1	2½	1	<2	28	
Northern Ghana	About 1100		Andropogoneae	3–4	7–10			29	
Mali	1000–1300		Short bunch grass	3	12–15			18	
Zambia	About 1000		Miombo woodland	2	up to 25			7	Pallid sandy soils

[a] Fallow farming in our terminology. *Source*: Nye and Greenland (1961, p. 128).

The frequency of movement and the distances covered seem to increase with rainfall. In semi-humid Africa shifts are gradual, as described by Allan. In humid parts of the Philippines, for example, Wallace (1970) found that an average household moved once every 5 years and usually a distance of 5–10 kilometres (p. 73). In the Lower Rio Negro (Amazonas Basin, Brazil) shifting cultivators move into a primary forest, cultivate their plots for 2 years, and shift their hut site at 10-yearly intervals. During their lifetime, they may move about 500–600 kilometres, using water-ways for transport (Weidelt 1968, p. 28).

By the direction of the migration we can distinguish, according to Spencer (1966, p. 177), between (1) random shifts, (2) linear shifts onto newly cleared adjoining lands, which result in the steady and progressive clearing of mature forests, (3) cyclic shifts within a given area claimed by the group or family, and (4) reduced cyclic shifts, brought about by increasing population densities, which might finally lead to sedentary living.

Another important distinction is the one between cultivation by indigenous cultivators who aim primarily at covering subsistence needs, and cultivation by immigrant settler communities with cash cropping as the major objective. The first group usually thinks in terms of cycles of land use. Their plots are small and they work with hand-tools, so their capacity to clear and crop has narrow limits. The latter group, which is important in Latin America, is usually equipped with motor saws. Larger plots are cleared, more leaching and erosion occurs, and the intention is often to mine the land for a few years only.

3.2.3. *Rotation systems*

Shifting farming is practised not only by migrating cultivators but also by sedentary cultivators. In each case, however, cropping and fallowing alternate, and this alternation can have an irregular or regular character. In the case of a regular sequence, a definite number of fallow years follows a definite number of years of cultivation. A further distinguishing characteristic of the rotational pattern is whether land is selected for use in small plots (where, for example, cultivation is restricted to termite mounds, moist valley bottoms, etc.) or whether continuous tracts of land are cultivated.

The rotation systems in rain-forest agriculture are mostly simple. 2–4 years of cultivation are followed by one to three decades of fallowing. Miracle (1967) calls such patterns 'classical long-fallow systems'. In the savannas of Africa, on the other hand, we often find complicated rotation systems in which short-term fallow periods of 1–2 years, medium-term fallow periods of 3–5 years, and long-term fallow periods of 6 years or more alternate in a single cycle of land use. The short-term fallow periods can usually be attributed to unexpected lack of labour during the cultivation period, whereas increasing weed growth and decreasing fertility lead to medium-term and long-term fallow periods (see § 4.3.2).

An example will illustrate this type of situation. In the Kilombero Valley, Tanzania, where shifting of cultivation rice is pursued in an area where elephant grass dominates the natural vegetation (Baum 1968, pp. 32–4), the following rotation systems have evolved according to the settlement density.

1. *Rotation cycle 45 years.* When there is an abundance of land, cultivation takes place within two rotations, one shorter and one longer. As a rule, the shorter rotation runs as follows:

3 years of rice-growing (with one annual harvest);
3 years of grass-fallow.

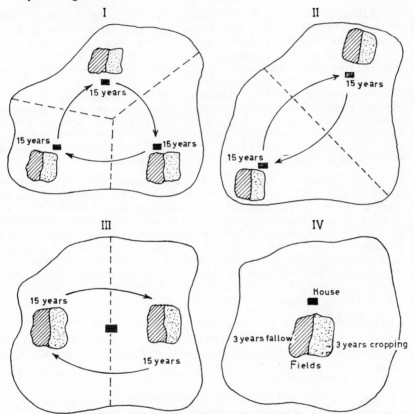

FIG. 3.1. Rotations of shifting cultivators in the Kilombero Valley, Tanzania.

   I = short rotation: 3 years cropping, 3 years fallow
       long rotation: 15 years occupation, 30 years unoccupied
       total duration of rotation cycle: 45 years, frequent changing of huts
  II = total duration of rotation: 30 years, semi-stationary housing
 III = total duration of rotation: 30 years, stationary housing
 IV = short rotation only: 3 years cropping, 3 years fallow
                  (From Baum 1968.)

Several families put their rice plots together into a block, in order to obtain better protection against vermin. They alternate between two blocks, one carrying rice, the other one lying fallow for about 15 years. Thereafter, the families move into a neighbouring area where the 'short rotation' is likewise practised. In a few out of the way parts of the inner valley, families even move into a third area. Then they move back to the first area, which has by then been fallow for 30 years.

The short rotation of 6 years is thus supplemented by the long rotation over a period of 45 years. This results in a total R value of 17, i.e. 17 per cent of the cultivated land is used annually (see Fig. 3.1, part I). The total area composed of the three zones farmed by each family is occupied by these families in accordance with traditional land rights. Immigrants may not settle there without permission. As a rule, the household members know their plots in each of the three areas and cultivate them in regular cycles.

II and III. *Rotation cycle of 30 years.* When there is not enough land available for the families to move between three areas, the 'long rotation' is limited to two areas. The families live in each one for 15 years and rotate 2–3 years of rice with 2–3 years of fallow. This results in a total R value of 25, i.e. 25 per cent of the occupied land is cultivated per annum in contrast to 17 per cent of the land with three cropping areas (see Fig. 3.1, part II).

Fig. 3.1, part III represents a modification of this type of shifting cultivation associated with a higher degree of stationariness. The settlers have a permanent dwelling-place, as, for example, near the road. The fields, however, are still rotated in 15-year cycles, i.e. the land on one side of the settlement is used for 15 years, alternating 3 years of cultivation and 3 years of fallow. Subsequently the fields on the other side of the settlement are cultivated.

IV. *Fallow systems* (see Chapter 4). Growing population density induces the abandoning of 'long rotations'. Where land becomes increasingly scarce, 2–3 years of cultivation are followed by the same number of fallow years. The dwelling-place is permanent, either next to the road or in some other favourite location. Fixed and recognized boundaries are established for holdings and fields. Permanent and generally acknowledged rights of use are derived from traditional land laws. The typical value for this intensity of land use is $R = 50$, which results when three years of cultivation are followed by the same number of fallow years.

3.2.4. *Clearance systems*

The type of clearance work and the order it takes vary according to the vegetation that is to be cleared, the distribution of rainfall, the crops grown, the available tools, and the cultural background of the population. Following Miracle (1967, pp. 41–7), we classify the forms clearance takes as follows:

(a) *Burn and plant.* One of the simplest forms merely involves burning off a thick and dry secondary vegetation. Immediately after burning, maize is planted, and matures before the secondary vegetation has recovered from the burning sufficiently to produce brushwood and leaves. Cook (1921) describes shifting cultivation of this kind in his 'Milpa agriculture' in Central America.

(b) *Burn, hoe and cut, plant.* The burn, hoe and cut, plant system is reported only in savanna areas, where fire effectively eliminates most of the vegetation.

Cutting is limited to the remaining trees and bushes. Except for the reversal of the burning and hoeing operations, this method of agriculture is much like the cut, burn, plant system.

(c) *Cut, burn, plant.* The cut, burn, plant method of agriculture is the most common one reported. Typically, vegetation is cut—usually towards the end of the dry season—allowed to dry for a while, and then is burnt as the rainy season approaches and the soil must be prepared for planting.

The cut, burn, plant method is divided by Miracle into ten major subdivisions, according to the amount of cutting and burning done and whether further cultivation operations are employed.

(d) *Cut, plant, burn.* The cut, plant, burn technique is the only system in which the vegetation residue left in the fields is burned after crops are growing, and is reported only in forest areas, or areas near the forest belt which would possibly still be forest except for the activities of man. Usually the crop occupying the field when it is fired is bananas, plantains, or manioc. In such systems, clearing is often done over a relatively long period, and crops are planted as soon as the smaller vegetation has been cut. By the time all trees to be removed have been felled, the planted crops have established themselves, and the field is given a burning, which apparently inflicts little or no damage upon the growing crops.

(e) *Cut, bury refuse in mounds, plant.* Various sorts of composting operations are reported as an integral part of field preparation in some of the African savannas. After vegetation is cut, it is buried in mounds in the field, or perhaps the debris is first burnt and the ashes are buried. In other cases, the debris is buried in mounds and then ignited, possibly because of the beneficial effect of heat on soil fertility (see Nye and Greenland 1961, p. 71).

(f) *Cut, add extra wood, burn, plant, hoe.* The cut, add extra wood, burn, plant, hoe method, generally referred to as *chitemene* in Zambia, involves cutting bush from a greater area than is to be planted in order to obtain a hotter fire and a large amount of ash on the land that is to be sown.

(g) *Cut, wait one season, plant (forest).* Several tribes in the Congo Basin partially cut vegetation, plant bananas, and then wait a season before completing the clearing operations and planting other crops among the established bananas.

(h) *Killing trees by ringing, ridging, planting.* This series of operations, additional to those mentioned by Miracle, is practised in the dry forests of southern Tanzania. Ringing replaces clearance work. When the trees have lost their leaves the light can reach the ground, and the leaves that have fallen off and vegetation on the ground are dug into the ridges. The trees are never felled; they are left as dead, standing timber.

Shifting systems may also be classified according to how complete the removal of the vegetation has been (Spencer 1966, p. 181). Occasionally the farmers restrict themselves to ringing or creating clearings by felling some trees. If fire does not consume the trunks and the larger branches, these are left lying on the ground, and cultivation is carried on among the debris. In other cases, however, the vegetation is thoroughly burnt off, and in some cases a second fire may be necessary. Differences in clearing technique appear to be fairly well explained by variations in ecological conditions. The simplest and least intensive techniques are found in the humid forest belt. The number of operations and amount of labour required are greater in the savanna areas (Miracle 1967, p. 159).

3.2.5. *Cropping systems*

Shifting cultivation is almost exclusively carried on as farming with annual and biennial crops. In humid climates we find above all rice economies or root-crop economies where the emphasis is on manioc, sweet potatoes, yams, and taro. Widespread, and extending into the drier climates of the savannas, are maize and millet economies. In between are numerous mixed economies, which are based on root crops as well as grain crops. There is ample evidence to show that shifting cultivators are not static or traditionally fixed in their cropping patterns (Miracle 1967; Baum 1968; Spencer 1966). Most of the crops or varieties grown by African shifting cultivators have been introduced in the last 300 years by traders, settlers, and administrators. According to Spencer (1966, p. 169), a change from cropping with root crops to grain cropping with rice can be observed in Asia. Where shifting cultivation is practised by sedentary farmers with permanent homesteads, complementary activities are usually to be found: the growing of fruit trees, a permanent garden, plots with irrigated rice, a fish-pond.

3.2.6. *Tool systems*

Shifting cultivation in the rainforest is still occasionally practised without cultivation implements; after burning off, seed is sown in the ashes. The axe and the matchet are the main tools. Where the ground is prepared, as is usually the case, tools for cultivation are required. Following the suggestion of Spencer (1966, p. 131), we can distinguish between digging-stick systems, hoe-systems, and plough-systems. Digging-stick systems are found in the humid rainforests. Hoe-systems predominate in the savannas, particularly where the cultivation period stretches over several years, since in these circumstances soil cultivation and hoeing of weeds are necessary. The plough-system can be employed only in shifting cultivation where an unused grass vegetation grows in the fallow periods.

3.3. General characteristics of shifting systems

The various shifting systems, from one continent to the next, are characterized by some common organizational features. The relative importance of these features change according to the local land conditions, but they are recognizable as tendencies in almost all cases. Of special interest in this respect are the spatial organization of cropping, the cropping principles, and the organization of the fertilizer economy, of animal husbandry, and of the labour economy.

3.3.1. *Spatial organization of cropping*

The farmer with a stationary home and land that is permanently cropped tries to create favourable growing conditions for his crops: i.e. he tries to control nature. Shifting cultivators, on the other hand, are usually highly skilled at adapting their cropping practices to the environment in which they are working. Four aspects of adaptation are important.

1. The choice of the plot of land to be cultivated. The main determining factors are soil fertility, the effort of clearing the vegetation, the amount of weed growth, the danger of vermin, and location in relation to the road and a source of water. On fertile soils, a smaller plot is more able to cover food needs than on poor soils. Fertile soils, however, are by no means preferred in all cases. Expenditure of labour per unit of production is a much more decisive factor. If, with the same effort, three times as much poor land as good land can be cleared and cultivated, even with only half the yield per unit, there is still a net gain (Allan 1967, p. 94).

2. The choice of crops.

3. The organization of cropping in mixed cropping, phased planting, and crop rotations.

4. The arrangement of short-, middle-, and long-term fallows.

The decisive factor in balancing these requirements is in most cases the very precise knowledge of soils and plants held by the shifting cultivators. Allan (1967, p. 5) writes on this subject, about central Africa:

The 'shifting' cultivator has an understanding of his environment suited to his needs. He can rate the fertility of a piece of land and its suitability for one or other of his crops by the vegetation which covers it and by the physical characteristics of the soil; and he can assess the 'staying power' of a soil, the number of seasons for which it can be cropped with satisfactory results, and the number of seasons for which it must be rested before such results can be obtained again. His indicator of initial fertility is the climax vegetation and his index of returning fertility is the succession of vegetational phases that follows cultivation. In many cases his knowledge is precise and remarkably complete. He has a vocabulary of hundreds of names of trees, grasses, and other plants and he identifies particular vegetation associations by specific terms. This fund of ecological knowledge is the basis of 'shifting' cultivation.

Conklin (1957) tells us the same about the rainforest in the Philippines, where the Hanunóo of Mindoro distinguish more than 1600 different plant types, which is a finer classification than that employed by systematic botanists, and includes the astounding number of 430 cultivated species. Conklin's vivid description of what a Hanunóo *swidden* in full swing looks like (pp. 18–19) gives an excellent picture of the degree to which this agriculture apes the generalized diversity of the jungle that it temporarily replaces:

> Hanunóo agriculture emphasises the intercropping of many types of domesticated plants. During the late rice-growing seasons, a cross-section view of a new (plot) illustrates the complexity of this type of swidden cropping (which contrasts remarkably with the type of field cropping more familiar to temperate zone farmers). At the sides and against the swidden fences there is found an association dominated by low-climbing or sprawling legumes (asparagus beans, sieva beans, hyacinth beans, string beans and cowpeas). As one goes out into the center of the swidden, one passes through an association dominated by ripening grain crops, but also including numerous maturing root crops, shrub legumes and tree crops. Poleclimbing yam vines, heart-shaped taro leaves, ground-hugging sweet potato vines, and shrublike manioc stems are the only visible signs of the large store of starch staples which is building up underground, while the grain crops fruit a metre or so above the swidden floor before giving way to the more widely spaced and less-rapidly maturing tree-crops.

Besides the systematic adaptation to variations of soil within a plot described by Conklin, we find in most shifting systems that cultivation is often different on those plots that are especially suited to individual crops. Miracle (1967, p. 13) describes this type of cultivation outside the rotation systems on the main plots:

> Cultivated fruits, vegetables, rice, yams, taro, condiments, and tobacco are outside the crop sequences and each of these are in a small environment especially favourable to them. Oil-seed, watermelons, tomatoes, mock tomatoes (*Solanum* spp., grown only for their leaves), roselle, egg plant, red pepper, mangoes, and papayas are grown in the courtyard where they are free from weeds and fertilized by ash and refuse. Bananas are grown on small mounds; and okra, yams, peanuts (as a vegetable), and various vegetables are grown on broken ridges in the courtyard. Rice, bananas, taro, sesame (as a vegetable), maize and others are grown on old refuse heaps. Moist, shady spots under the kitchen eaves are used as tobacco nurseries. Ash accumulations found outside the courtyard, especially those resulting from the burning of stumps during the clearing of a field, are used for tobacco and various vegetables. At the base of trees that have been killed, climbing varieties of cowpeas, lima beans, and oil-seed gourds plus bottle gourds and calabashes, are planted. Living trees are used to support yam. Uninhabited termite mounds are flattened and planted with white sorghum, rice, and cowpeas.

The attempt to combine cash cropping with a regular supply of foods for household use leads, in most shifting systems, to diversified production units. A shifting holding usually consists of a number of distinct components:

1. The dwelling-place consists of the huts and the surrounding yard. The

materials used in constructing the huts depend on the degree of permanency in housing. If the hut is to be used for less than about 6 years, then wood and grass are used. If it is to be used for more than 6 years, then clay is the main building material. Farmers who expect to live permanently in a given place usually endeavour to build stone houses.

2. Usually a garden plot near the dwellings is cultivated with vegetables, gourds, bananas, etc., as long as the family lives in one place. The longer the family lives in one place, the greater is the number of fruit trees grown near the yard.

3. The differentiation of production on the outlying fields ensures that each is cultivated with the crop best suited to it, partly by mixed cropping and rotations, and partly by having several plots on different soils (Guillot 1970a, pp. 47–96).

4. In addition to cropping, shifting cultivators gather 'wild' products from the surrounding fallow. These are usually not truly wild plants, but the remains of domestic plants within a 'tumbledown' fallow.

5. If the farmers live in permanent dwellings, or move only within a limited area, the shifting cultivation is generally supplemented by other major food-producing activities, including planting of tree crops, permanent gardens, permanent irrigation fields, and fish-ponds.

3.3.2. *Cropping principles*

The adaptability of shifting systems is closely related to three principles of cultivation which are linked in various ways. These are: mixed cropping, phased planting, and crop rotations, i.e. alternating crop mixtures which are regarded by de Schlippe (1956) as *pseudo-rotations*.

(a) *Mixed cropping.* Mixed cropping, which means the simultaneous growth of two or more intermingled useful plants on the same plot, is an important characteristic of most shifting systems. It usually involves the relay-planting technique, in which a maturing crop is interplanted with seeds or seedlings of the following crop. Mixed cropping can be found in a great number of farming systems but with differences in technique. In shifting systems crops are rarely grown in rows. The plot is usually uneven, and tree stumps, roots, termite moulds, and low areas with poor drainage would interfere with the rows. A worthwhile labour economy is not achieved, since the farmers rarely work with equipment drawn by animals. Row cultivation, moreover, would not allow the mixture of plants grown on each piece of land to be adapted to the changing soil conditions. In shifting cultivation, therefore, we find predominantly mixed cropping, which is not organized with the help of a formal principle, but is based on the best possible adaptation to the surface condition.

Mixed cropping is usual practice in shifting systems but also in a great number of more intensive farming systems and this for the following reasons (Norman 1973, p. 138), some of which have been verified in particular conditions and some of which are the result of hypotheses drawn from knowledge about tropical plant production:

1. *Physical–technical reasons*

(i) Mixed cropping allows a fuller use of light, nutrients, and water. Most crops have widely differing leaf canopies, and their combination permits a more efficient use of incident light and a favourable distribution of carbon dioxide (Suryatna and Harwood 1976). Nitrogen uptake is higher. The degree of root interpenetration seems to increase with the degree of crop dissimilarity as different crops use different soil layers. There is therefore less run-off and leaching. More transpiration occurs through plants than directly from the soil, and the time with leaf cover is extended.

(ii) Mixed cropping tends to reduce the adverse conditions in the ecosystem. Disease and insect damage may be less and is likely to be more evenly distributed. Suryatna and Harwood (1976) found a reduction of corn-borers in mixed stands. Also damage to one crop may be compensated to some degree by the more vigorous development of another one (Banta and Harwood 1973, p. 6).

(iii) Mixed cropping tends to produce less erosion than sole stands under the same circumstances, and so contains an element of resource maintenance (Norman 1973, p. 138). Where there is a vertical arrangement, the rain, for example, falls from the bananas onto the manioc and then onto the beans, and only then does it reach the soil. Mixed cropping has a protective effect similar to that of the original forest or bush vegetation.

2. *Socio-economic reasons*

(i) The above physical–technical reasons often reflect themselves in higher returns per hectare and per year; the following contributing elements can be distinguished: the growing season is more fully utilized, fuller use of the environment factors is made at any time, symbiotic effects occur (nitrogen fixed by groundnuts is used immediately by maize), and disease and insects may produce less damage. The increase in total output is probably the most important reason why farmers prefer mixed cropping (Norman 1973, p. 138).

(ii) Mixed cropping generally requires more labour per hectare than sole stands, and very intensive forms may run into steeply increasing labour inputs per unit of output. Intermediate forms, however, may require less labour per unit of output than sole stands because the dense vegetation tends to suppress weeds and relay-planting techniques save cultivation work. Particularly interesting is the fact, established by Norman (1973,

p. 138), that a more even distribution of labour over the season can be obtained.

(iii) A farmer may reduce risks by having several plots planted at different times, but under the condition of fallow systems mixed croppings is the traditional and perhaps the more effective way of reducing risks.

(iv) In smallholder farming, a varied food supply is usually sought. Over and above this, the householder relies on a continuous supply of fresh food, since there are insufficient storage facilities and the storage losses are high. Both aims naturally lead to phased planting and mixed cropping. Table 3.2 shows, with an example of the consumption of a shifting cultivator's family, the variety of products that a tropical smallholding produces, and the frequency of consumption, which should coincide where possible with the harvest, because of the limited storage facilities.

Mixed cropping is more difficult to mechanize, and innovations in crop production are usually crop-specific and more easily applied to sole stands. In the course of economic development farm systems tend to specialize and mixed cropping is increasingly replaced by sole stands. However, these conditions do not yet apply for most shifting and fallow systems. It should also be noted that mixed cropping may be changed into intercropping (two or more useful crops in proximate but different rows), and inter-cropping, associated with modern techniques, fits well into the condition of a developing economy with high rural population densities.

(b) *Phased planting.* Another technique which is generally applied by shifting cultivators, but which is to be found in other systems as well, is the phasing of planting, i.e. parts of the plot producing the same crop or crop mixture are planted in a time sequence stretching sometimes over months in order to distribute labour requirements and risks and to obtain a more even supply of food for the household (compare Fig. 4.5).

(c) *Rotations.* Shifting cultivation with short-term cycles is rarely connected with crop rotations. Usually one crop or a given mixture of crops is grown for several years. However, when shifting cultivation is carried on with cultivating periods lasting for several years, or even a decade, crop rotations are evolved—a logical continuation of the tendencies that can already be seen in phased cropping. De Schlippe (1956, p. 207) calls the change in the composition of mixed cropping pseudo-rotation, but as a rule it is a matter of carefully considered cropping sequences, which can justifiably be considered as genuine rotations. Amongst the great number of forms rotations take, several important features can be observed.

1. In the crop rotations of shifting cultivators, there is rarely a sequence of monocropping. Usually, various mixed crops follow one another, with often only a gradual change in the composition of the mixture. Immediately after clearing, crops that require more fertile soils are chiefly planted, like

TABLE 3.2

Consumption chart of a shifting cultivator's family in Manhaua, Mozambique

Product	1	2	3	4	5	6	7	8	9	10	11	12
Crops produced by the family												
(1) Staple foods containing starch												
Manioc	×	×							×	×	×	×
Maize in milk-ripeness			×	×								
Maize as ripe corn					×	×						
Rice						×	×					
Sweet potatoes							×	×	×			
Sorghum								×				
Sorghum-corn (ecununga)								×				
Sorghum-cane (maele)							×	×				
(2) Staple foods containing protein												
Beans (boer boer)									×	×		
Beans (jugo)					×							
Beans (manteiga)					×							
Green beans (boer boer)							×	×				
Green beans (nyemba)			×									
Green beans (jugo)				×	×							
Manioc leaves	×	×	×	×	×	×	×	×	×	×	×	×
Sweet potato leaves							×	×				
Bean leaves of all kinds		×	×	×	×	×						
(3) Additional foods and spices												
Onions											×	×
Tomatoes						×	×					
Gherkins			×									
Aubergine (2 kinds)										×	×	×
Quiabo (*Hibiscus esculentus*)		×										
Groundnuts						×	×	×	×	×	×	
Sugar-cane			×									
Pumpkins	×	×	×									
Sorghum-cane (ecununga)					×	×						
Collected food												
(4) Staple foods (animal protein)												
Grasshoppers	×	×	×						×	×		
Mice and other rodents						×	×	×				
Caterpillars				×	×							
Larvae of butterflies											×	×
(5) Additional foods												
Fungi (9 kinds)	×	×										×
Wild fruits (7 kinds)	×										×	×
Honey					×	×	×					

Source: Pössinger (1967, p. 199).

yams in the rainforest, and cotton, maize, groundnuts, and sesame in the savannas. As the cultivation period proceeds, crops that make little demand on soil fertility are preferred. Biennial or semi-perennial crops like manioc or bananas are frequently grown last of all, because these can survive longer than other crops against the competing weed and bush, on account of their height and the shade they create, before gradually disappearing in the tumbledown fallow.

2. Nearly every plot is different from the point of view of soil-type, degree of incline, and distance from the hut, and nearly every field is suited, therefore, to a different crop rotation. The various crop sequences that Richards (1961) found in the fields and gardens and on old hut sites of the Bemba demonstrates this point (Fig. 3.2).

3. It is rare for one rotation to be applied persistently on a big plot. Division of the plot into several smaller sections is more typical. Each small piece of land then supports a different sequence of crops or mixture of crops. This again is clearly demonstrated in Fig. 3.2.

4. In shifting systems, it is common for the same mixture of crop-types or the same crop to be grown on a suitable plot for several consecutive years. When the cultivation period increases, growing 'at intervals' some-times arises, a system typical of the African savannas, when, for instance, 2–3 years of grain crops follow 2–3 years of cotton.

'Much of the most effective human utilization of the natural habitat,' writes Geertz (1963, p. 17), 'consists of changing generalized communities into more specialized ones . . .'. This tendency also can be seen in the shifting cultivators' farming principles. As land use becomes increasingly intensified and commercial, organizational principles like monocropping and row cultivation are increasingly preferred to the various types of mixed cropping.

3.3.3. *Characteristics of the fertilizer economy*

(a) *The regeneration of soil fertility in long-fallow systems.* The various features of the organization of cropping that are applied by shifting culti-vators aim primarily at an effective use of the existing soil fertility. In humid and semi-humid climates, where vegetation tends to be ample, the main store-house of nutrients is not the soil but the standing vegetation.† The amount of nutrients stored in the vegetation is certainly less in semi-arid climates, but the general principle remains: forest and bush vegetations reduce leaching and store nutrients, which are made available to crops to a significant degree by fire clearance and the resulting temporary increase of the pH. The basic feature of shifting cultivation is therefore the cost-free, effortless regeneration of soil productivity during the fallow period, especially when the fallow consists of forest or bush vegetation. The damper and warmer the climate,

† The following percentages of the total store of cations in tropical forest ecosystems (soil and biomass) was found in the biomass: (1) mature secondary forest Ghana, 42 per cent; (2) lower montane forest at 2500 m altitude, New Guinea, 19 per cent; (3) Alluvial lowland forest, Brazil, 73 per cent. In all three cases almost all of the phosphorus was stored in the biomass (Greenland and Herrera 1975, Table 6).

and the poorer the soils, the more rapidly the organic substances break down and nutrients are leached, the more rapidly the crop yields tend to sink. The first crops are comparatively ample, because nutrients are mobilized and the ashes increase the pH, but the harvests in the subsequent years become more and more meagre. On unstable soils of the rainforest, for example, the third crop will usually yield only half, or less than that, of the first. An illustration

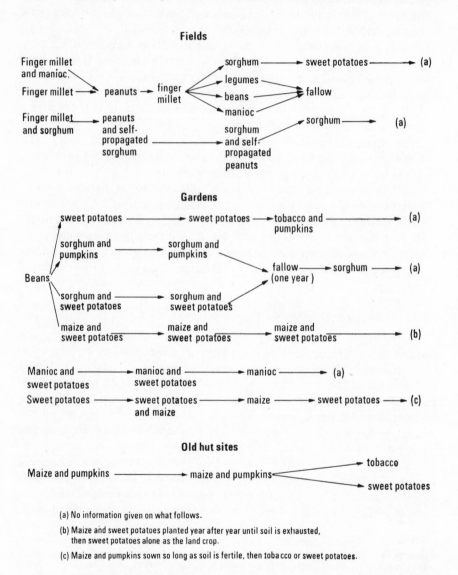

(a) No information given on what follows.

(b) Maize and sweet potatoes planted year after year until soil is exhausted, then sweet potatoes alone as the land crop.

(c) Maize and pumpkins sown so long as soil is fertile, then tobacco or sweet potatoes.

FIG. 3.2. Crop sequences of the Bemba, Zimbabwe (Richards 1961; Miracle 1967, p. 110).

of this is provided by Nye and Greenland (1961) (see Fig. 3.3). If further utilization of the plot no longer promises to be rewarding, the shifting culti-vator allows it to lie fallow for a fairly long period, in the course of which the humus and nutrient content of the soil tends to regenerate but not, however, the pH (Watters 1971, p. 40). Shifting cultivation allows soils poor in nutrients regularly to produce relatively high and certain yields in a climate in which arable farming is always a difficult struggle against nature. This is usually achieved without manure, terracing, or any other means of maintaining

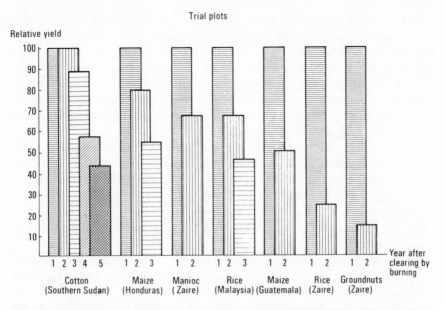

FIG. 3.3. Decline in yields under prolonged cropping without fertilizer in a humid tropical climate (data from Nye and Greenland 1961, pp. 73–74, drawing from Andreae 1977, p. 129).

fertility that involve a high expenditure of labour. If the fallow periods are long enough, this type of land-use system represents a balanced exploitation of the resources available for farming.

The regeneration of soil fertility in the upper soil layers is largely due to the effects of a tree or bush fallow. The deeper tree and bush roots ensure that nutrients are brought up from the subsoil. The fallen leaves enrich the surface, and the minerals that rain water has leached down are brought up again. The fallow vegetation covering the soil and the comparatively cool temperatures of the covered soil facilitate the regeneration of the soil. Non-symbiotic nitrogen fixation is substantial and may amount to 75 kg ha^{-1} of nitrogen per year in a secondary rainforest, 45 kg ha^{-1} of nitrogen in a highland forest, and 15 kg ha^{-1} of nitrogen in a savanna forest (Dart and Day 1975, p. 35).

Provided the fallow periods are long enough, a slash-and-burn system proves to be in no way harmful to the soil. In fact, the yields after a fallow period with secondary forest are in some cases higher than those after the first clearing. The smallness of the plots, the considerable shading of the soil during the cultivation period that arises from mixed cropping, the presence of tree stumps in the field, and other factors mean in addition that balanced shifting cultivation involves little risk of erosion damage. Another advantage of field shifting is the fact that losses through plant diseases remain comparatively slight.

(b) *Fertilizer application in shifting systems.* Since shifting cultivation is associated with a low degree of commercialization, the export of crops over and above the local sale of produce is not a major cause of the loss of nutrients. There is, however, a movement of nutrients from the field to the dwelling-place, and therefore some change in the location of soil fertility. Secondly, nutrients are lost during the years of cultivation because of leaching and erosion. Particularly important is the loss of the nutrients that are washed away together with the ashes.

The long-term fallow period is the most common method of regenerating the fertility of soil. In addition, however, there are various forms of fertilizing.

1. Utilization of hut refuse. The first step in the development of a fertilizer economy consists usually of applying household refuse. It is normally put where it can be useful to plant growth. Thus, for example, fertilization of bananas near a hut with household refuse is common among shifting cultivators.

2. Establishment of gardens on old hut sites. A general practice when the hut has been shifted is to use the old hut site in order to grow crops that demand fertile soil. The movement of the huts means, in fact, that heavily fertilized plots are systematically obtained. This applies particularly where cattle are reared and kept overnight in a kraal near the hut. This practice may be considered as the first step in the development of folding systems (see § 4.3.3).

3. Green manuring and use of compost. Where ridge or mound cultivation is practised, it is common to clear the vegetation and to allow it to dry before the soil is worked. It is then dug right into the centre of the ridge or the mound as a green manure. In the Porka system in Ghana, the cleared vegetation is thrown on heaps and later used as compost (Manshard 1968, p. 90).

4. Working with pit systems. In the pit systems, and the hole system in Tanzania, the fallow vegetation and weeds are collected in pits or in heaps. These are covered with earth, cultivated, and dug over several times. In these shifting systems, the cropping period lasts for up to 12 years, and is followed by a long-term fallow period.

5. Working with heated mounds. With a number of African tribes, the fallow vegetation is covered with earth and then set alight and charred, which results in cultivation taking place on little mounds of carbon. This is done in order to increase phosphate availability (Allan 1967, pp. 73–4; Guillot 1970b, p. 315; Miracle 1964, p. 84).

6. Ashes as fertilizer. Where fire-clearance methods are used, ash in any case serves as fertilizer. In some regions an ash concentration on small plots is achieved with considerable labour costs. In the chitemene system, as practised in some parts of Zambia and southern Tanzania, branches are transported from the forests in the vicinity to the farmland, piled up a metre thick, and burnt to increase the ash fertilizer.† Millet is sown in the ash. This method produces comparatively high and certain yields of 1.5t ha^{-1} on very poor soils. The main disadvantage is the size of the area from which the branches are taken. The forest requires 15–20 years to regenerate. For the annual cultivation of one hectare of finger millet, a nutrient supply of 30–100 ha of forest is required.

Richards (1961), reporting on the ash system practised by the Bemba, writes:

Cut branches are trimmed enough so that they will lie flat in a pile for burning, and are laid in rows, each man taking care to keep his branches separated from those of his neighbours. Women later carry the branches to the plot to be cultivated and place them radially, with the smallest diameter to the centre. At the time of Richards's study, branches were cut from an area up to six times as large as that to be planted. Pieces of wood are piled one on the other until the pile is about 65 centimetres high, care being taken to keep the thickness of the pile uniform. The Bemba are said to often single out barren patches of a field as spots where branches were not properly stacked before burning. The last task in piling branches is to outline the edge of the circle with small branches, in order to keep the fire from spreading into the bush, and to provide an extra thickness of ashes at the edge for growing gourds. [Taken from Miracle 1964, pp. 13.–18.]

Whether, or to what extent, the yields of these systems are to be attributed to the effect of fertilizer or to the yield-promoting effect of heat is an open question:

It may be that this is the mechanism through which the heat of the chitemene burn exercises its main effect, but there is as yet no direct proof. Whatever the explanation, it seems clear that the ash-planting peoples, with no livestock and only the simplest equipment, have devised an effective means of conferring a high if transient fertility on soils of very low intrinsic value. [Allan 1967, p. 74.]

7. Application of mineral fertilizer. The use of mineral fertilizer in shifting systems is rare, limited to those with a higher degree of cash cropping. Mineral fertilizer has been adopted especially where tobacco and cotton

† The ratio of cleared land to that of field acreage varies from 10:1 to 20:1 in Small Circle Chitemene to 5:1 and 10:1 in Large Circle Chitemene, and even less in Block Chitemene (Schultz 1974, p. 68; Allan 1967, p. 86). The number of crop years tends to be the shorter the smaller the ratio.

are grown. The majority of shifting systems do not involve the importation of purchased nutrients. They depend on the nutrients present in the soil.

3.3.4. *Characteristics of animal husbandry*

In the rainforests, the cultivators have few domestic animals. Mostly there are a few goats, sheep, or chickens. In south-east Asia and West Africa, pig-keeping is often practised. As a supplement we find fishing and collecting caterpillars and other minor sources of animal food. In the semi-humid African savanna, in which there is often plenty of grass for fodder, the tsetse fly has hitherto prevented cattle-rearing, apart from in limited areas in West Africa and Zaire where tsetse-tolerant cattle breeds are kept. The situation is different in Latin America, particularly in Brazil, where shifting cultivation and a large herd of cattle are often combined, and the cattle eat the grass on the cleared open spaces. In the drier savannas, animal-rearing among the shifting cultivators is usually on a large scale, and is closely connected with arable cropping. Under these conditions slash-and-burn agriculture may convert bush-covered land into grazing. The aims and characteristics of cattle-rearing, as they are connected with slash-and-burn agriculture, are discussed in more detail in Chapters 4 and 9.

3.3.5. *Characteristics of the labour economy*

The operations carried out by shifting cultivators can conveniently be considered separately as follows:

(a) *Clearance of wild vegetation.* From the standpoint of the labour economy, clearance is the most strenuous activity. The area that can be cleared by a household without a motor saw is often less than cultivators would wish (Thornton 1973, p. 70). Clearing is frequently, but by no means in all cases, carried out within a system of fire-farming. Labour input into clearance varies widely (between 200 and 1400 h ha^{-1}) depending on the vegetation and the thoroughness of the work. In Latin America, where saw felling is common, labour inputs are significantly less than in Africa (Watters 1971, p. 51). Table 3.3 gives some examples of the labour requirements for fire clearance for upland rice in West Africa which are between 200 and 400 h ha^{-1}.

(b) *Land preparation and planting.* Immediately after burning, planting can often be carried out without cultivating the soil. However, most shifting cultivators do in fact disturb, loosen, and move small amounts of surface soil as part of the operation of planting (Spencer 1966, p. 31). In the savanna, or on grassland, or where cultivation lasts for several years, the soil has to be thoroughly loosened, because it is usually much harder, but on the whole the shifting cultivator needs to prepare the soil comparatively little. He replaces, as it were, cultivation with the hoe by clearance work with the axe and matchet. Thorough deep-soil preparation with the hoe is less characteristic of shifting cultivators than of cultivators in fallow systems.

(c) *Weeding.* The fallow suppresses the growth of weeds, and after fire clearance the soil is often weed free. Many shifting systems manage, therefore, with little weeding. Only as the period of cultivation increases does the effort spent on hoeing and weeding increase. Scaring away birds may, however,

TABLE 3.3

Some examples of input–output relations with upland rice in African shifting systems (per hectare of first crop after clearance)

Country	Sierra Leone	Liberia	Ivory Coast	Cameroun
Location	Bum	Gbanga	Man	Begang
Year	1971–2	1972	1974–5	1976–7
Technique	Traditional	Traditional	Traditional	Traditional
Method	Sample	Sample	Sample	Sample
Technical data				
Labour input (man-hours)[a]				
Slashing and felling	135	254⎫	301	115
Burning and clearing	79	164⎭		184
Planting[b]	127	107	142	181
Weeding	145	37	292	416
Bird-scaring[c]	256	44	222	294
Harvesting	268 ⎫	164	218	277
Threshing	59[d] ⎭		84	224
Total	1069	770	1259	1691
Yield (t/paddy)	1·23	0·97	1·74	0·77
Economic analysis ($)				
Gross return	112	122	203[e]	152
Material inputs				
Seeds	4	4	8	8
Tools	3	3	4	1
Income before interest and overheads	105	115	191	143
Income per hour of work	0·10	0·15	0·15	0·08

The definition of terms and comments on the presentation of the tables are given in Note (4) of the Appendix.

[a] 6 hours per work day. [b] Including minor land preparation works. The land is not cultivated before planting. [c] Including hut building, fencing, etc. [d] Estimate from other surveys. [e] 1973–4 prices. *Sources:* Sierra Leone: Spencer (1975); Liberia: Van Santen (1973); Ivory Coast: Lang (1976); Cameroun: Fotzo (1977).

demand much time and may compete as an activity for children with going to school.

(d) *Harvest, transport of harvest, and processing.* Together with clearing, harvesting is the most labour-demanding task. This applies particularly to root economies, in which the transportation of harvested produce—there are usually no roads or carts—and the processing of the crop tend to be time-consuming, as, for example, with manioc.

In addition, various other activities have to be considered. Tending livestock may be very time-consuming where shifting farming is practised by people with a pastoralist background, and where men spend most of their time with their cattle. Getting water often takes much of the wives' time. Visiting local markets is also very time-consuming. Work in the fields usually absorbs not more than half the working time. It is important in this connection to realize that the seasonality of labour demand in shifting systems is usually less pronounced than in more intensive arable systems. Clearing, harvesting, and processing are the major time-consuming activities, and there is more latitude for doing these tasks than there is with cultivation or weeding.

Table 3.4 contains two examples of the sequence of operations in shifting systems. The first year involves the clearance work. In subsequent years labour input is less, because manioc has been established and covers the land. In years 2–4 most of the work is harvesting. Material inputs are negligible. Most remarkable is the efficiency of the commercialized production of manioc flour in Brazil. Human labour is the only support energy invested, and the relationship of energy input to food energy output is one of the most favourable to be found in agriculture (§ 3.4.6).

In the traditional forms of shifting cultivation, the execution of the labour operations is usually governed by the customs of labour division. A number of main types of labour division can be recognized:

(a) *Division of labour with respect to different jobs.* A clear division of jobs between men and women is usually found wherever forest and bush vegetation must be cleared. In fire-farming in the forest, clearing is the man's job, while planting, weeding—if done at all—and harvesting are carried out by the women.

(b) *Division of labour with respect to plots.* Short-term cycles of cropping and fallowing give rise to a fallow vegetation consisting of grass and bush. The input for clearing is comparatively low. Consequently, work is divided between men and women in terms of areas. Both men and women tend to cultivate their own plots. Men do most of the heavy work (clearing) and the women perform the tasks requiring more dexterity (weeding, harvesting). Fig. 3.4 is taken from de Schlippe's book (1956) about the Zande, who live on the borders of the Sudan and Zaire. It provides an example of the distribution of plots to the members of the family.

(c) *Division of labour with respect to specific crops.* When larger plots are planted with cash crops, then the women are usually responsible for cultivating the food crops, whereas the men farm the plots with cash crops. Housework and collecting firewood are almost always women's and children's tasks, while the men are responsible for cutting the wood used for building and for erecting the houses.

Although the implements of shifting cultivators as a rule are limited to hand-tools unassisted by domestic animals, the productivity of labour is by

TABLE 3.4

The economics of manioc–maize production in two shifting systems and of maize–sesame–manioc production in a fallow system (per hectare)

	Ghana	Brazil	Kenya
Country	Ghana	Brazil	Kenya
Location	Begoro, Eastern Region	Castanaha, Para	Kilifi District
Rainfall (mm)	2000	2800	750
Year	1972	1977	1977
System	Shifting	Shifting	Fallow
Technique of cultivation	Hoe	Hoe	Hoe
Method	Interview (cases)	Interview (cases)	Interview (cases)
Labour input (man-hours)			
1st Year			
Land clearance	665	136	489
Planting	207[a, b]	124[a, b, c]	148[a]
Weeding	277	296	311
Interplanting sesame	—	—	10
Harvesting maize (including shelling)	280	80	221
Harvesting rice or sesame	—	160[c]	44[d]
2nd Year			
Land preparation	—	—	296
Planting maize	—	—	148
Weeding	277[b]	—	311[a]
Interplanting manioc	—	—	40
Harvesting	933[b]	672[e]	221[a]
3rd Year			
Weeding manioc	277	—	210
Harvesting manioc	605	—	750
4th Year			
Weeding manioc	138	—	—
Harvesting manioc	121	—	—
Total labour	3780	1468	3199
Yields (t)			
1st Year maize	1·481	0·60	0·650
1st Year rice or sesame	—	0·32	0·100
2nd Year maize	—	—	0·550
2nd Year manioc	7·407	10·00	—
3rd Year manioc	4·815	—	5·600
4th Year manioc	0·925	—	—
Gross return ($)			
Maize	110·00	52·00	146·00
Rice or sesame	—	50·00	35·00
Manioc	426·00	519·00	214·00
Total	536·00	621·00	395·00
Material inputs			
Seed	8·00	20·00	6·00
Tools	9·00	28·00	5·00
Income before interest and overheads	519·00	573·00	384·00
Income per hour of work	0·14	0·39	0·12
Number of fallow years after cropping	6	8	3
Length of rotation cycle	10 years	10 years	6 years
Gross return per ha and year ($)[f]	54·00	62·00	66·00
Income per hectare and year ($)[f]	52·00	57·00	64·00

[a] Maize. [b] Manioc. [c] Rice. [d] Sesame. [e] Including the processing of tubers into flour. [f] Average per year of total hectarage. *Sources:* Ghana: Rourke (1974); Kenya: Thorwart and Reeves (1978, personal communications); Brazil: personal information.

no means lower than in more intensive traditional land-use systems. However, the number of hours of agricultural work per available family labourer is usually low. Most of the labour capacity remains unutilized, though this is obviously not because of lack of employment possibilities. The more obvious explanation is that there is neither the need nor the incentive to invest more labour.

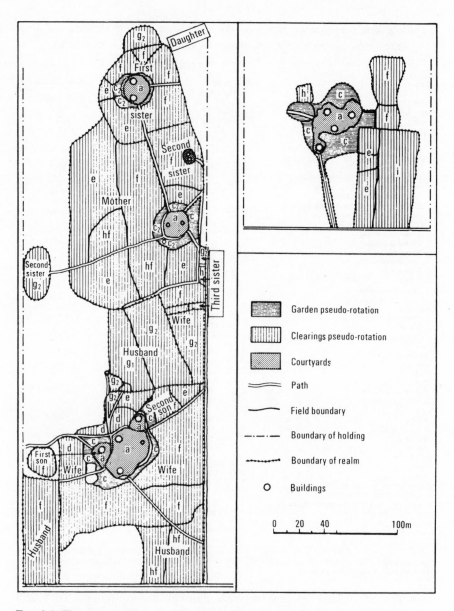

Fɪɢ. 3.4. The responsibilities of different members of a household for the individual plots in a Zande holding (from de Schlippe 1956, p. 109).

(a) courtyard; (c) maize–sweet potatoes; (d) maize–oil seed–gourds; (e) groundnuts–*Eleusine*, (f) main *Eleusine*-grass; (g) *Eleusine*-grass; (g₁) sesame-grass; (g₂) sesame-grass on fire-preserved area; (h) bean patch; (hf) beans–*Eleusine* or sorghum; (i) cotton field; (j) manioc fallow.

3.4. Examples

3.4.1. *Commercialized shifting cultivation: rice holdings in the Magdalena Valley, Colombia* (see Table 3.5)

Shifting cultivators in the Magdalena Valley, Colombia, are of the settler type (Colonos). They move into virgin forest on ferralitic soils with a very low pH (3.8–4.8), clear some land and broadcast rice. They sell poles for fencing and cleared land to large-scale ranchers. The farm consists of the home plot with some maize and manioc and two types of rice production. After forest clearance rice yields about 0·9 t ha^{-1}. Thereafter the land turns into a bush fallow which is cleared again after 3–5 years and planted with a second crop of rice yielding about 0·5 t ha^{-1}. This time rice and grasses are seeded together. After the harvest the land is sold to cattle owners for a price of about $62 per hectare and the Colono moves further into the virgin forest. Farmers do not aim for a permanent land use. They practise an effective approach to turn forested land into pastures. The pastures tend, however, to move rapidly to a level of low productivity. The carrying capacity of the pasture declines from 2 ha per livestock unit after clearance to 3–4 ha in later years.

3.4.2. *Shifting cultivation combined with lowland rice: holdings in Sierra Leone and Liberia* (see Table 3.5)

The data from Sierra Leone depict the classical traditional shifting cultivation pattern with subsistence-oriented upland rice only, while rice holdings in Bong County, central Liberia, combine permanent hut sites with shifting cultivation, and the result is a system with three fairly distinct parts (van Santen 1974):

1. The average holding comprises about 1.6 ha of upland rice. The rice is broadcast by women into the ashes, and is slightly covered with soil by scratching. There is no weeding. Clearance performed by men, and sowing and harvesting, mainly carried out by women, are the major operations. Most of the plots are at a 30–90 min distance from the house, and field huts are built on the more remote sites. Most of the rice is followed by bush fallowing to re-establish soil fertility. Between one and three crop years follow 4–20 fallow years. On average about 9 ha of fallow support one hectare of cropping, i.e. the R value is 10.

2. Upland cropping is supplemented by lowland cultivation. Fire clearance of the grassy vegetation is followed by broadcasting pre-germinated seeds. Wet plots are planted. Most of the rice is followed by sugar-cane, mainly to produce rum, the major cash activity. Fallowing is less frequent, and the main reason is weed control.

3. Each farm has a permanent home plot of about 0.4 ha, with plantains, bananas, 4–5 different fruit trees, some root crops, and vegetables.

The Liberian holdings already indicate the development path this farming

TABLE 3.5

Farm management data of shifting cultivation with upland rice in Colombia, Sierra Leone, Liberia, and Ivory Coast

Country	Colombia	Sierra Leone	Liberia	Ivory Coast
Location	Middle Magdalena Valley	Bum	Gbanga, Bong	Man
Rainfall (mm)	3000	2500	1900	1750
Year	1970	1971–2	1972	1974–5
Method	Sample of 210[a]	Sample of 22	Sample of 20	Sample of 30
Persons per household	7·0	8·4	7·2	6·9
Labour force (ME)	2·0	5·2	2·7	3·2
Land claimed (ha)	120·00	25·00	16·00	12·20
Upland rice	6·00	1·74	1·34	1·74
Swamp rice	—	—	0·40	—
Other food crops	—	0·16	0·68	0·40
Sugar-cane	—	—	0·12[b]	—
Coffee and cacao	—	—	0·20	3·30
Total crop area	6·00	1·90	2·74	5·44
Cropping index (per cent)	5	8	17	24[c]
Yields (t ha^{-1})				
Upland rice	0·80[d]	1·23	1·12	1·74
Coffee	—	—	0·20	0·22
Economic return ($ per holding)				
Gross return				
from crops[e]	516	220	480	1580
from sale of timber and cleared land	331	—	—	—
Total	847	220	480	1580
Purchased inputs[f]	183	15	12	62
Income[g]	664	205	468	1518
Productivity				
Gross return ($ ha^{-1} total land)	7	9	30	130
Gross return ($ ha^{-1} crop land)	141	116	175	290
Gross return ($ ME^{-1})	423	42	178	494
Income ($ ha^{-1} crop land)	111	108	171	279
Income ($ ME^{-1})	332	39	173	474
Labour input[h] (man-hours)	1980	1919	1975	2340
Income per man-hour ($)	0·34	0·11	0·24	0·65
Man-hours per family ME	990	369	731	731
MJ per man-hour[i]	17	14	17	n.a.

[a] Information from project personnel. [b] Production of rum for sale. [c] Arable land only. [d] 0·9 t ha^{-1} on newly cleared land and 0·5 on second clearing, see text. [e] Paddy prices: $109·80 in Colombia and Liberia, $94 in Sierra Leone, and $254 in the Ivory Coast. [f] Including farm-produced seeds. [g] No wage labour is employed in these systems, and income is almost identical with farm family income, before income from non-farm work, taxes, and interest. See note on methodology in the Appendix. [h] Total agricultural work at 5 hours per day in Liberia, 6 hours per day in Sierra Leone and the Ivory Coast. [i] Mega Joule. *Sources:* Colombia: Arnhold and Lindemann (1974, p. 38); Sierra Leone: Spencer (1975); Liberia: van Santen (1974); Ivory Coast: Lang (1976).

system is likely to take over time: with increasing soil-fertility problems in arable upland farming, more emphasis will be given to valley-bottom farming and the planting of tree crops.

3.4.3. *Shifting cultivation with upland rice of sedentary coffee farmers in the Ivory Coast* (see Table 3.5)

Shifting cultivators around Man, Ivory Coast, tend to plant coffee into their rice clearings. A relatively large area of 3.3 ha per holding is occupied by low-input robusta coffee mixed with shrubs and some plantains. Much of the upland rice is interplanted with maize and manioc. After the rice crop which comes first after clearance, groundnuts, yams, and sweet potatoes are grown in small plots. Coffee planting led to permanent homesteads, and land use is in the process of change from 'shifting' pattern to a fallow system.

3.4.4. *Shifting cultivation in the semi-humid tropics: manioc–cotton holdings in the Central African Republic* (see Table 3.6)

Shifting cultivation in the humid tropics is mainly practised with rice and root crops. In semi-humid areas cropping is often more diversified. In the Ouham District of the Central African Republic land use is changing from shifting—which still predominates—to fallow farming. Cotton, the cash crop, is planted first after clearance, and is followed in the second year by an association of sorghum, groundnuts, and several legumes and grain legumes. Most of the second-year plot is interplanted with manioc, which is the third year's crop, and at the same time some kind of tumbledown fallow. On average 3 crop years are followed by 6.5 years of bush fallow.

Land clearance is done by men, often in groups. Subsequent crop production is strictly individual. Both men and women work with cotton, while food crops are the domain of women. Tools are the only purchased inputs. Yields are low, but labour productivities (returns per hour of work) are higher than is usual in more intensive systems. The employment content of the farm is low, and much time is devoted to hunting, collecting, and fishing.

At the time of the survey, prices in the area were such that the use of fertilizer did not pay. The expansion path of the system is indicated by the introduction of trypano-tolerant cattle (Baoulé) and the substitution of hoe- by ox-plough cultivation.

3.4.5. *Shifting cultivation in the semi-arid tropics: maize holdings at Malindi, Kenya* (see Table 3.6)

Shifting cultivation in the Magarini area occurs under decidedly marginal conditions. Rainfall is low and highly erratic. Only four humid months per year can be expected, and rainfall distribution is bimodal. The short rains are insufficient for a crop and during the long rains even short-term crops suffer moisture stress in three years out of four.

Farmers adapt by practising an extensive type of shifting cultivation. The bush fallow vegetation is cut and burnt, and the area is cropped for 2–3 years, almost entirely with maize, intercropped with sesame, green gram, and cowpeas. The crop area per family is rather large and varies between 2·5 and 5 ha. This is related to a low maize yield of 400–600 kg ha^{-1} and a relatively low labour input of 400–600 hours. Clearance, weeding, and harvesting are the major activities. The soil is normally not cultivated. Phased planting, low plant densities (15 000 plants in 5000 plant holes per hectare), interplanting, and the manioc plot (as food reserve in case of crop failure) are typical adaptions of husbandry practices to a highly risky setting. The area is infested by tsetse flies and cannot be grazed by cattle. Farmers keep some goats which show tsetse tolerance.

The natural conditions of production at Magarini are such that a change in system is very difficult. Fertilizers are hardly economical. Timely planting could increase yields significantly but requires traction power, oxen or tractors. Ox-power implies tsetse eradication by complete clearance of larger blocks of land. Tractors are too expensive in relation to expected yields. Ploughing would imply the complete clearance of the bush-fallow vegetation, and this would lead to even more depleted soils and lower yields. Substituting the bush fallow by a ley system requires the establishment of intensive animal husbandry, and again tsetse eradication. It is doubtful, therefore, whether change towards a more intensive system is economically viable, and, if so, it would probably not be made successful through a smallholder system.

3.4.6. *Commercialized shifting cultivation: manioc–flour producers at Castanaha, Para State, Brazil* (see Table 3.6)

Shifting cultivation in Para State occurs under the conditions of ample rainfall and ample land, but fragile soils. Rice, maize, and manioc are planted after the clearance. Rice and maize are minor products. The major objective is to obtain manioc which is processed on the farm into manioc flour, a marketable product. 8–9 years of fallow follow 1–1½ years of crops. Harvesting and processing occur continuously throughout the year. The farmer works with family and hired labour. Production is fully commercialized and more than a century of routine in the process has led to high labour productivities. Production occurs almost without purchased inputs. In terms of the efficiency in the use of support energies, this is one of the most productive systems of the tropics with an energy ratio of 65 units of food energy, ready for consumption as 'farinha' manioc flour, for each MJ energy input, in the form of human labour (compare Leach 1976, p. 375). Returns per hectare of land, however, remain low because of the lengthy fallow period.

3.5. Problems of the system

3.5.1. *Difficulties in the introduction of innovations*

The shifting cultivator has been, and still is, a primary agent in the development of productive cultural landscapes out of forested regions. He begins the process of transforming the forested landscape, and he is followed by other settlers who continue the process (Spencer 1966, p. 5).

In temperate and subtropical climates, in the tropical high altitudes, and in the drier savannas of the tropics, the gradual intensification that creates an increase in the value of R has led to a gradual transition from shifting

TABLE 3.6

Farm management data of shifting cultivation in the Central African Republic, Kenya and Brazil

Country	Central African Republic	Brazil	Kenya
Location	Ouham	Castanaha, Para	Malindi
Rainfall (mm)	1500	2800	850
Year	1969	1977	1978
Type of farming	Manioc–cotton	Manioc–maize–rice	Maize–sesame
Method	Survey	Case study	Model
Number of holdings	n.a.	1	1
Persons per household	6·0	n.a.	10·0
Labour force (ME)	2·7	3·5	4·0
Size of holding (ha)[a]	8·0	50·0	13·00
Cotton	1·1	—	—
Food crops	1·4	4·50	4·00
Total crop area	2·5	4·50	4·00
Fallow	5·5	45·50	9·00
Cropping index (per cent)	31	9	31
Livestock (number)			
Goats	2	—	15
Chicken	3	—	5
Mule	—	1	—

Output of holding	tonnes	dollars	tonnes	dollars	tonnes	dollars
Cotton	0·36	56	—	—	—	—
Manioc	2·80	52	45·00	2334	0·30	10
Maize and sorghum	0·49	48	2·70	281	1·48	183
Groundnuts	0·19	16	—	—	—	—
Sesame and rice[b]	0·04	10	1·44	225	0·17	65
Grain legumes	0·10	9	—	—	0·12	38
Livestock	—	16	—	—		34

Economic return ($ per holding)			
Gross return	207	2840	330
Purchased inputs	2	125	11
Income	205	2715	319
Wages and taxes[c]	9	938	n.a.
Family farm income	196	1777	319
Sales as percentage of gross return	37	90	28

Productivity			
Gross return ($ ha^{-1} total land)	26	57	25
Gross return ($ ha^{-1} crop land)	83	631	83
Gross return ($ ME^{-1})	77	811	83
Income ($ ha^{-1} crop land)	82	603	80
Income ($ ME^{-1})	76	776	80
Labour input (man-hours)[d]	1122	7267	4154
Income per man-hour ($)	0·18	0·37	0·08
Man-hours per family (ME)	415	n.a.	1038
MJ per man-hour	n.a.	48	8

[a] Estimate of land claimed. [b] Rice in the Brazilian case only. [c] Taxes in the case from Ouham and wages in the Brazilian case. [d] Based on 6 hours of work per day. *Sources:* Central African Republic: Seminar für Landwirtschaftliche Entwicklung (1977/78); Kenya: Reeves (1978, personal communication); Brazil: personal information.

cultivation to fallow systems and finally to permanent cultivation. The warmer and damper, in other words the more tropical, the climate, the greater the difficulties that this development encounters. The weakness of shifting cultivation lies in the fact that the productivity of labour and soil can hardly be increased within this system. Most shifting systems are not in a position to absorb a growing population, to allow a steady broadening of cash production, or to derive continual and accumulating benefit from yield-increasing and labour-saving technical innovations. The problem obviously does not lie in any antipathy to innovation on the part of the shifting culti-vators. Past experience shows that they are willing to include new crops and varieties in their agricultural programmes and do so very quickly.

The Portuguese and Arabs, as well as the colonial governments of the nineteenth and twentieth centuries in Africa, can take the credit for intro-ducing a series of important crops into the indigenous farm economies. The introduction of cash crops like cotton, manioc, tobacco, and sesame leads to increased income. Sales from shifting cultivators' holdings are already of considerable importance for the supply of home markets, and in some cases also for export. The yields per hectare can be raised by denser planting, timely weeding, and plant protection. By relieving families of the arduous work required for such tasks as shelling groundnuts and peeling and drying manioc, local centres for processing release labour for work in the fields. The same applies to the introduction of the bicycle, which makes the transport of the harvest considerably easier and quicker. But all measures that increase the cropped area or the yield per hectare shorten the fallow period or make the regrowth of a fallow vegetation more difficult; and the fertility of the soil is closely related to its content of nutrients and organic matter, and thus to the length of the fallow period and the vigour of the fallow vegetation.

In the forest regions and in the tsetse-infested savannas, there is a lack of livestock to provide animal manure. Mineral fertilizer is usually not worth-while, because of unfavourable cost–return ratios, which can hardly be avoided in shifting systems. Shifting cultivation is limited to the hoe; ploughs cannot be used unless roots and tree stumps are cleared, and that would prevent the growth of the desired fallow vegetation. In shifting cultivation proper tracks are almost unknown; there are only footpaths. The frequent changes of field would necessitate an exceptionally large network of tracks. The absence of tracks, however, prevents the introduction of carts. The harvest must for the most part be carried or transported by bicycle, and those people who have animal manure have to carry it to the field in baskets.

The shifting of fields leads to the change of hut sites, and much time is used in building huts. Those people whose fields and huts are moved at intervals have little inclination to invest work in permanent improvement of the land, and a countryside with shifting cultivation is therefore distinguishable by the lack of permanent improvements, such as irrigation, drainage, and tree plantations. In general, the family units are scattered. The lack of stationary housing and the long distances between fields create difficulties in storing food, collecting surplus produce, and division of labour between farming and non-agricultural pursuits. Shifting cultivation is thus a hindrance to the development of villages and towns, central political bodies, and advanced societies. The land-use system moulds the economic attitude of the farmers who practise it. Shifting cultivators are in the habit of regarding the soil as a free gift of nature, not as capital that has to be maintained. Where soil fertility decreases and land is in the process of being eroded, shifting cultivators are much more inclined to mine the land and to look for virgin territory than to invest in labour soil conservation.

3.5.2. *The tendency towards 'degraded' shifting systems*

With increasing population and growing production for sale, a number of distinct stages of development can be distinguished.

1. The oldest form is probably the shifting of cultivation and of the farming group in one direction within a primary vegetation.

2. According to the growing number of other claimants to land, the groups share the total area among themselves, and then each family shifts in a circle in the area allotted, which is covered by secondary vegetation, and moves on after one or several decades. Relatively intensive cultivation is temporarily practised near the dwelling ($R = 30$–50).

3. With a further increase in population, instead of circular shifting with long-term intervals of fallow and cropping there often occurs the shifting of fields within a fairly limited area, with the tendency to stationary housing, where possible along a road. On the permanent site the hut position is sometimes changed.

4. A continuous increase in population and the development of cash cropping involve an intensification of shifting cultivation by shortening the fallow (R rises to 30–60). At this stage, the basis of land utilization has changed to such an extent that the system can no longer be regarded as shifting cultivation, but is a fallow system. Such systems are considered in the next chapter.

These processes usually lead to a situation of relative overpopulation on the land, which then loses soil fertility. Without changes in technology and inputs, only a small number of people can be supported compared with other land-use systems.

Naturally the crop yields of shifting systems vary greatly according to climate and soil-type. For example, whereas on the acid soils of the Amazonas Basin 10 or more fallow years follow $1\frac{1}{2}$ years of cultivation, with the same rainfall on volcanic soils in Zaire there are only 2- or 3-year fallows for the same period of cultivation. In any case, however, the shortening of the fallow or the lengthening of the cultivation period beyond a certain point disturbs the equilibrium of the land-use system. Shifting cultivation, which can be called 'balanced exploitation' where the fallow period is sufficiently long, becomes soil mining. Geertz (1963, p. 26) argues that in the rainforests of south-east Asia, in an ecologically-balanced shifting economy, no more than 20–50 people can live per square kilometre with guaranteed subsistence.†

Fig. 3.5 shows these developments in the form of a graph. The yield of the soil drops with the number of cultivation years and recovers in the fallow period. In (a) the fallow period lasts longer than the regeneration of the soil requires. R is low (11). This is shifting cultivation with production reserves. The situation in (b) corresponds to that of shifting cultivation without production reserves ($R = 29$). The fallow period is, however, long enough to restore soil fertility to

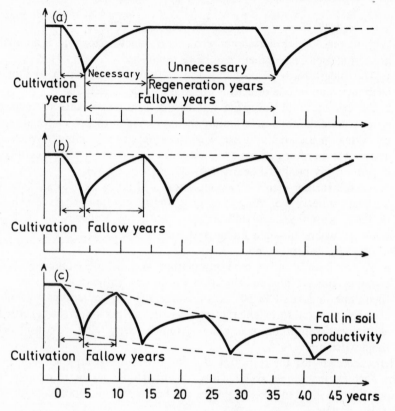

Fig. 3.5. The relation between length of fallow and soil productivity in shifting cultivation (from Guillemin 1956).

its original level. (c) shows what happens as the fallow is shortened ($R = 46$). The fallow is no longer sufficient to restore soil productivity and the yields per hectare fall, though since the shortening of the fallow period means that a greater part of the total area is cultivated, the fall in yields per hectare may well be accompanied by a rising total production. However, the result is a continuous degrading process.

† Assuming an even use of land, 3 crop years and 15 fallow years, 0·3 ha of crops per person, not more than 56 persons can be supported per square kilometre.

The increase in cultivation involves a more thorough clearing, more intense hoeing, and, at the same time, the death of roots and tree stumps. It means more erosion and more leaching; soil structure deteriorates, and soils become more compact. After the years of cultivation, the fallow ground does not produce bush quickly enough, especially on soils with low fertility, so that only arable weeds, grass, bamboo, or wild bananas establish themselves. Without ample fertilizer no herbaceous cover has the same ability to regenerate fertility as a woody one. Trees and bushes act as 'nutrient pumps' (Obi and Tuley 1973, p. 107), but bush regeneration is made difficult by more intensive cropping. Where cattle are kept, grazing is an additional impediment to forest or bush regeneration. Spencer (1966, p. 127) reports that most shifting cultivator groups are conscious of the threat of the spread of grassland as shifting systems are intensified. Nevertheless, almost everywhere in areas with shifting cultivation, the burning of vast bush and forest areas takes place in the dry season, partly from habit, partly to make hunting easier, and partly to get early fresh grass for the cattle before the beginning of the next rainy season. Regular grass fires also help to prevent the growth of forest and bush.

The interaction between grass and fire transforms the countryside of unbalanced shifting systems from a natural forest or bush vegetation into a derived savanna.† In large areas of Africa and Latin America this process is deliberately encouraged because grazing for the cattle is desired. The result is a decline in biological activity. Grass fallows may be assumed to produce a third or less dry matter than a bush or tree fallow. Less nutrients are stored in the biomass, and there is more leaching. A balanced and complex forest system is replaced by a less complex and unbalanced grass system.

The spread of more productive systems of land use is, as a rule, the result of necessity and incentives. When shifting systems are modified, the incentives come from cash cropping, and the main pressure tends to be the relative shortage of land. Increasing cash production, as well as the increasing need for subsistence crops, and both together in interaction with the introduction of technical advances, lead to efforts to regulate and improve shifting systems or to replace them by more efficient patterns of land use.

3.6. Development paths of shifting systems

Shifting cultivation is obviously not essential in the tropics, but is rather an expression of a certain stage in population density, technology, and price relations. However, in circumstances of relatively mild population

† An example: Aerial photographs of the Jagua–Bao region of the Dominican Republic show that from 1948 to 1966 the forested area declined from 227 to 147 km², while the area under pasture increased from 98 to 163 km². Shifting cultivation has been the main agent in turning forested land into grassland (Antonini, Ewel, and Tupper 1975, p. 67 ff.).

pressure, shifting cultivation has been widely suited, ecologically and socio-economically, to the simple needs of man and the potentialities of the environment. Its replacement by some more satisfactory land-use system is therefore by no means easy in the light of our present knowledge.

3.6.1. *Improved shifting systems*

(a) *Shifting systems with planted fallow vegetation.* The replacement of a natural fallow vegetation by planted shrub legumes has proved a technical success in a number of cases. Nitrogen fixation is high, fallows can be shortened, and the soils that were under cover yield highly, but the practice has rarely been adopted by farmers. Exceptional cases are the planting of Leucaena as a fallow crop on sloping land in Cebu, the Philippines, and of pigeon peas in Kwale District, Kenya. The main reason for non-adoption probably is that the establishment of a planted fallow is expensive in terms of labour, and the additional benefits obtained from the shrub legumes are low.

(b) *Shifting systems in forestry reserves (Taungya system).* Forest developers in several areas (India, Burma, Indonesia, Bangladesh, East Africa) success-fully combine shifting cultivation with planned reforestation. They allow and control the cultivation of food crops in the forest reserves by a limited number of shifting cultivators. The cultivators are allowed to cut down, burn, and farm a certain area of unimproved forest allotted to them, on condition that the land is vacated after 2 or 3 years of cropping and that the forest trees (mahogany, teak, pines, etc.) which are planted in the second year are properly weeded. Then they receive a new plot. This procedure is advan-tageous to both sides: the farmers have the chance of farming within a restricted area, while there is no danger of forest destruction; the forestry department saves the clearing costs of reforestation, because these are borne by the shifting cultivator. The only, but considerable, disadvantage in this process is that relatively few shifting cultivators can be absorbed. The felling cycle of the timber is much longer than the usual fallow of 15 years, reaching 80 years with teak and mahogany and about 30 years with pines. This method is thus well adapted to forest areas with little population pressure, but it is unsuitable for more densely populated areas that are inhabited by shifting cultivators.

(c) *Shifting systems in plantations.* The production of the famous Deli tobacco in Sumatra, close to Medan, is one of the few cases where shifting cultivation has been maintained within the framework of a large-scale enterprise. Traditional peasant farming and company farming were integrated. Bush-fallowing for 7–8 years was interrupted by a tobacco crop, produced by several companies, and followed by a single upland crop produced by peasants, who claimed the area. The integration of company and peasant farming on about 200 000 ha lasted for roughly a century, but increasing land pressure resulted in a separation of peasant and company activities. By 1975 the system operated with a tobacco crop only and no peasant farming. Between 2500 and 3500 ha are cropped annually with tobacco, out of a total of 59 200 ha under the control of a state-owned company.

After 7–8 years of bush-fallowing a careful seedbed is prepared by fire-clearance and several ploughings and harrowings. The production process lasts for a few months only. Thereafter the land is quickly reoccupied by bush, which (due to deliberate efforts in the past) contains a high percentage of various *Mimosae*. The production process is highly labour-intensive. Forty thousand permanent workers are supported by 140 000 seasonal workers. Yields oscillate between 0·7 and 1·0 t ha^{-1}. But high yields are not the main objective—the production process is geared to obtaining a quality produce, mainly wrapping leaf, selling at prices of about $30 kg^{-1}. Gross returns of the tobacco crop amounts to about $8000 per hectare. Prolonged cropping apparently affects the quality of soil conditioning by fallowing. Substituting the fallow years by permanent farming or by ley farming—as has occurred with tobacco farming in Zimbabwe—has been found to be un-rewarding. A bush-fallow of 7–8 years has so far remained unsurpassed as the most efficient cultivation method for obtaining a wrapping leaf with the desired qualities. There are thus very special circumstances which maintain the profitability of a shifting system under the conditions of farming with modern technologies (K. Schneider, personal communication).

(d) *The 'couloir' system.* A large-scale experiment to regulate and improve shifting systems took place under the Belgian colonial administration in the Congo. In 1960, about two million hectares belonging to 200 000 shifting cultivators were cultivated under the *couloir* system. The system consisted of a sufficiently large and fertile area being selected, declared a *paysannat* in agreement with the population, and divided into straight strips. The individual families each received a strip of forest of 8–12 ha, of which 1–2 ha were cleared and cultivated whilst the remainder was fallow.† A rotation of cultivation years and fallow years was established on the plot. The plots cleared by each family were next to each other, so that normally a cleared corridor ran through the forest. As clearance continued, this corridor shifted in one direction and left behind incipient secondary forest (Fig. 3.6).

At Yangambi the rotation was practised that is illustrated in Fig. 3.6. The strip for cropping in a particular year—say 1958—is felled and prepared for cultivation. The following year (1959) the next strip is felled and prepared, and the year after that (1960) the next strip, and so on. When the strip for 1961 has been opened up, the one to be exploited next is at the other end of the

† An example: 1600 ha are available for 200 families. The rotation consists of 3 cultivation years and 16 fallow years. The annual clearance amounts therefore to 0·4 ha, the area of cultivation to 1·2 ha (Tondeur 1956, p. 77). For the numerous variations in the organization of the *paysannats*, see Pinxten (1954).

block, next to the strip that was cleared for 1953. The clearing of alternate strips then proceeds through the block for 10 years, each strip being cropped for only 1 year. When the twentieth year after the start of the work is reached, felling takes place again on the strip first cleared in 1953, where a 19-year-old regenerated forest exists. All subsequent strips have also had a 19 years' fallow before they are cleared again for cropping.

No matter how beneficial some features of the *couloir* system may have been, it has not survived in Zaire after independence. The shifting cultivator who is used to freedom of movement obviously finds the compulsory rotation of land oppressive. Maintenance of the rotation probably required the administration of the *paysannat* by an official from an outside institution, with

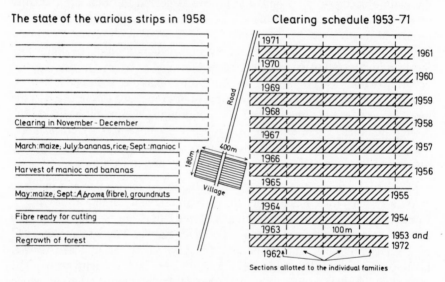

FIG. 3.6. Controlled shifting cultivation in the *couloir* system (from Dumont 1966, p. 40).

sufficient 'power keys' (like project rules, control of credit and sales) to be able to make the farmers adhere to a system that they would not otherwise have preferred. The planning and supervision of a *paysannat* thus makes considerable staffing demands. It is a system which requires (1) willingness of the households to co-operate, (2) reasonably uniform land quality, and (3) reasonably uniform family labour capacities; and it usually fails because these prerequisites do not exist.

An approach involving land reform seems to be more promising. As part of the Zande Scheme, Sudan, 12–16 ha on a catena were allotted to about 25 farmers in the early 1950s. No legal titles were given. In 1971 the farmers still shifted within their area and recognized the others' boundaries (personal inquiry).

3.6.2. *The change from shifting systems to other farming systens*

(a) *From shifting to permanent upland farming.* The obvious development of shifting systems lies in a progressive shortening of the fallow period, which leads finallly to permanent cultivation systems. In high-altitude areas, on fertile soils and in the dry savannas, the development towards permanent farming has taken place to a considerable degree in a relatively short period. In the humid and semi-humid tropical lowlands, on the other hand, the introduction of systems with permanent arable farming proves to be a difficult problem from the standpoint of agronomy and farm economy. The first experiments with this kind of system, which were carried out in Yangambi, were an unequivocal failure. Dumont (1966, p. 37) describes them:

> On the first experimental plots at Yangambi, the forest was cut down, burnt, and uprooted. The soil was ploughed deep and, as a precaution, sown with *Pueraria, Calopogonium*, and other leguminous crops which had proved their worth in plantations. These were used as green manure and various grasses (*Pennisetum purpureum*) were sometimes sown as well. The land was ploughed a second time and then cultivated on a 2-year rotation. As the region has rain throughout the year, with only a slight abatement from December to April, it was possible to introduce four breaks into this rotation. A crop of upland rice was followed by two of manioc and one of groundnuts. This was followed by a leguminous cover crop. In this manner it was hoped to begin continuous cultivation on the pattern of temperate agriculture. The ant-hills, which occupied about a third of the area, were not cultivated. The soil was exposed to the harmful effects of chemical photo-oxidation for more than half the time, for 13 months out of 24, to be exact. Deep ploughing, expensive as it was, impaired the soil structure. The absence of a vegetation cover for so much of the period increased the amount of percolation, and the removal of plant foods from the topsoil was thus accelerated. The cover crops themselves did not have any apparent effect in improving the soil and their woody stems decomposed too slowly. Bacterial activity was hindered by excessive isolation and was unable to keep up an adequate supply of humus.
> The yield of every crop fell rapidly. The ears of rice did not swell. The 150 acres on which the experiment had been conducted had to be abandoned after a few years. Too far from the edge of the forest to be reseeded quickly and with the micro-climate of its soil reduced almost to aridity, the area was still only thinly wooded 10 years later. The Belgians frankly recognized their mistakes. Their notions of farming in temperate regions were obviously out of place in the Congo. In 1940 they decided to start again, but this time they took the Bantu system as their only sure starting-point and tried to improve it.

More recent experience shows that permanent cultivation in a humid tropical climate is technically possible and that balanced systems can be achieved, but only at a high cost, and the question is whether these systems are economic from the point of view of the farmer (see § 6.6).

(b) *From shifting to regulated ley farming.* An alternative possibility of evolution is the replacement of the bush fallow that occupies the land for several years by plants that are themselves usable, regenerate the soil productivity, and allow use of the plough. The obvious way to achieve this is by transition to the ley system. However, arable crops, grasses, and leguminous crops in a rainy climate do not have the capacity of the forest or of tree crops to keep a reasonable balance between leaching and the remobilization of nutrients. In a rainy climate ley systems can take the place of forest and bush fallow only with the help of heavy applications of mineral fertilizer. However, the right economic conditions for ley systems demanding much fertilizer are found only rarely (see § 4.6.2).

(c) *From shifting cultivation to perennial crops.* The humid and semi-humid tropics are by nature forested, and the natural environment is well suited to perennial crops like bananas and to tree crops, a fact that is widely recognized and used by the shifting cultivators. The slowing down of the shifting cycle leads to stationary housing with more or less permanently cultivated gardens. Permanent banana groves develop, offering soil protection as well as high yields of calories per hectare. Wherever remunerative markets are available, shifting cultivators tend to plant tree crops like rubber, coconuts, and cocoa in their plots, since these are less choked by weeds and demand less of the soil than annual crops. The cultivation years are thus not followed by regeneration of fallow vegetation, but by a stand of tree crops whose effects on the soil resemble those of forest and bush vegetation. Evolution of this kind has taken place on a large scale. Nowadays in tropical Africa and Asia hundreds of thousands of hectares that were previously utilized by shifting cultivators support cacao, oil-palms, and coffee. The evolution from shifting systems to perennial crops is obviously suitable for the natural conditions of the wet tropics in particular; it is also a change that is relatively simple and cheap.

(d) *From shifting cultivation to irrigation farming, in particular to wet rice.* The experience of tropical Asia shows that the best system of arable cropping in a warm, humid climate is the concentration of production in the relatively fertile, hydromorphic valley bottoms, i.e. replacement of shifting cultivation over large areas by irrigation farming, especially with rice, in small, intensively cultivated plots. The transition from the shifting system to the wet-rice system, or the combination of both methods, can be observed in large areas of the tropics (Madagascar, East Africa, Sri Lanka, Indonesia). It depends, as a rule, on the initiative of the indigenous population, and obviously represents an evolution that is ideally adapted to the special environmental conditions (see § 7.4).

(e) *From shifting cultivation to artificial pastures.* Artificial pastures are well suited to the natural conditions of extensive areas in the humid and semi-humid tropics. The output in terms of dry matter is very high, and the problems of producing fodder of sufficient quality are manageable. High inputs of fertilizer seem to be required in most areas and the approach is therefore a reasonable one only under the condition of favourable price relations between beef and fertilizer (Crowder 1971; Ruthenberg 1974). In Para State, Brazil, however, cleared land is increasingly planted with the grass *Bracharia humidicula* which—on not too acid soils—seems to be an effective producer with low inputs of phosphate, and bush regrowth is apparently controlled by the vigour of the grass.

(f) *From shifting cultivation to ranching.* In some areas in Africa, and to a greater degree in Latin America, especially Brazil, shifting cultivation serves as a method of obtaining cleared grazing land. The shifting cultivator clears a piece of land and cultivates it for several years. He then sells the crop land and clears a new plot. The period of crop farming is not followed by the regeneration of a fallow vegetation of bush, but by the transition into grassland, facilitated by the heavy grazing, supplementary bush clearing, and planned burning. Huge areas of formerly forested land have been transformed by shifting cultivation and burning into savannas which on better soils seem to be stable ecologically and support extensive grazing (2–9 ha per livestock unit). In other cases, as they are to be found in Panama, the sequence is: forest—2 crop years—natural pasture—bush regrowth, and cultivation shifts within a larger ranch area. Little is known, however, about the long-term effect on the leaching of nutrients and on the pH, and it may be assumed that such a system is heading towards a low-level equilibrium, which, however, produces some meat with very low inputs. Similar methods of using shifting cultivation to turn bush into grazing land are widely practised by cattle-owning cultivators in the African savannas.

4. Fallow systems

4.1. Definitions and genesis

EXPANDING cash production and the growing subsistence needs of an increasing population lead to a gradual extension of arable farming at the expense of the fallow, and short-fallow systems replace long-fallow systems. *Cultivateurs avec jachères* (Faucher 1949, p. 9) cultivate areas by alternating cropping and bush or grass fallows. Brinkmann (1924, p. 959) used the term *Umlagewirtschaft* to indicate that stationary farmers alternate between cropping and long-term fallows with a natural fallow vegetation. In this type of land use the value of R is between 33 and 66. An R value of 50 can be regarded as typical, and arises when, for example, 3, 5, or 10 fallow years follow 3, 5, or 10 years of cropping. In such a situation 50 per cent of the arable land is cultivated annually.† The length of fallow is mostly insufficient for a fallow vegetation of forest to regenerate. Bush or grass fallows are typical. In areas not infested by the tsetse fly, the fallow vegetation is grazed by cattle, sheep, and goats. The characteristics of these farming systems are distinct from those of shifting cultivation, and we classify them as fallow systems. Where the fallow vegetation is systematically grazed we speak of unregulated ley systems. Table 4.1 illustrates this development amongst the Bailundu of Angola. Within a few decades, the reduction of fallows has

TABLE 4.1

Development of the land-use system of the Bailundu, Angola

Traditional farming system, which has almost disappeared $R = 17$	Hoe-systems predominant in the 1960s $R = 29$	Plough-systems which are expanding $R = 54$
6 years arable farming	6 years arable farming	6 years arable farming
5 years pasture	5 years pasture	5 years pasture
10 years bush fallow	10 years bush fallow	
15 years bush and forest		

Source: Pössinger (1967, p. 114).

† A further criterion is that several fallow years succeed several cultivation years. In semi-arid climates of the subtropics the rotation cereal–fallow is widespread. From the point of view of farm management, these dry-farming systems are fundamentally different from the fallow systems discussed here.

increased the R value from 17 in the traditional system to 31 in hoe systems and 54 in plough systems.

Fallow farming is usually characterized by clearly defined holdings with largely permanent field divisions. Quasi-stationary housing predominates, since the changing of hut sites is a matter of moving short distances only. Families generally have *de facto* or registered ownership of the land, in contrast with most shifting systems, in which the holding boundaries are not usually clearly defined, housing is more or less of a migratory nature, and land rights are even less precisely defined.

Most fallow farmers practise hoe-cultivation, often in the form of a soil-conserving and yield-maintaining, but labour-demanding, ridge or mound culture. They usually cultivate a larger area than shifting cultivators in the same environment, with a fairly large cash-crop area, which may take half the total land cropped. In Africa in particular the economic importance of fallow farming is far greater than that of shifting systems. Most of the population and areas that would normally be associated with shifting cultivation are in fact part of relatively intensive fallow systems.

In general terms, a reduction of the fallow period causes a reduction of the yield per hectare, unless there is fertilizer application or manuring, which is not normally the case in these systems. The warmer and rainier the climate and the poorer the soils the more necessary is the long-term forest fallow and the greater the loss in yield due to a reduction in fallowing. The extension of cropping at the expense of the fallow, however, may lead to a rising overall production from the total area available, in spite of decreasing yields per hectare cropped. The lower yield is usually more than balanced by cropping larger acreages. The process of increasing output by consuming soil fertility may last for decades or centuries, and even the low-level equilibrium yield of the total area may surpass the output of a balanced shifting system on the same piece of land. There is consequently a general tendency to reduce fallowing as the need for more output arises, and this tendency is particularly strong where it has been found that a few years of fallowing are sufficient to have a marked effect on yield levels. For the West African lowlands, Nye and Greenland established that over half of the mobilized nutrients, except phosphorus, in an 18-year forest fallow had been accumulated during the first five years (1961, p. 32). At Ukiriguru, Tanzania (842 mm mean annual rainfall), 3–4 years of grass fallows are reported to have been as effective as 1.5 t of manure per hectare (Peat and Brown 1962, p. 313). With increasing cash cropping and subsistence demands farmers tend therefore to remain as long as possible at an intermediate stage, about half-way between balanced shifting cultivation and permanent cultivation at a low-level equilibrium: in bush and grass fallow systems, which are usually very diversified in their cropping pattern, but where the emphasis is on fields with an R value of about 50.

4.2. Types of fallow system and their geographical distribution

Fallow systems can be found under a wide range of natural and social conditions, but the underlying reasons are generally the same: the extension of cash cropping and the increasing subsistence demand of growing populations compel farmers to move away from shifting to fallow systems, but more intensive systems are not as yet either necessary or profitable. We classify the various types of fallow system according to fallow vegetations and altitudes.

4.2.1. Bush-fallow systems

Bush-fallow systems are widespread in the humid and semi-humid tropical lowlands of West Africa and in extensive parts of Latin America (Watters 1971). Fallow systems in the humid and semi-humid climates are often related to tree-crop development, and two types can be distinguished.

(a) *The evolutionary type.*† Fallow systems evolve out of shifting systems if the planting of tree-crops gives rise to stationary housing. Cultivators are restricted to an area close to the house. More and more land is taken by permanent crops and by food crops; less and less remains for fallows. Farmers shy away from cultivating remote fields, because of the transport involved. Fields closer to the hut are preferred even if this means lower yields, more weeding, and the growing of root crops instead of maize and millets for home consumption. A further evolution of the system would mean permanent upland cultivation, with possibly wet rice in the valley, based on an intensive fertilizer economy of the Asian type.

Valley-bottom development tends to induce similar changes on the R value of upland cropping as the planting of tree-crops. Paddy-rice cultivation requires permanent homes and creates a relative shortage of upland in the vicinity of lowland. Where shifting cultivators change to paddy-rice cultivation in valley bottoms, short-fallow systems usually replace long-fallow systems on the surrounding upland. Especially clear examples of developments of this kind can be seen in Madagascar and Indonesia (see Chapter 7).

(b) *The involutionary type.* This type develops out of the evolutionary type where subsistence cropping increases rapidly without adequate fertilization. The competition for scarce land between tree crops (for cash) and food crops becomes acute. Subsistence is usually given preference to cash. More land is devoted to root crops, mainly manioc. Crops and bush fallows alternate under a stand of trees (often oil-palms), which is the thinner as the need for food is greater. The final state is then mainly manioc under a sparse stand of poor oil-palms.

† *Involution* is the attainment of higher total income in the area accompanied by a lower income per head, because of population increase. *Evolution* is the attainment of a higher income per head.

An example is the land use of the Krobo in Ghana (Benneh 1970, p. 201). It began with shifting cultivation. Oil-palms were the first cash crop. Cacao followed and later almost disappeared because of disease and soil-fertility problems. The present state is one of bush fallowing under a sparse stand of oil-palms, with manioc and maize as the major crops. People are moving out of the area because there is more productive employment elsewhere.

4.2.2. *Savanna-fallow systems*

Fallow-systems are predominant in the African savannas and in Brazil's northeast. Again two types have to be distinguished.

(a) Where alluvial or volcanic soils of high fertility are found in tropical regions with heavy rainfall, the cultivator's problem is less one of conserving soil fertility than of controlling weed growth. Consequently, fallow years may be replaced by more labour input, and cultivators change from shifting to fallow systems wherever cash cropping or concentration of population demands a more intensive use of land. Table 4.2 shows a course of development that is typical of the fertile areas of the semi-humid savannas in Africa, the particular example being the fertile plain covered with elephant grass in the Kilombero Valley in Tanzania.

Soil fertility permitting, people prefer to enjoy the advantages afforded by the accessibility of markets, water and neighbours. The easily recognizable tendency to move into areas already somewhat densely populated is an indication that these last advantages are considered more important than the higher labour input per unit of yield required in fallow systems as compared with shifting systems. Baum (1968, p. 35) writes:

> The intensification of rotation cycles and the tendency towards stationary living bring about changes in farming pattern. Stationary dwellings are accompanied by permanent gardens where more and more beans, maize, sweet potatoes, manioc, and fruit trees are planted. Whoever has the means builds himself a solid house with a tin roof. Where fallow or even permanent farming is being practised, technical innovations are spreading. New crops are planted. Interest in selling the produce grows in direct correlation to the degree of permanency in farming. Here and there tractor ploughs are used and cotton is treated with insecticides.

(b) Distinct from these systems, which rely on favourable edaphic conditions are the fallow systems on badly leached soils in the derived savanna (northern part of eastern Nigeria, Korhogo, Ivory Coast, Zande, southern Sudan, etc.). The fallow vegetation is mainly *Imperata cylindrica*, and cultivation usually means obtaining a crop from a plot which remains full of *Imperata* even during the cropping years. Here farmers arrive after one or two decades of cropping at a low-level steady state.

4.2.3. *Fallow systems and unregulated ley systems in semi-arid areas*

In semi-arid areas, as cultivation spreads at the expense of fallows, forest or bush vegetation gives way to a *cultivation steppe* with large areas of grass

TABLE 4.2

Intensity of fallow systems and farming patterns in the Kilombero Valley, Tanzania

System of land use	Extensive shifting cultivation	Intensive shifting cultivation	Fallow systems	Permanent farming
R value	15	30	50	70
Permancy of the dwelling-place	Frequent shifting of dwelling-place and garden	Infrequent shifting of dwelling-place and garden; sometimes stationary housing	Stationary housing and garden	Stationary housing and garden
Housing material	Tree poles	Tree poles and clay	Clay/tin roof	Stone/tin roof
Main input	High input for clearing; low input for weeding	High input for clearing; low input for weeding	High input for weeding; low input for clearing; increasing application of agricultural innovations; increasing investments	
Typical agricultural innovations	New annual crops and new crop varieties	New annual crops and varieties; some new permanent crops	Introduction of insecticides; introduction of mineral fertilizer; increasing planting of perennial crops	

Source: Baum (1968, p. 36).

fallow, which is often dotted with patches of bush. In Africa this process is causing the tsetse fly to be displaced, so that cattle can be kept, and the grass fallows and harvest residues can be used as fodder. The concentration of cattle dung in the kraals where the animals spend the night provides the beginnings of an organized folding system. Fallow systems of this nature may be considered as a stage in the transition to ley farming proper and are therefore called unregulated ley farming.

The economic importance of this kind of land-use system is considerable, especially in Africa. The cotton holdings of the savannas in western and eastern Africa, and the groundnut-millet holdings in western Africa, may be classified wholly or partly as a form of unregulated ley farming.

4.2.4. *Unregulated ley systems in high-altitude areas*

Greatly reduced dependence on fallow is the outstanding feature of the agricultural systems in temperate, high-altitude areas of Africa, Asia, and Latin America (Miracle 1967, p. 143). Short-fallow systems, with R values that are much higher than in long-fallow shifting systems, are consequently widespread in tropical high altitudes. The fallow vegetation consists as a rule of grass which is used for grazing; in other words, we find wild unregulated ley systems, as in the drier savannas. This kind of farming is widespread in, for example, the highlands of Burundi, Tanzania, Kenya, and Ethiopia. Comparatively large areas of unregulated ley systems exist in the mountainous parts of Latin America, where the grass fallow is grazed by sheep, llamas, and alpacas, and, to a lesser extent, by cattle (Dion 1950; Fals-Borda 1955; Sick 1969, p. 164; Watters 1971, pp. 96 and 154). Some unregulated ley farming is still to be found in mountainous areas of northern India (Himachal Pradesh). In these areas, double cropping with rice and wheat is carried out for some years, and followed by several years of grazing.

In most cases we still do not find the permanent field divisions that characterize regular ley systems. Instead, suitable parts of the grazing area are cropped for a few years and then left to go back to grass.

4.3. General characteristics of fallow systems

Fallow systems are particularly widespread in tropical Africa. The following description of the characteristics, therefore, refers principally to African conditions.

4.3.1. *Spatial organization of cropping*

Countrysides used by fallow cultivators are substantially modified by man's effort to maintain and improve his living. Each topographical zone, each segment in the catena, each soil type has a specific function and the vegetation, whether planted or natural, is modified in order to serve man's purposes.

However, like the shifting systems from which they have usually developed, fallow systems are characterized less by efforts at controlling nature than by a complex and skilful adaptation to the natural conditions of production. Each holding is usually a combination of plots (1) on different soil-types, (2) with different numbers and kinds of interplanted trees, (3) at different distances from the hut, and consequently (4) with different fallow and cropping patterns. Two principles for the ordering of this complex situation offer themselves: the differentiation of cropping (i) according to a catena (Fig. 4.1), and (ii) differentiation according to the distance from the hut.

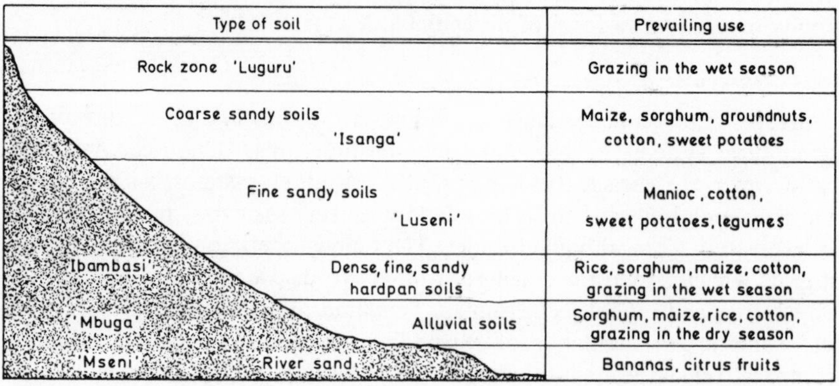

Type of soil	Prevailing use
Rock zone 'Luguru'	Grazing in the wet season
Coarse sandy soils 'Isanga'	Maize, sorghum, groundnuts, cotton, sweet potatoes
Fine sandy soils 'Luseni'	Manioc, cotton, sweet potatoes, legumes
'Ibambasi' Dense, fine, sandy hardpan soils	Rice, sorghum, maize, cotton, grazing in the wet season
'Mbuga' Alluvial soils	Sorghum, maize, rice, cotton, grazing in the dry season
'Mseni' River sand	Bananas, citrus fruits

FIG. 4.1. The catena and the principal uses of the different types of soil in Sukumaland, Tanzania (from von Rotenhan 1968, p. 58).

Fig. 4.1 shows the catena of various soils in Sukumaland, Tanzania. The most suitable crops are grown on each type of soil. A Sukuma holding consists, where possible, of a strip of land running through all the soil levels. At the top of the slope are the fields with maize and sorghum, followed by plots chiefly planted with cotton and manioc. The hardpan soils serve principally for grazing. Finally, in the valley bottom, we find plots with rice and sweet potatoes.

Fig. 4.2 illustrates, with an example from Senegal, the 'ring' cultivation that has developed in western and eastern Africa with the introduction of cash crops into arable farming (Pelissier 1966; Dumont 1966, p. 66; Maymard 1974; Milleville 1974; Savonnet 1970, p. 34; Lahuec 1976, p. 150; Morgan 1969, p. 301). The intensity of cropping decreases in concentric circles from the domestic gardens in the vicinity of the huts to the often widely scattered peripheral fields (see Fig. 4.2). Close to the house soil fertility is high because of trees, household refuse, and cattle droppings. The next circle is largely degraded in fertility, but the distant plots are more fertile because of longer fallow periods. All this, however, is usually no static situation. Fallow cultivators tend to move their huts at long-term intervals, thus creating new

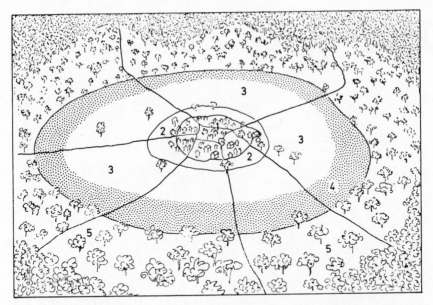

Fig. 4.2. Spatial organization of land use in N'Gayène, Senegal (from Pelissier 1966, p. 474). (1) Houses and gardens; (2) permanent cultivation; (3) intensive fallow systems; (4) intensive shifting cultivation; (5) bush and extensive shifting cultivation.

circles of land use overlapping the old ones, and their sons tend to move into the bordering bush or forest and to develop 'extension' areas as shown in Fig. 4.3 (Gilg 1970, p. 186). Holdings of this type usually include a great number of various subsistence and cash activities carried out on different plots within the holding area, so much so that fallow systems, particularly those in semi-arid areas, may be considered as among the most diversified farming systems in the tropics (Pradeau 1970, pp. 91 *et seq.*; Rouamba 1970, p. 137). The main features of these holdings are as follows:

1. Permanent gardens with fruit trees and perennial crops like bananas and papayas are found in the immediate vicinity of the hut, which is either no longer shifted or is moved only a short distance at long intervals.

2. Near the huts or villages, what is called the *dung-land* is mostly culti- vated. It is used for crops that require more fertile soils, and should guarantee a good yield. The intensity of land use frequently corresponds to garden culture.

3. Adjoining this we find intensive fallow systems in concentric circles of varying size. These fields are often used for growing the staple food and cash crops. The fallow is mostly used as pasture.

4. The intensity of the crop cycle decreases proportionately to the distance from the farmstead. Between the cropping years the ground is left fallow for long periods; in other words, shifting cultivation is practised.

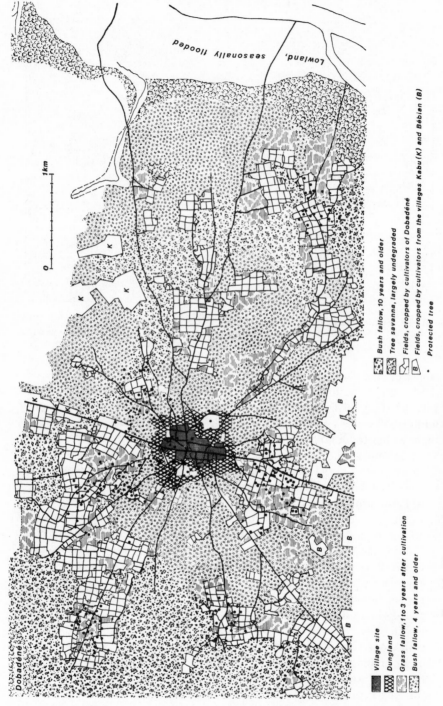

FIG. 4.3. The spatial organization of land use in a semi-arid savanna; Dobadéné, Tschad. *Source:* Gilg (1970).

Lowland, seasonally flooded

1 km

0

Dobadéné

K

Village site

Dungland

Grass fallow, 1 to 3 years after cultivation

Bush fallow, 4 years and older

Bush fallow, 10 years and older

Tree savanna, largely undegraded

Fields, cropped by cultivators of Dobadéné

B Fields, cropped by cultivators from the villages Kabu (K) and Bébian (B)

• Protected tree

5. In some areas the holdings include plots in valley bottoms, which are used for cropping sweet potatoes, taro, or sugar-cane. On the valley bottoms rice cultivation is also widely practised. In addition millet or maize is cropped on moist plots in the dry season, thus providing the household with an off-season supply of fresh food.

6. Trees are often found scattered over the area: fruit trees (e.g. mangoes in eastern Africa), fodder trees (e.g. *Acacia albida* in Senegal), or sisal hedges. In humid and semi-humid areas of West Africa much cropping occurs under oil-palms.

7. Where cattle are kept, a calf enclosure is occasionally found near the hut or inside the kraal.

8. The fallow and communal pasture are grazed by cattle, sheep, and goats.

9. Livestock-rearing in a stationary homestead is supplemented here and there by a certain amount of nomadic herding: some of the herds belonging to the cultivators graze on remote grazing areas watched over by herdsmen.

By no means all holdings that are classified as belonging to fallow systems incorporate all the activities cited above. In most cases, however, there is a pronounced spatial differentiation within the holding which is often not yet stable. Hut sites are moved and land permanently cropped may be abandoned after several decades of use (Benoit 1977, p. 104).

4.3.2. *Cropping principles*

The main source of cultivators' sustenance in fallow systems is arable farming, with animals playing a supplementary role only. In the savannas, crop yields vary considerably from year to year, more so than those obtained by shifting cultivators in the humid zones. Consequently cultivators try above all to secure a reliable supply of food. Another aim is to obtain tasty food: millets and maize are usually preferred to manioc or sweet potatoes. In addition there is as a rule a dominating cash crop like cotton, groundnuts, or tobacco in each farming system. These are the usual objectives on which the organization of cropping is based, and they require a gradual change in cropping principles from those that are typical of shifting cultivation.

(a) *Sole stands and mixed cropping.* In most fallow systems mixed-cropping practices are applied. The cultivation of such crops as cotton and groundnuts, however, and new varieties of high-yielding maize have led to a considerable number of sole stands, particularly in semi-arid regions. In humid and semi-humid climates, we find the change from plant associations covering the land for most of the year to the growing of mainly maize and manioc, in which the percentage of maize is the higher the more ample the available land. In West Africa manioc and maize are frequently grown under oil-palms, and the number of plant species in the crop mixture is higher the closer the plot to the house and the higher the degree of land scarcity (Okigbo 1974*a*, p. 15).

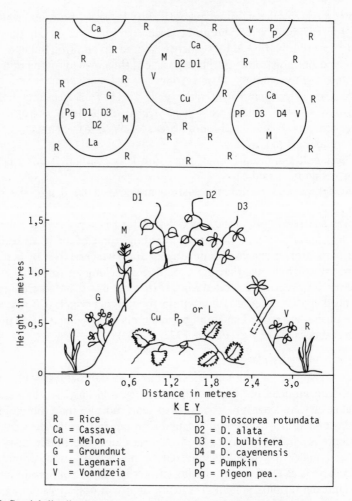

FIG. 4.4. Spatial distribution of crops on a mound in Abakaliki, Anambra State, Nigeria (from Okigbo 1978, p. 18).

Mixed cropping usually represents a careful adaptation of the cropping pattern to topographic features. Fig. 4.4 shows an example of crop production on hydromorphic soils at Abakaliki, Nigeria. 'The crops are located on the mounds according to their root system and in relation to their tolerance to high water table' (Okigbo 1978, p. 18).

(b) *Phased planting.* Phased planting is perhaps even more widely applied in fallow systems than in shifting systems, particularly in semi-arid areas, even though losses due to not planting at the optimum times are considerable. The point is illustrated by the practice followed in Sukumaland, Tanzania (see Fig. 4.5). The cultivators are aware of the unreliability of the rain, particularly in the drier regions. To allow for this uncertainty, they distribute the planting of their crops over a long period, preferring to spread the risk instead of aiming at maximum yields. They also prefer to spread the demand for labour over several months by phased planting. Household requirements offer an

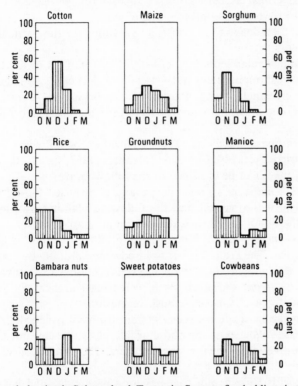

FIG. 4.5. Phased planting in Sukumaland, Tanzania. Seventy-five holdings in three locations were studied, each with several hundred plots, and with only minor differences at the beginning of the rainy season. The percentage of the total area used for each crop sown in different months is shown. (From von Rotenhan 1968, p. 62.)

additional explanation, planting being so timed that the farmers can get food from the garden or field for consumption periods of several months. Thus the planting of subsistence crops in particular, such as beans, sweet potatoes, groundnuts, and maize, is spread over several months.

(c) *Crop combinations*. Cash crops compete with subsistence crops for land and labour. Consequently cultivators tend to adopt those crops which yield more calories per unit of land and labour. With increasing land shortage millets tend to be replaced by maize, and grain crops generally by root crops.

(d) *Crop rotation.* As short-fallow systems replace long-fallow systems, we can see the pseudo-rotation (de Schlippe 1956) being increasingly displaced by proper crop rotations. When compared with the crop mixtures and pseudo-rotations, of shifting cultivators, the rotations of fallow farmers show several distinct tendencies.

1. Whereas the rotations of shifting cultivators tend to include a wide range of different crops, as cultivation becomes more permanent cropping is restricted to fewer crops on each type of soil.

2. There is a tendency to use different rotations for the particular soils of different plots.

3. Fixed and regular sequences of crops like rotations of the kind found in permanent cropping have not yet evolved, but crops demanding fertile soil are increasingly being planted at the beginning of the period of cultivation, and crops demanding less fertility are planted at the end.

(e) *Organization of fallowing.* The organization of fallowing is as a rule especially varied. According to von Rotenhan (1968, p. 64), the variety of forms that it takes can be classified in the following way.

1. There is a *half-year's fallow* if the land is not used during a 6-month period between the harvest and the following planting. In most wet-dry climates, the 6-month fallow is the rule on all arable land.

2. There is a *proper fallow*, if no crops at all are grown on the land during the whole year. The ground is left to be covered naturally by grass or bush. Subject to the influence of climate and soil in most African savanna areas, a *short-grass steppe* develops. With increasing length of the fallow, this short-grass steppe changes into bush vegetation.

A distinction can be made between different forms of proper fallow:

1. *Permanent fallow* exists on those arable plots that are not likely to be used for cultivation in the foreseeable future. They usually serve as grazing areas.

2. If the yields from a plot that has long been cultivated have fallen so much that further cultivation appears uneconomic, then it is given a *long-rest fallow* of 5–20 years. Frequently several fields at once cease to be cropped, so that we can speak of *fallow blocks* as opposed to *cultivated blocks*.

3. There is a *medium-term fallow* if a plot of arable land within a cultivated area is left uncultivated for 2–5 years. The aim of this form of fallow is to combine the regeneration of the soil fertility with the elimination of weeds. Medium-term fallows are particularly widespread.

4. The *short-term fallow* of 1–2 years has generally not the aim of restoring soil fertility, but is caused rather by other circumstances, such as the illness of a farmer or unfavourable weather, which may make the planting of the field difficult.

The vegetation during the fallow changes according to the type of fallow. In humid and semi-humid climates the long-rest and medium-term fallows would be occupied by bush, and short-term fallows would consist mainly of grasses. In a semi-arid climate all types of fallow would mainly be covered by grasses and serve as pastures.

4.3.3. *Characteristics of the fertilizer economy*

The higher intensity of land use necessitates more evolved manuring techniques, although it must be recognized that in most fallow systems, as in shifting systems, the fertilizer economy is still rudimentary. The continuity of cropping depends on either:

1. high soil fertility, which is not visibly impaired by frequent cropping, in which case fallow periods are not interpolated so much to regenerate fertility as to eradicate weed growth; or
2. the ability of the soil to regenerate during the fallow period; or
3. stabilization of yields at a low level.

Generally, repeated cropping with short, unfertilized grass fallows brings about a long-term deterioration of soil fertility. Repeated cultivations favour sheet and gully erosion. Bush fallows are too short, and the root systems of grasses are not enough to prevent leaching. There is not only a loss of organic matter in the soil. Portères (1952), for example, has demonstrated that in some areas of Senegal the soil eventually loses its structure: clay components are washed out, the soil becomes increasingly pervious, the proportion of sand in the topsoil increases, and the ability to retain water is reduced.

In so far as cultivators have recognized the problem of soil mining and practise manuring, a number of fertility-restoring methods can be identified.

(a) *Shifting of the hut and kraal.* It is in any case traditional for the home and farmstead to be shifted at fairly long intervals.

> The huts become dilapidated after some years, the site foul and eroded, the kitchen gardens reach the point of exhaustion, and it is better to move to a new site than to attempt to rehabilitate the old dwellings. These rebuilding moves are generally short, sometimes a mile or more but often no more than a few hundred yards . . . [Allan 1967, p. 7].

The wish to obtain an especially fertile plot that can serve as a permanent garden for some time also motivates the shifting of the holding and the kraal.

(b) *Systematic folding.* In some cases we also find a systematic folding economy (Froelich, Alexandre, and Cornevin 1963, p. 71; Dumont 1966, p. 65; Pelissier 1966; Westphal 1975, p. 102), fallow systems are combined with livestock-rearing, the cultivators can use the manure from the kraals. Transporting the manure to the fields, however, and cultivating it into the soil present considerable problems where there are no tracks, carts, or draught animals. The cultivators therefore prefer to move the kraal and, in the case of the Wasukuma (see § 4.4.3), for example, they do so irregularly, thereby obtaining a series of manured plots. In some areas of Senegal, and on the west coast of Madagascar and Ethiopia, the kraals are moved regularly, so that fairly large plots are systematically manured. In these cases, it is the area around the farmstead which is treated in this way, with the consequence that it can be cropped more or less permanently (see § 4.3.1).

(c) *Fallow vegetation.* Bush-fallow systems depend on a vigorous fallow vegetation with the double objective of re-establishing soil fertility and suppressing undesired weeds. In a few cases farmers establish nitrogen-fixing shrubs such as *Leucaena* (Philippines), *Anthonotha* (Eastern Nigeria), or pigeon peas (East Africa). In Eastern Nigeria farmers observe with satisfaction the growth of *Eupatorium odoratum*, a vigorous leafy shrub which effectively eliminates *Imperata* in few years of bush fallowing.

(d) *Weeds.* Weeds are probably the most important yield-depressing factor in fallow systems, but they have also to be considered as a useful element in the land-use system. De Schlippe (1956, p. 212) indicates that weeds are useful in maintaining soil fertility in the plots of the Zande, Sudan. In hundreds of interviews in various parts of Africa and Asia I have tried to establish that poor weeding may be a deliberate part of the farmer's husbandry practices. Farmers, however, invariably argue that this is not so and that shortage of labour or low return for the additional work are the only reasons for poor weeding. It may be assumed nevertheless that, in a number of traditional systems, weeds have been effective in reducing erosion, shading the soil and preserving soil moisture, providing mulch material, fixing some nitrogen, and maintaining a balance between harmful insects and their predators (Moody 1975). There are numerous systems where farmers only weed in the early stages of plant development. Sometimes they only weed a few centimetres around the useful plants (pepper at Batangas, Philippines), and grassy weeds often provide a dense cover under a maturing crop, so that seasonal fallowing comes close to having a green manure crop.

In the derived savanna of Eastern Nigeria crop production occurs within a 'crop–*Imperata*' complex. Weeding is shallow, so that *Imperata* is not eradicated. The objective is to subdue the growth of *Imperata* long enough so that the leaf canopy of maize or manioc can establish itself above the grass. The result is a two-storey physiognomy with the crop leaves on top and a dense cover of *Imperata* below.

(e) *Green manuring.* A widespread practice in hoe cultivation, especially in ridge cultivation, is for the fallow and weed vegetation to be dug systematically into the soil. Sometimes the gathering of green manure from further afield is organized as well. Pössinger (1968, p. 117) reports, for example, that farmers in Luanda, Angola, artificially increase the size of termite mounds, which are preferred for cropping. The cultivators pile up maize straw, manioc stalks, and other plant waste in a circle around the mound, and the termites break down this plant material on the spot.

(f) *Mineral fertilizers.* Where cotton, groundnuts, maize, tobacco, and other annuals are grown as cash crops, more and more cultivators apply mineral fertilizers. Mineral fertilizers—when it is economically worthwhile— can be introduced comparatively easily in fallow systems, because it can be applied efficiently on small plots without requiring tracks, carts, or much labour. At the same time, it can be stored without appreciable deterioration, is easy to apply, and produces an increased crop after a relatively short period. For these reasons, it is usually easier to introduce the application of mineral fertilizers than to introduce manuring with available cattle dung.

Even though the various stages of incipient fertilization may be worthy of note, it must nevertheless be remembered that they are mostly inadequate. Fallow systems often represent nothing more than degraded forms of previously balanced systems of shifting cultivation.

4.3.4. *Characteristics of animal husbandry*

Whereas shifting cultivation is largely practised with little or no livestock, in fallow systems of the semi-arid climates in particular large stocks of cattle are found. In extensive areas of the African savannas, it is the interaction of arable cropping and cattle-keeping that keeps the tsetse fly at bay. Slash-and-burn agriculture thins out the forest or bush vegetation and destroys the breeding grounds of the tsetse fly; intensive cropping prevents bush regeneration and therefore prevents the insect from re-establishing itself.

The aims of stock-keeping are varied.

1. Cattle are kept to cover the risk of harvest failure or sickness. When land is not privately owned, cattle are kept as a means of support in old age.

2. In close relation to this there are social functions. Cattle acts as bride-price, while a large herd is a status symbol.

3. Farmers want a supply of meat and milk for the household; sale of this produce is not very important.

4. A factor of increasing importance is the provision of the traction power for ox-plough cultivation.

5. Only in a few cases do fallow cultivators consider the contribution of manure as an essential purpose of stock-keeping, although it might be of considerable relevance to their farming.

Harvest residues, fallow grazing, and natural grazing provide the fodder. Communal use of grazing land is customary: everybody has the right to allow any number of animals to graze on the fallows, pastures, and stubbles.

Fodder cropping is practically non-existent. Occasionally, balanced feeding is achieved by moving the livestock seasonally to grazing areas some distance away, which have not been used earlier in the season.

The composition of the livestock herds depends above all on the availability of fodder. Where there is good grassland with sufficient watering places, the number of cattle per family is high, and a high proportion of the stock consists of male animals. With increasing shortage of grazing and

TABLE 4.3

Structure and returns of livestock in Shinyanga, Sukumaland, Tanzania, 1976

Method	Cattle	Sheep	Goats
		Models built on surveys	
1. *Herd size*			
Average herd size of cattle-owning households[a]	26·0	12·7	11·3
2. *Herd composition* (per 100 head)			
Females, 0–1 years	10	19	21
1–2 years	7	17	19
2–3 years	5⎱	54	49
over 3 years	41⎰		
Males, 0–1 years	10	4	3
1–2 years	5	1	2
2–3 years	3	—	—
bulls, bucks	6	5	6
oxen	13	—	—
3. *Technical coefficients*			
Calving/lambing/kidding rate (per cent)	57	73	68
Calf/lamb/kid mortality (per cent)	25	12	13[b]
Weaning rate (per cent)	43	—	—
Yearling and adult mortality (per cent)	10	24	21
Calving interval, days	640	—	—
Average lactation period	350	—	—
Average production per cow in milk (after calf requirements) 0·6 kg per day			
Average annual milk production per cow in milk: 220 kg			
4. *Outputs* (per 100 head)			
Live offtake, heads sold	8	9	9
kg carcass	577	70	95
heads consumed	2	19	18
kg carcass	110	95	90
Value of total live offtake ($)	343	123	138
Mortalities consumed, head	13	18	14
kg carcass	693	171	133
value (half the life price) ($)	173	64	50
Hides, value ($)	37	27	27
Milk, kg	5300	—	—
Milk, value ($)	662	—	—
Total value of output ($)	1215	214	215
Draft availability	13 oxen		
5. *Labour input* (hours per average herd, including children-hours)			
Watering in rainy season (hours)	237	n.a.	n.a.
Watering in dry season (hours)	964	n.a.	n.a.
Grazing rainy season (hours)	1456	n.a.	n.a.
Grazing in dry season (hours)	1939	n.a.	n.a.

[a] Half the households own cattle. [b] Under the semi-humid tropical lowland conditions of West Africa mortality rates are around 40 per cent (de Haan 1978, personal communication). *Source:* Pudsey (1977, personal communication).

higher cattle densities, primarily female stock is kept, the male calves being slaughtered soon after birth. Where fodder is not sufficient to keep large herds in one place, the lending out of cattle is common.

Table 4.3 gives information about the technical coefficients, the outputs, and inputs of livestock production in a typical grass–fallow farming system in a semi-arid African savanna.

An important sociological feature is the fact that the ownership of live-stock is much more concentrated than the cropping of land. In Sukumaland, for example, 25 per cent of the families crop 55 per cent of the cultivation area, but 25 per cent own 80 per cent of the animals (see Fig. 4.6).

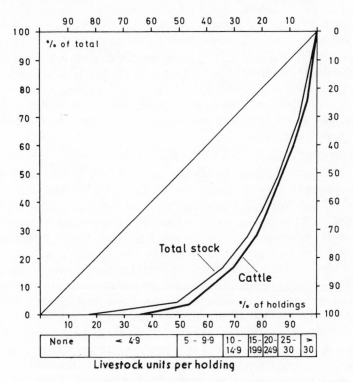

FIG. 4.6. Distribution of livestock property in Sukumaland, Tanzania (from von Rotenhan 1966, p. 57).

4.3.5. *Characteristics of the labour economy*

Important differences between shifting and fallow systems may also be found in the labour economy. The increasing degree of permanency in land use, and the increasing amount of cash cropping, induces rather striking changes.

(a) *Technical change*. There is a change in work pattern. Tree-felling and clearing are the main jobs of fire-farming shifting cultivators; the more tedious tasks of cultivating with the hoe and of weeding are the most time-demanding activities of the fallow cultivators, particularly where these processes cannot yet be carried out with ploughs and weeders.

(b) *Economic change*. A great amount of data about labour use in fallow systems has been collected and ordered by Cleave (1974, p. 32). His information and the examples in this book tend to indicate that fallow farmers work more days per year than shifting cultivators and indeed than most permanent cultivators, first because more hours per hectare and per unit of output are required (compare Tables 4.5 and 6.3) and secondly because a greater cash-crop area has to be taken care of. In those fallow systems cited by Cleave (Northern Nigeria, Uganda, Zimbabwe), hours of field work per man per year vary between 700 and 1000, while permanent farmers (North Central Nigeria) work about 500–700 hours. The working day in fallow systems usually varies between 5 hours and 7 hours, while it is 3–6 hours in permanent systems.† In fallow systems there are already more incentives for work than in shifting systems, and there is still more land available to employ labour productively than in permanent systems.

(c) *Sociological change*. We find a change from the rather rigid traditional division of labour between the sexes, which is still typical of many shifting systems, towards a more flexible use of the available labour force. The demands of cash cropping seem to be the driving force behind this. An important feature is the increasing involvement of men in field work such as weeding and harvesting (compare Tables 4.4 and 5.2). This in turn indicates a new tendency in farm management. Traditionally, several persons in the household produce separately the food they need for themselves and their dependants on a plot that might be considered a subfarm within the family holding. With increasing permanency and commercialization, land and labour use within a holding seem to become more centrally organized. Shifting systems usually show labour exchange between families, while partly commercialized fallow farmers increasingly rely on hired labourers or take up paid off-farm work.

Typical of fallow systems in the African savannas is the cultivation with hand tools of comparatively large areas of 2–8 ha per family. The labour input per hectare is rather high. A lot of hard work, mainly cultivation, weeding, and harvesting, has to be done in the growing season. Table 4.5 gives an idea of the input–output relations for some major crops in fallow systems. It should be noted that the labour input is comparatively low in the Ghana example from a semi-humid area, where fire-farming substitutes some part of the

† The hours of work per working day vary widely with the type of work and whether or not there is a peak season in labour demand. Two hours to four hours per day are typical for slack seasons and 5–7 hours for peak seasons (Cleave 1974, pp. 55–99; von Rümcker 1972, p. 26).

TABLE 4.4

Use of time in shifting and fallow systems
(as a percentage of daytime, 84 hours per week)

System Time Method	Uélé, Zaire Shifting (cotton–rice) 12 months 1959–60 Daily observation of 9 selected households		South-east Ghana Fallow (maize–manioc) 13 months 1970–71 Sample 27 households[a]	
Number of persons	12 males	22 females	12 males	53 females
Agriculture (per cent)	19	30	38[b]	30[b]
Collecting, hunting, fishing (per cent)	14	9	1	—
Household construction, etc. (per cent)	15	28	5	26
Social activities (per cent)	11	6	12	7
Meals, leisure, etc. (per cent)	35	20	25	20
Illness (per cent)	4	4	4	3
Absence from home (per cent)	2	3	15	14
Total (per cent)	100	100	100	100
Field work per person and year (hours)	832	1300	1659	1267

[a] Including two non-agricultural households. [b] Including 2·6 per cent off-farm work, and including marketing. 0·7 per cent for males and 7·1 per cent for females. In the Zaire study marketing is shown as part of the social activities. *Source:* Zaire: de Smet (1972, p. 297); Ghana: Thornton (1973, p. 38).

hoe cultivation which is required in semi-arid areas. The labour input per hectare is high when compared with Asian (Indian) plough systems (see Table 6.3). Traditionally, labour, supplemented by some seeds and tools, is the only input. Modernization not only implies the use of fertilizer but also requires a significantly higher labour input, as the data from Malawi show. Where the price for the produce is low, as in the cases cited from Malawi, returns per hour of work are so low that cash cropping and modernization were not an interesting proposition in 1972.

A significant change in the work situation is brought about by the introduction of the ox plough, which is technically possible wherever areas with a fallow vegetation of grass have been de-stumped. First and foremost, it eases the burden of work by relieving the hard toil of land cultivation. At the same time, fields of mostly regular rectangular shape are formed, the size of fields increases, and the cropped area is expanded at the expense of the fallow (see Table 4.6). Where the ox plough is not combined with the ox-drawn weeder and row cultivation, as, for example, in cotton cropping in Sukumaland, the labour-saving in cultivation may easily be cancelled out by the greater work required for weeding. For the full success of animal traction, to obtain the greatest possible reduction in the workload, the ox-drawn weeder and cart, as well as the plough, are necessary.

A further feature of this land-use system is the seasonal nature of labour

TABLE 4.5

Some examples of input–output relations in fallow systems with hoe cultivation (per hectare per crop)

Crops	Maize	Maize	Maize	Finger millet	Sweet potatoes	Manioc	Cotton	Cotton[b]	Tobacco
Country	Ghana	Kenya	Malawi	Uganda	Uganda	Colombia	Tanzania	Malawi	Uganda
Year	1971	1974	1972	1972	1972	1974	1976	1972	1972
Technique[a]	Trad.	Mod.	Mod.	Trad.	Trad.	Trad.	Trad.	Mod.	Trad.
Method	Sample	Sample	Sample	Sample	Sample	Sample	Model	Sample	Sample
Technical data									
Labour input (man-hours)									
Clearance	210	531	144	512	558	140	150	152	1277[c]
Land preparation	24	80	13	374	390	77	192	150	371
Planting	90								
Weeding	186	297	342	339	17	408	192	432	475[d]
Fertilizer and pesticide application	—	30	174	—	—	4	—	115	n.a.
Harvesting	90	147	208	426[e]	375[e]	116	432	710	391
Processing		72	368				222	566	1467
Marketing, etc.			153				22	315	193
Total	600	1157	1402	1651	1340	745	1210	2440	4174
Yield (t)	1·91	3·60	2·32	1·05	8·39	7·90	0·50	1·28	0·77
Economic analysis ($)									
Gross return	166	317	85[f]	147	388	181	114	256	650
Material inputs									
Seed	3	8	3	1	—	—	—	—	n.a.
Fertilizer and pesticides	—	15	29	—	—	1	—	58	60
Tools, mechanization, etc.	1	1	1	1	1	1	1	10	79
Income before interest and overheads	162	293	52	145	387	179	113	188	511
Income per hour	0·27	0·25	0·04[f]	0·09	0·29	0·24	0·09	0·08	0·12

[a] Modern = use of fertilizers and improved seeds. [b] Average of ten 'progressive' farmers. [c] Half the farmers use tractors for ploughing. [d] Including desuckering. [e] Estimate from other surveys. [f] At official price. Farmers actually sell at higher prices. *Sources*: Ghana: Thornton (1973, p. 27); Kenya: Frey (1976); Malawi: Dopieralla (1974, p. 77); Uganda: Kyeyune Sentongo (1973a); Colombia: Diaz, Pinstrup-Andersen, and Estrada (1974); Tanzania: Pudsey (1977, personal communication).

TABLE 4.6

Characteristics of hoe- and plough-regions in Sukumaland, Tanzania

	Hoe-region Ukerewe, Kwimba	Plough-region Shinyanga
Basic method	Ridge cultivation	Flat cultivation
Shape of fields	Irregular	Regular
Average size of fields (ha)	0·5	0·9
Cultivated area per farm (ha)	2·4	3·3
Area cultivated per worker (ha)	0·7	1·3

Source: von Rotenhan (1968, p. 56).

demand. In shifting systems, the task of clearing does not have the same rigid timetable as has cultivation in fallow systems. The fallow farmer must wait until the beginning of the rainy season to cultivate, whereas the shifting cultivator can carry out clearance work in the dry season. Weeding and harvesting, the two activities which require most of the work, are fixed in time and produce pronounced peaks in labour demand (Cleave 1974, p. 120). Such is the case, for example, with groundnuts. As Fig. 4.7 shows, both in

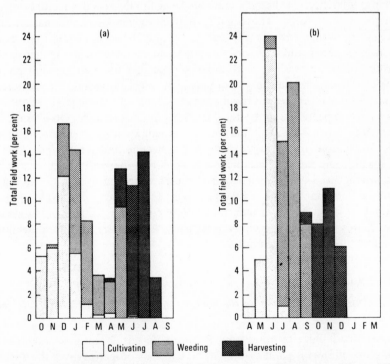

FIG. 4.7. Seasonal distribution of field work in fallow systems of Tanzania and Senegal (from Lacombe-Orlhac 1967, annex 1; von Rotenhan 1968, p. 63). (a) Cotton-millet holdings in Sukumaland, Tanzania. (b) Groundnut-millet holdings in Senegal.

Sukumaland and Senegal 90 per cent of the field work is concentrated into 6 months, and 50–60 per cent in only 3 of those months. Since field work, apart from herding of cattle, is almost the total work of the holding, considerable underemployment occurs during the remaining months. In this connection, we must bear in mind that the number of working hours per day does not give a satisfactory idea of the different labour demands. In seasons of peak demand on labour, performance per hour is considerably higher.

Finally, it should be pointed out that the labour economy of partly commercialized fallow systems is usually far removed from the uniformity of traditional subsistence farming. Differences in the man-to-land ratio, differences in the choice of crops and the adoption of innovations, different access to development services, and different opportunities for cash cropping—these factors all combine to bring about a wide variation in the size of enterprises and in the smallholders' performance within a given location. In most fallow systems, only a few hectares are cultivated and a few animals kept, i.e. the size of the agricultural enterprise is rather small. This, however, does not mean that the average area cultivated by each family or each worker is about the same. Within each type of farming, even where land is still ample, we find striking differences in the area cultivated per family and per member of the available labour force. The equality that characterizes traditional shifting systems, in which each family cultivates only what is required to cover subsistence needs and everyone uses the same practices, no longer exists. Within fallow systems, we find larger and smaller units, richer and poorer families. The differentiation in cropping is the more noticeable the greater the extent of land shortage and the higher the degree of commercialization.

Even more pronounced are the variations in performance. On average the hours of field work per available man-equivalent are low. There are, however, wide fluctuations in the input of work per hectare and returns per hour of work within the range of production possibilities offered by a fallow system. Farm-managerial ability, particularly in cash cropping, is one of the major factors differentiating progressive producers from others. The explanation is an obvious one: the performance of fallow cultivators differs widely, much more than that of shifting cultivators, because there are more opportunities for effort, drive, and ingenuity.

4.4. Examples

4.4.1. *A bush-fallow system in the humid tropical lowland of Eastern Nigeria* (Table 4.7)

Most smallholder farming from Benin to Eastern Nigeria is organized in root-crop–maize–oil-palm systems. The cover of oil-palms gives the appearance of an open forest to the countryside. Under the oil-palm cover cropping with manioc, yams, and maize alternates with several years of bush fallowing.

Fallow periods of more than 6 years and a predominance of maize or yams indicate an ample availability of land. Less than 6 fallow years and an increasing reliance on manioc indicate land shortage. Land shortage is particularly pronounced in Eastern Nigeria. The area around Umokile and Owerri had 300–500 persons per square kilometre in 1963. Decades ago farmers reacted to the increasing land shortage in a way that is classical for Africa:

1. The crop area was extended at the expense of the fallow. Farming changed from shifting cultivation to bush fallowing.
2. Extensive areas were planted with oil-palms, which became a major industry in the 1920s and 1930s.

Both developments proved to be insufficient to support a growing population. Traditional oil-palms are an extensive type of land use in terms of employment and income. Under the condition of sandy soils in a humid tropical lowland climate, increasing land use by the reduction in fallowing runs into serious soil fertility problems. Manioc in the area, grown on almost identical soils, produces in farmers' fields 10.8 t ha^{-1} in 18 months after $5\frac{1}{2}$ years of fallowing and 3.8 t after $3\frac{1}{2}$ fallow years. The yield per hectare and year (crop period + fallow period) amounts to 1.54 t and 0.76 t respectively. The loss in manioc is partly compensated by the return from other crops grown in association with manioc. Total output of all crops per hectare and per year is 0.65 t dry matter ($156.00) in the system with $5\frac{1}{2}$ fallow years and 0.39 t dry matter ($94.00) in the system with $3\frac{1}{2}$ fallow years. These figures indicate substantial negative marginal returns to land use intensification (Lagemann 1977, p. 67). Farmers, by adhering to their traditional practice of reducing fallowing in case of need of food, apparently tend to act irrationally, because with less cropping and more fallowing total output would be higher and less labour would be required. The increase in output by turning towards permanent cropping at a low-level steady rate, which is usually achieved under favourable edaphic conditions or in semi-arid climates, is thus not possible in the area under consideration. Farmers had to adapt differently (Okigbo 1974a; Flinn 1974).

1. Off-farm employment and out-migration gained in importance. In the example, the greater part of the family income originates from off-farm income. Farming is expected to supply the food base, particularly a continuous flow of starchy roots, palm-oil, and vegetables.
2. The land belonging to a household experienced a very pronounced spatial differentiation: (i) a small compound plot is farmed with great intensity; (ii) a field adjacent or close to the compound, usually covered by oil-palms, produces arable crops within an intensive fallow system; (iii) a distant field, larger in size and usually with a low-density oil-palm cover, produces food in an extensive bush-fallow system.
3. Tree-crop growing and arable farming became increasingly integrated.

In the compound we find a multi-storey spatial organization of useful palms, bushes, bananas, standing arable crops, and creepers. The integration of arable cropping and tree crops induced a change in cropping pattern. Preliminary surveys indicate that maize is more sensitive to shade than yams (a reduction in yield of 30 per cent compared with only 9 per cent with yams) (Okigbo 1974a). Similar relations are expected with cocoyams and manioc. Root crops consequently gain in comparison with maize. Some of them effectively produce starch even under the shade of palms (Flinn 1974, p. 14).

The *compound plot* in the examples has only 400 m². At the height of the vegetation period it carried the following number of crops: 1 oil-palm, 2 coconut palms, 30 other useful trees and bushes, 307 yams, 230 cocoyams, 80 manioc plants, 46 maize stalks, and 566 vegetables. Altogether 60 different plant species were found on compound plots in the survey area (Okigbo 1974a, p. 17). Trees and bushes are trimmed before the major planting season to provide light for crops lower down the storey. Seedlings are watered where necessary. Mounds are covered with mulch material. All household refuse, mulch materials, goat manures, etc. are collected in pits and used for manuring. The goats are stabled during the main-crop season and are fed with palm leaves, crop residues, and household remainders, and their manure is highly appreciated. Latrines (deep pits) are occasionally shifted and provide nutrients to bananas mainly. Nutrients and organic matter are imported into the compound from the fields and by household purchases. The result is a very high output and much employment. In 1974–5 the compound in the example produced goods worth $225 at local market prices. On a per-hectare basis this would amount to the staggering sum of $5625. The compound occupies only 2 per cent of the farm land and 9 per cent of the crop area, but produces about half of the farm output.

The farming of the fields is very extensive compared with the compound. The *homestead field* (less than 5 minutes walk from the compound) shows a rotation of 1½ crop years and 3½ fallow years. Yams, which grow well under the shade of the oil-palms, are interplanted with manioc and then with maize. In the second year the manioc takes over, and during the year the field turns into a tumbledown fallow. The *distant fields* are much larger, and there are fewer oil-palms if any. Fallows tend to increase with the distance from the compound. Traditionally 6–8 years of fallowing was the practice, but in the example fallowing has already been reduced to 3½ years. After fire clearance, manioc is grown and interplanted with groundnuts and maize. No mulching or manuring is practised on the fields. Some farmers deliberately plant *Anthonotha*, a nitrogen-fixing plant, as fallow bushes, but this is the only observable effort to maintain soil fertility in the fields. Yields are low and amount in the example to not more than 3 t of produce per hectare, compared with 28 t in the compound. The value of the output on a per-hectare basis is

TABLE 4.7

Farm-management data of bush-fallow systems in Eastern Nigeria and Uganda

Country	Nigeria	Uganda	
Location	Umuokile/Owerri	Kigezi	
Rainfall (mm)	2300	940	
Altitude (m)	100	1200	
Year	1974	1971–2	
Type of farming	Root-crops–oil-palm	Tobacco–millets–plantains	
	Traditional hoe-farming	Traditional hoe-farming	
Method	Case study	Sample	
Number of holdings	1	88	
Persons per household	9·0	6·1	
Labour force (ME)	2·0[a]	2·6	
Size of holding (ha)	1·40	2·52	
Compound plot	0·04⎫		
Tree crops	0·19⎭	0·48	
Homestead field (manioc–yams–maize)	0·06	Short rains, millet	0·38
Distant field	0·36	Long rains, tobacco	0·24
		Long rains, food-crops	0·43
Total arable cropping (ha)	0·46	1·05	
Cropping index (per cent)	38	52	
Livestock (number)			
Goats	6	3	
Chicken	24	4	
Yields (per hectare)			
Compound plot (t produce)	28·4	Millet (t)	1·05
Compound plot ($)	5625	Sweet potatoes (t)	8·30
Fields (t produce)	3·1	Plantains (t)	4·60
Fields ($)	439	Tobacco (t)	0·77
Economic return ($ per holding)			
Gross return of which	679[b]	278	
Root crops (per cent)	53	23	
Other food crops (per cent)	7	23	
Tree crops (per cent)	27	8	
Tobacco (per cent)	—	43	
Livestock (per cent)	13	3	
Purchased inputs	45	37	
Income	634	241	
Wages for hired labour	46	10	
Family farm income	588	231	
Off-farm income	559	n.a.	
Total family income	1147	n.a.	
Productivity			
Gross return ($ ha^{-1} total land)	485	110	
Gross return ($ ha^{-1} crop land)	1045	182	
Gross return ($ ME^{-1})	340	107	
Income ($ ha^{-1} crop land)	975	158	
Income ($ ME^{-1})	317	93	
Labour input (man-hours)	1675	2358	
Income per man-hour ($)	0·38	0·10	
Man-hours ME^{-1}	837	907	
MJ per man-hour	14	n.a.	

[a] The remaining household labour force has off-farm employment. [b] Valued at prices on local markets. These prices are very high. Fresh manioc is valued at $120 per tonne and fresh yams at $190 per tonne. *Sources:* Nigeria: Lagemann, Flinn, Okigbo, and Moormann (1975); Uganda: Kyeyune-Sentongo (1973a).

only 1/12 of that of the compound. The fields contribute, however, to the output of the compound by providing most of the starch (root crops) and fat (oil-palms) for the household members, and they supply most of the organic matter for mulching.

There are several possible avenues for the improvement of the system. (1) In some locations undeveloped valley bottoms are available which could produce wet rice, vegetables, and fish in ponds. (2) In the past the extension of a multi-storey cropping pattern was limited due to lack of nutrients. This restraint no longer applies. Mineral fertilizers are available. (3) Livestock industries (chicken, pigs, and goats) relying on purchased and farm-produced fodder could provide employment, income and manure. (4) High-yielding, oil-palm varieties could replace the low-yielding palms. There is thus much scope for increasing the output of the system, but little is as yet known about the yields, input requirements and labour productivities of a modernized multi-storey cropping system. Increasing the yields of arable upland cropping is apparently the most difficult proposition. The higher the cropping intensity, the lower the base saturation of soils and the lower is the return to modern inputs (Flinn and Lagemann 1976, p. 7).

4.4.2. *A fallow system in the derived savanna: tobacco–millet holdings in Uganda* (Table 4.7)

The example from Uganda shows the characteristics of fallow farming in an area with two pronounced rainy seasons, each averaging 450 mm per annum. Here clearly the tendency is for perennial crops and much plantains are grown. Roughly half the arable land is under crops and the other half is a bush–grass fallow which cannot be utilized by cattle because of tsetse flies. Much millet is grown and serves for making beer. Tobacco is the cash crop of the area. Tobacco is competing with millet for labour and so the emphasis in food production is on plantains and sweet potatoes.

4.4.3. *Unregulated ley farming in semi-arid Africa: cotton–sorghum holdings in Sukumaland, Tanzania and groundnut–millet holdings in Mali* (Table 4.8)

An example of arable farming in the drier savannas is the Wasukuma's land use in the 'cultivation steppe' to the south of Lake Victoria which receives a rainfall of 700–900 mm during a two-peaked six-month rainy season. A typical Sukuma holding used to be a strip of land stretching from the top of a hill to the bottom of a valley which includes different soil-types carrying different crops (see Figs. 4.1 and 4.7). In 1976 the population was villagized and individual farmers cultivated their plots in 2–3 ha blocks of land with an average distance of 2.7 km from the village. The average walking time per day from the village to the field and back amounted to 1.5 hours (Pudsey 1977, personal communication).

TABLE 4.8

Farm-management data of grass-fallow systems (unregulated ley farms) in Tanzania, Mali, and the Sudan

	Tanzania	Mali	Sudan	
Country	Tanzania	Mali	Sudan	
Location	Sukumaland (Maswa)	Kayes	Damazine	
Rainfall (mm)	800	500[a]	732	
Year	1976	1977	1976	
Type of farming	Cotton–sorghum–cattle	Groundnut–millet–cattle	Sorghum–sesame	
Technique	Traditional hoe-farming	Traditional hoe-farming	Partly mechanized	
Method	Model	Case-study	Model built on survey[b] of 43 farms	
			year 2	year 7
Number of holdings	1	1	year 2	year 7
Persons per household	8·5	10·5	n.a.	n.a.
Labour force (ME)[c]	3·7	3·5	n.a.	n.a.
Size of holding (ha)	6·90	19·00	n.a.	n.a.
Cotton	1·25	—	—	—
Groundnuts	—	1·00	—	—
Sesame	—	—	84	84
Sorghum, millet, maize	1·25[d]	2·80	302	302
Rice	0·60	—	—	—
Manioc	0·70	—	—	—
Total crop area	3·80	3·80	386	386
Grazing (fallow)	3·10[e]	15·20[e]	n.a.	n.a.
Cropping index	55	35	n.a.	n.a.
Livestock (numbers)				
Cattle	21	9	—	—
Sheep and goats	10	6	—	—
Yields				
Cotton (t ha^{-1})	0·50	—	—	—
Groundnuts (t ha^{-1})	—	0·70	—	—
Sesame (t ha^{-1})	—	—	0·19	0·11
Sorghum (t ha^{-1})	0·60	0·60	0·73	0·23
Economic return ($ per holding)				
Gross return				
Cotton	148	—	—	—
Food crops	327	228	47 272	18 382
Livestock	276	138	—	—
Total	751	366	47 272	18 382
Purchased inputs	40	9	20 743[f]	11 643[f]
Income	711	357	26 529	6739
Wages for hired labour	48	—	8736	9658
Family farm income	633	357	—	—
Net return before land rent and interest	—	—	17 793	−2919
Productivity				
Gross return ($ ha^{-1} total land)	109	19	n.a.	n.a.
Gross return ($ ha^{-1} crop land)	198	96	122	48
Gross return ($ ME^{-1})	203	105	n.a.	n.a.
Income ($ ha^{-1} crop land)	187	94	69	17
Income ($ ME^{-1})	192	102	n.a.	n.a.
Labour input (man-hours)	8238	3640	32 000:	36 000[g]
Income per hour ($)	0·09	0·10	0·83	0·19
Man hours per ME	2226	1040	n.a.	n.a.
MJ per man hour	n.a.	14	112	34

[a] Unimodal pattern of rainfall. [b] Inputs and outputs in year 2 and 7 after clearance. [c] On-farm labour force only. [d] Mostly interplanted with grain legumes. [e] Additional communal grazing is available. [f] Including farm-produced seeds. [g] Guestimate of the orders of magnitude. *Sources:* Tanzania: Pudsey (1977, personal communication; Mali: Lagemann (1978, personal communication); Sudan: Simpson (1979).

Sukuma holdings are typical subsistence-cum-cash-crop enterprises. The food-crop area, with predominantly maize in the more humid and sorghum in the drier parts, amounts to 0·25 to 0·30 ha per person and 1·6–1·8 ha per family. The cotton-crop area varies with the objectives and possibilities of the farmer and is in the example, with 1·25 ha, somewhat above the average. Rice and sweet potatoes are cropped in valley bottoms (Fig. 4.8). A gradual increase of manioc cultivation over time has been observed in the area, which

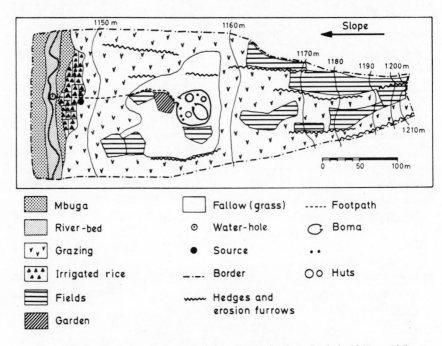

Fig. 4.8. Land use in a Sukuma holding, Tanzania (from Ludwig 1967, p. 104).

may be taken as an indication of growing land shortage. Traditionally 6–8 crop years were followed by 6–8 years of grass-fallowing (von Rotenhan 1966; Collinson 1962–5). In much of Sukumaland, however, fallowing has in the last decade significantly been reduced and occupies no more than 20–30 per cent of the potentially arable land. The example (Table 4.8) is taken from an area where the traditional pattern of fallowing is still maintained. Observations in the various surveys indicate that the reduction in fallowing has been accompanied by a gradual decline in cotton yields. Manure, however, is used in exceptional cases only. It usually remains in the boma and the boma site is moved in long-term intervals, thus producing highly fertilized plots for garden crops.

An important characteristic of Sukuma agriculture is the intensity and quality of hoe-cultivation. Crops are planted on ridges, 50 cm high and 1·3 to 1·5 m apart. The ridges are organized in such a way as to reduce erosion. Labour inputs per hectare are high and the return per hour of field work is low, as in most hoe-cultivation systems. Roughly half of the holdings own cattle and the average herd size is 21, plus some sheep and goats. Household surpluses tend to be used for cattle purchase, to establish a buffer against crop failures. Also livestock is an asset which does not lose its value in inflationary periods. Cattle are also used as dowry, but most families also accept other forms of payment.

The livestock is individually herded, in the rainy season mostly on fallows, and in the dry season on wide valley bottoms (mbuga) which then lose their swampy character. The value of livestock products amounts to roughly a third of total gross returns and a third of the offtake is sold. The labour input for watering and herding a herd of cattle amounts to 4500 hours for a holding with cattle, and thus surpasses the labour requirements of arable cropping with 3414 hours. However, most of the labour need falls into the dry season. Also ox-plough cultivation expanded rapidly since the 1950s and has become the predominant mode of cultivation in the drier, less densely populated areas.

Sukuma farming experienced the rise from subsistence orientation to cash orientation within a few decades, but the system clearly tends to encounter serious constraints. The land resources are increasingly utilized and soil fertility declines. The labour capacity is fully utilized by labour-demanding types of cropping and livestock husbandry. The return per hour of work is low. The effectiveness of mineral fertilizer is low because of low pH and much leaching on the mainly sandy soils. Most farms have become too small for a lay system proper. The system is heading towards a low-level equilibrium situation with permanent cropping on depleted soils.

The groundnut-millet holdings in Mali receive one-third less rainfall than the holdings in Sukumaland, but the rainfall pattern is unimodal and allows groundnut and millet cultivation. Land use is decidedly more extensive. Crops are grown with a minimum tillage technique and not with the labour-demanding ridge cultivation practised by the Wasukuma. Food and animal production are subsistence-oriented. Most of the cash is received from household members who work in higher rainfall areas of West Africa.

4.4.4. Partially mechanized fallow farming in a semi-arid climate: sorghum–sesame farms in the Sudan (Table 4.8)

The Sukuma data contrast with those of mechanized farming in the Central Rainlands of the Sudan where about 1·4 million hectares are operated by large-scale tenants (Simpson 1979, p. 6). Partial mechanization prevails, where cultivation and harvesting are mechanized, while great numbers of migrating labourers weed the land. The gross return to the land in these large-scale farms is a third or a quarter of that of smallholder farming in Sukumaland. The income per hour of work, however, is much higher under the condition of ample land.

In most cases production is not yet stationary. Virgin land is cleared, cropped for a number of years, and then abandoned. Permanent arable cropping is tried by a number of farmers, but runs into major weed and soil-fertility problems. Production then occurs at a low-level equilibrium which is insufficient to cover operating costs (Simpson 1979, pp. 29–32). A growing number of farmers tend towards fallow systems where several crop years are followed by a lengthy fallow with a natural vegetation of bush and grass.

Large-scale sorghum–sesame farming, as practised in the Central Rainlands, is not yet a balanced system of land use. Crop yields are relatively high after clearance averaging 0.7 t ha^{-1} with sorghum and 0.23 t ha^{-1} with sesame. After 6 to 8 years they are down to a half or a third of that level due to weeds and declining soil fertility. Crop production is highly profitable in the initial years. Overtime operation costs tend to surpass revenues (Simpson 1979, p. 51). Farmers tend therefore to 'mine' the land for several years and then to move to virgin land. Arable production thus expands at the expense of traditional grazing areas of pastoralists who object. There is social unrest in the area.

For the system to become viable and stable, herbicides and mechanized weeding will have to be introduced. At the same time there will have to be a change from crop production only to a ley system with an alternation of crop years and several years of grazed leys—which might be solutions of the weed and soil-fertility problem.

4.5. Problems of the system

4.5.1. *Soil-fertility maintenance*

Partly commercialized fallow systems are distinguished from shifting systems both by more intensive land use and by the fact that technical innovations can be more easily applied. People live in stationary housing with permanent gardens. New varieties and production methods can make head-way, especially where a major cash crop is grown, like cotton or groundnuts. Cultivation with ploughs is possible where the fallow vegetation consists of grass. At the same time, the first steps are being taken in investing labour in plantations of perennial crops, permanent roads, and other long-term capital works.

Despite these development advantages, there are a number of basic draw-backs which are inherent in fallow systems of this type, and are not just shortcomings in husbandry. A central problem is the maintenance of soil fertility. Manuring with animal dung and fodder cropping are exceptional, and the effectiveness of mineral fertilizer is usually marginal, due to interactions between the low cation-absorption capacity of most soils and bottle-necks in the labour economy which allow fields to be crowded with weeds. Other things being equal, the yields per hectare are often less than in shifting systems. At the same time, especially where the plough has replaced the hoe, erosion damage is on the increase. Fallow systems exist by soil-mining, and efforts to prevent it are usually not economic given the price relations in the location and the preference of the people concerned. The return to soil-preserving measures (green manuring, compost, terracing, etc.) is too low in relation to the disutility of effort, and there is not yet the scarcity of land which would bring about a change in preferences.

However, fallow systems may supply a high output for a very long time and may be economically very acceptable during the early stages of industrializa-tion. Developments which move shifting systems into more commercialized fallow systems with extended cash cropping (with cotton, groundnuts, rice, etc.) are usually regarded as important successes in agricultural policy.

But fallow systems are rarely balanced systems. They are transient systems which usually move more or less slowly towards a low-level steady state. Observations indicate that the decline in yields per hectare is not linear or a smooth function of the land-use intensity (Flinn, Jellema, and Robinson 1974, p. 8). In Eastern Nigeria, for instance, yields which are relatively high after clearance of long-term fallow decline as the R value increases, but tend to stabilize at an intermediate level with an R value of about 50. Farmers tend to remain at that level of land-use intensity as long as possible, because the difference in yield between such a fallow system and permanent cultivation at a low-level equilibrium seems to be very pronounced (see Fig. 4.9).

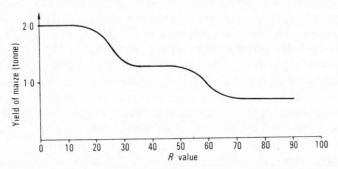

Fig. 4.9. Yield levels with growing R value: a hypothesis for Eastern Nigeria. Source: Personal enquiries.

Farmers in fallow systems usually elect to follow a number of alternatives before changing to permanent upland cultivation.

1. There is strong pressure for moving into expansion areas with lower population densities and for taking up off-farm work.

2. Intensification occurs earlier in the valley bottoms than on the upland because of higher marginal returns to additional labour (see § 6.4.2).

3. Farmers tend to develop dualistic holdings with a small permanent compound farm, planted with trees and bananas and fertilized with household wastes, and a larger outfield which remains in a fallow sequence.

These three strategies certainly prolong the lifetime of a fallow system, but the change to permanent cultivation occurs as a necessary consequence of population increase and more cash cropping, and within the traditional technology this means a gradual movement towards a low-level steady state (see § 6.5).

For the West African Sudan Zone the process is described by Funel (1978) as follows:

1. The initial stage is fallow farming with extended fallow years.

2. The introduction of folding around the village establishes a belt of permanently cropped land.

3. Permanent cropping surpasses the manure supply. Land use is no longer balanced.

4. Finally all suitable land around the village is permanently cropped. There is not enough manure; fertility and yield decline. 'Le système se trouve totalement bloqué' (p. 4).

4.5.2. *Degeneration of the livestock economy*

Another conspicuous aspect of fallow systems is the degeneration of the livestock economy once grazing becomes scarce. Static animal-husbandry practices within a changing farming system almost inevitably lead to problems. As long as there is sufficient grazing land available, the herds per family are of a good size and the cattle are well fed. An increase in the area needed for arable farming necessitates a reduction of the grazing area. Simultaneously, the proceeds from crop farming are partly being invested in cattle, to obtain security and status. The results are an increase in the cattle population and a reduction in the grazing area, causing complementary damage. The condition of the cattle deteriorates and erosion increases. On over-grazed land, a high proportion of the rain runs off. Over-grazing encourages the thickening of bush, particularly those species on which livestock do not browse. Thus in the long run over-grazing reduces the carrying capacity of the grazing area, reduces the length and the effectiveness of the fallow period, and increases the amount of clearance work required.

The practice of unorganized communal grazing is at the root of the problem. Usually everybody has the right to graze all his cattle on the available grazing, including the fallows. No improvement of the grazing economy is of value, however, unless the stock numbers are limited to the carrying capacity of the grazing land, and unless the movement of stock within any given area is controlled by a rotational grazing system. A change in the land-tenure systems is a prerequisite for a change in husbandry practices at the farm level.

4.5.3. Bottlenecks in the labour economy

In most fallow systems only about one-third to two-thirds of the available labour capacity is absorbed in field work (see Cleave 1974, p. 32). Nevertheless cultivators almost everywhere, regardless of the degree of underemployment, unanimously feel that labour shortage is the most important factor limiting output. When asked about this, they almost invariably argue that they would like to produce more and earn more, but they are not in a position either to cultivate large areas, or to change to labour-demanding crops, or to cultivate, plant, weed, and harvest in time, or to increase the labour input per hectare of a given crop, because there is not enough free family labour available during the relevant weeks, and there are not enough funds to pay hired labour. Much, if not most, of the cultivated land is neither planted nor weeded according to a timetable that would optimize yields.

The fact that important labour shortages within family holdings and a high degree of underemployment go hand in hand may be traced back to a number of major factors.

1. It is well known that labour peaks in agriculture, particularly with arable crops, occur seasonally. The Wasukuma, for instance, have a great deal of work to do between December and April. In the second half of the summer there is no field work at all.

2. Agricultural work is not the only work that has to be performed in smallholdings. In 14 cotton–maize holdings in Machakos District, Kenya, Heyer (1966) established that on average throughout the year the ratio of working hours on the land to household work is 1 to 1·7, i.e. general household work is much more time-consuming than field work.

3. The labour capacity of the various persons in the household is not utilized equally, because of traditional concepts of the way in which work is divided between men and women. It may be, for instance, that the women cannot weed in time because the work conncected with the household demands too much attention. On the other hand, the men who are unemployed at this time do not consider that tending the food crops should be part of their work.

4. Another important factor is the low efficiency per hour of work. This is often caused by unsuitable implements. In addition, smallholders are accustomed to making a concentrated, sustained effort only in connection with a few procedures, such as felling, clearing, and hoe-cultivation. The other jobs have traditionally been done in a leisurely manner.

5. Finally, the extent of underemployment is calculated according to norms of working hours per year and per man-equivalent. The smallholders who refer to difficulties in the labour economy are naturally not

familiar with such norms. Many of them do not accomplish more work because they regard a 5-hour day, even at planting time, as a complete utilization of their working capacity.

Many of the problems of the labour economy arise because these systems are a transitional phase between shifting and permanent land use. The advances in the labour economy that are characteristic of permanent arable farming—row cultivation, ploughing, mechanical weeding, proper farm layout, and so on—have not yet been introduced on a large scale and often do not fit into fallow systems. Usually the transition to cash cropping, with its consequent labour demands, takes place within a fairly short period, before the farmers are accustomed to longer hours of work per day. On the other hand, the cropping areas are relatively large, and the labour requirement is seasonal. Consequently, there is a lively interest in labour-saving innovations, even more than in yield-increasing innovations. Farmers in Sukumaland do not generally change from hoe work to ox-ploughing to gain higher returns, but to replace hard work with the hoe by easier work with the plough. Particularly striking is the general interest in tractors, Wherever finance is available, and the return per hectare high enough to justify tractor ploughing, smallholders endeavour to buy tractors or to hire tractor services.

4.5.4. *Risk of drought*

Fallow farming is widely practised in semi-arid areas where droughts occur with certain probabilities. In a few highly developed farming systems— as for instance those of the Dogon in Mali—food-crop storage over several years is part of the traditional farming pattern (Gallais and Sidikou 1978, p. 16). Fallow farming expanded in extended medium-low potential areas of Kenya also, because farmers expect to receive famine relief in case of crop failure. In most cases, however, smallholders operate neither with appropriate storage strategies nor with any reliable famine-relief system. They generally prefer low-return and low-risk types of farming. Arable cropping is supplemented by drought-tolerant permanent crops such as tree-cotton (Sertao, Brazil) or sisal (Sukumaland), and animal production (Sanders 1979).

4.6. Development paths of fallow systems and unregulated ley systems

Production increase in fallow systems normally results from extending the crop area by clearing the land in neighbouring extension areas and from reducing the fallow area. The expansion of cash cropping in fallow systems is often opposed by an alliance of political economists, ecologists and rural health advisers, who argue that such an *économie de traite* exploits natural stock resources, produces higher incomes for a limited time only and that it is often accompanied by a decline in rural nutrition. It is usually overlooked in this context that the reduction in fallowing is mainly due to the food demand of a growing population. Once the land is fully occupied land-use intensification is the only answer to the challenge posed by increasing numbers of people living on the land. Cash cropping allows the purchase of manufactured inputs—in particular mineral fertilizers—which are essential for the development of more productive and stable land-use systems. The quicker the development of cash cropping the better the chances are of preventing the tendency towards involution, which is usually very strong in a smallholder subsistence economy. If the obvious tendency towards involution is to be replaced by evolution towards more productive and more balanced systems, then purchased innovations have to be introduced and cash cropping is required to buy them. We distinguish in this respect between improved systems and such innovations and new combinations that eventually create other, more productive systems.

4.6.1. *Improved systems*

In considering field cultivation, we must emphasize the supreme importance of better husbandry. Established routines of work, which might have been the best in shifting systems, are still practised in fallow systems, although the actual farming is more intensive. Consequently, there is much scope for better husbandry. The main shortcomings in husbandry that could be at least partially remedied are usually the following.

1. Bottlenecks in the labour economy often result in late planting.
2. Most farmers tend to use plant spacing that is greater than is advisable. They aim at robust individual plants and overlook the advantages of an optimum plant density.
3. No other factor is as much to blame for poor yields as late weeding. This can be only partly explained by labour problems. Probably farmers would harvest more if they planted smaller plots and devoted more care to weeding.
4. There is further scope for the introduction of new varieties. The cultivation of better-yielding types of cotton and groundnuts brought noticeable successes in a number of cases.
5. The lack of field roads and carts make the application of kraal manure difficult. This makes the introduction of mineral fertilizers all the more relevant in some locations.
6. Few other measures promise such great increases in returns in relation to the additional costs as the supplementation of the plough by an ox-drawn weeder, in conjunction with the introduction of row cultivation.
7. The quality of ploughing with oxen is usually very poor. Proper ploughing would not only raise yields, but would also significantly reduce the weeding task.
8. Where it is economic to keep tractors, very real improvements are

possible, especially timely cultivation, the better ploughing-in of weeds, deeper ploughing, and more effective use of mineral fertilizers.

9. In areas infested by the tsetse fly, the introduction of N'Dama breeds, which are tolerant to trypanosomes could be considered, in order to reap the benefits that cattle may bring.

Fallow systems as a whole represent a promising step towards introducing technical innovations into farming, and therefore towards agricultural development. The scope for application of innovations is incomparably greater than in shifting systems but any lasting improvement would require an intensive fertilizer economy and thus a change in the system.

4.6.2. *Change towards more intensive types of farming*

Fallow farmers, faced with increasing land shortage, may escape the low-level steady-state trap of traditional permanent farming under tropical lowland conditions by (1) developing valley bottoms with wet rice, and (2) by changing to permanent crops, in particular bananas (plantains), which can supply a great amount of subsistence food in a balanced land-use system (see § 8.4.3). However, these two avenues for evolution are not open everywhere. They require high investments and much time, and so few development efforts prior to the late-1960s aimed in either direction. Up to the mid-1960s most agriculturalists from temperate climates considered ley farming as the proper system to replace obsolete fallow systems.

(a) *Ley systems*. In many fallow systems, grass or bush vegetation grows in the fallow period and is then grazed. It is therefore natural to seek the solution to the problem through the changeover from unregulated to regulated ley systems. This involves replacement of the several years of fallow, with their natural regeneration of grass and bush, by deliberate cultivation of a plant or plant mixture which is more productive than the fallow vegetation, re-generates soil fertility, and allows plough cultivation. The techniques of ley farming in the conditions under which fallow farmers produce have been established by several decades of research. Fertilized leys are effective not only in semi-arid and semi-humid climates but also in humid climates, where 2 years of upland rice may be followed by 2 years of *Stylosanthes guyanensis*, grazed by trypano-tolerant N'Damas.

There is reason to assume that, given favourable price relations between agricultural produce and mineral fertilizers, ley systems are profitable pro-positions, but they require medium- or large-sized mechanized holdings. The problem is with the changing of existing smallholder fallow systems into ley systems. Ley systems call for considerable investment in clearing, livestock, fences, implements, carts, tracks, and buildings. The cost of draught oxen and their necessary accessories alone amounts to two or three times as much as the gross annual earnings from these holdings. Another pre-condition for regulated ley farming is land consolidation. Holdings have to be in one block, with fields large enough for plough cultivation.

The fact that a 'technological jump' must be made presents further obstacles. In the holdings in Sukumaland mentioned above a whole series of innovations have to be adopted in combination if success is to follow, and the same applies to most fallow systems. It is simply not enough to encourage the use of ox-carts to take animal manure from the kraals to the fields, since the tracks also have to exist. But since there are none, unpaid communal work is necessary to build them.

Farmers can scarcely be expected to establish leys and buy seeds and mineral fertilizer for them without obtaining a direct economic return. Keeping dairy animals again necessitates the harvest, transport, and storage of fodder, correct feeding, and, above all, keeping the animals healthy. The technological jump can scarcely be accomplished by an individual without the consent or the active support of the group, and this further complicates the issue. Profitable dairy farming involves keeping breeds with a high milk yield, maintaining their health, and making available suitable male animals or semen. The effectiveness of the necessary veterinary care on a single farm is doubtful unless neighbouring animals are also treated. Fodder cropping clashes with traditional rights because the fallow pasturage has been available for everybody's animals for generations. Thus, for anyone to crop fodder plants on the fallow it is necessary for the others to agree to renounce their grazing rights. It is generally the case that recommendations for a transfer to regulated ley systems with dairying, fodder cropping, dung and mineral manuring, etc., can be expected to succeed only where there has been a land shortage for a number of decades. Where there is no such pressure, there is little desire to change, especially as, even with the continuing loss of soil fertility, farmers can still gain a comparatively high income.

Another way of introducing stable systems of rain-fed farming is through settlements under close supervision. Farmers taking part in a settlement have to obey the project leader in matters concerning rotations, fertilizing, animal-keeping, fodder, veterinary supervision, and so on. The lack of knowledge and finance that prevent the individual from making the leap into proper ley farming are counteracted by the settlement rules and the credit afforded by the supporters of the settlement. In Africa, south of the Sahara, there are numerous examples of this kind of approach to the modernization of agriculture, but they usually reveal an unsatisfactory ratio between input and yield. Most settlements need continual financial support, or they are abandoned after a few years. The reasons are usually quite straightforward: on the one hand, the cost of personnel for supervision is high; on the other hand, the land that is mostly used is capable of producing only moderate yields. Close supervision, which can be rewarding in irrigation agriculture and with certain perennial crops, is normally not worth while with ley systems, unless cash crops with a high return per unit of area, like tobacco, can be cultivated. Given this situation, for the time being probably nothing can be done but to accept fallow systems. The long-term hope is that increasing commercialization will demonstrate more and more clearly the inadequacy of the system and create enough incentive for the farmers to remedy the situation. It is very possible that the judgement of Rounce (1949) applies generally to fallow systems. He wrote about Sukumaland: 'A situation of overpopulated villages and impoverished soils must be passed before better husbandry can spread' (quoted from von Rotenhan 1966, p. 52). By then, however, the countryside will be too crowded for ley systems.

(b) *Permanent cultivation.* The change from fallow systems to permanent cultivation is a difficult proposition in a humid and semi-humid tropical lowland climate, and the systems that do exist are either traditional 'island' situations or are in areas with fertile alluvial or volcanic soils.

Past efforts to introduce improved permanent cultivation systems to African lowlands failed, mostly with heavy losses. During the groundnut scheme (1947–9), about 90 000 ha were cleared and abandoned with a loss of about $70 million. Particularly conspicuous are the failures in the semi-humid or humid tropics. A group of about 40–50 French farmers settled in 1949 at Aubeville (Congo-Brazzaville) started farming according to European patterns and abandoned the site after a few years (Gourou 1969, pp. 101–12). Of Ghana's 30 large-scale mechanized State farms established in the early 1960s only a few survived, and losses have been heavy. West, East, and Central Africa are dotted with many more development projects which aim at substituting permanent arable cropping for fallow systems. None of them, to my knowledge, has become an economic success, and most failed within a few years, but nevertheless many more projects are being planned and financed and considered as worthwhile development efforts. The change from fallow systems to permanent cropping is still in the process of trial and error, at least in tropical Africa (see § 6.6).

The situation is different in the semi-arid and highland areas. Here the change from fallow systems to permanent cultivation has occurred in wider areas. Permanent cropping usually begins with the dung land around huts and with valley-bottom land, then gradually moving to upland. The long-term trend is clearly towards permanent arable systems, and it seems most likely that ley systems, as an intermediate stage in land-use intensity, will be by-passed.

(c) *Artificial pastures.* Another possible avenue for the development of land use in areas with fallow systems in semi-humid climates is the establishment of artificial pastures. The potential in terms of dry-matter and starch-equivalent production is very high, and acceptable levels of digestibility and protein for beef production can be provided by adequate management practices. Artificial pastures are much easier to establish and maintain in semi-humid or humid zones than fields with permanent cultivation. Tsetse-tolerant breeds may be kept in areas with Trypanosomiasis. However, heavy inputs of mineral fertilizer are required to maintain the pastures on most soil types, and price relations between beef and fertilizers have to be favourable to make artificial pastures economic (Ruthenberg 1974).

5. Ley systems and dairy systems

5.1. Definition

THE word 'ley' is used wherever several years of arable cropping are followed by several years of grass and legumes utilized for livestock production. A distinction is then made between unregulated and regulated ley systems. Unregulated ley systems in the tropics are characterized by a natural fallow vegetation of various grass species, a certain amount of bush growth on the pasture, communal grazing, and a lack of pasture management, all of which make such systems often more akin to short-term fallow systems than to regulated ley systems. Individual grazing, fencing, pasture management, and rotational use of the grassland are the usual characteristics of regulated ley systems. The seeding or planting of grass leys, the production of hay or silage, and the application of mineral fertilizer are indicators of intensive types. Ley systems show a more or less regular alternation between arable farming and leys. The sum of the arable years and the ley years tells us the rotation cycle. The leys may be used either for grazing or as meadows producing roughage for stabled cattle. However, the borderline between unregulated and regulated systems is often difficult to draw.

5.2. Types of ley farming and their geographical distribution

Regulated ley systems, which are widespread in the subtropics of the northern and southern hemisphere, are rare in the tropics. Those that do occur are more or less restricted to areas that show one or more special features, namely:

1. Those which lend themselves to the vigorous growth of palatable grasses.

2. Those where the dependence on the long-fallow system is not pronounced, i.e. where relatively few years of grass are sufficient to restore soil fertility.

3. Those where land becomes so scarce that intensification is required beyond unregulated ley farming, yet is not so scarce that permanent cropping is necessary.

4. Those where the health risks of intensive animal-keeping are comparatively low.

5. Those where price relations favour intensive forms of meat and milk production.

6. Those where sufficient draught animals or tractors are available to plough the leys.

7. Those where farmers have sufficient knowledge and capital to organize a proper fodder economy over the year.

These conditions are to be found in several tropical high-altitude large-farm areas. In addition, some systems with regulated ley farming have developed in semi-arid savannas, but ley systems in the semi-humid and humid tropics are still in the experimental stage.

The organization of ley systems is closely related to farm size. We can therefore classify the main forms into three principal groups.

5.2.1. *Regulated ley systems in smallholder areas*

There are but few cases where the evolution of indigenous farming leads to regulated ley systems. One of the most evolved traditional ley systems in the tropics is to be found in the highland of Ethiopia and involves a rotation including not only intensively grazed grass fallows but also 'teff' (*Eragrostis abyssinica*), a grass-type crop producing seeds for human consumption and valuable straw for cattle (Westphal 1975, p. 83 pp). Growing land shortage induced a change in system and we nowadays find mainly permanent cultivation (see § 6.4.1). Noteworthy are some modernized smallholder ley systems with intensive dairy production in Kenya (see MacArthur 1964 and Stotz 1977a).

5.2.2. *Regulated ley systems in settlement schemes*

The African savannas are dotted with settlement schemes, most of which aim at the introduction of regulated ley systems and modern animal husbandry. The organization of production within settlements is usually considered as an instrument for introducing the complete set of innovations that is required for the jump into a new and demanding technology. In the Kenya highlands, settlement of ex-European farms and estates led to the establishment of numerous smallholdings with ley farming and intensive milk production as major planning objectives, which achieved a certain degree of success. However, few of the numerous smaller settlement schemes in medium-altitude and low-altitude areas of East and West Africa, which were planned to incorporate mixed farming with leys, have proved to be successful in either a technical or an economic sense. They may be considered worthwhile pilot schemes, but cannot yet be regarded as economically viable and established farming systems.

5.2.3. *Regulated ley systems in large farms*

Most established ley systems are found on large, mechanized farms, primarily because the use of the tractor is possible only with a complete clearance of bush and stumps, i.e. it requires a fallow vegetation of grass. Furthermore, the profitability of the leys depends on modern forms of animal husbandry, which are much more easily introduced in large than in small farms.

The following large-farm ley systems are of some importance.

1. The tobacco farms in the drier African savanna of Zambia and Zimbabwe are largely based on leys. The prime objective of the ley is not to produce fodder but to free the soil from nematodes.

2. In the Kenya highlands, wheat, pyrethrum, sheep, and beef and dairy cattle are the main branches of activity in large modern ley farms (Mac-Arthur and England 1963). Less-evolved ley systems are to be found in the highlands of Angola and Mozambique (Pössinger 1967).

3. Extensive areas in the Cerrado, central Brazil, are cropped in a 5-year rotation. One crop of fully mechanized rice, yielding about $1·5$ t ha^{-1}, is followed by about 4 years of *Bracharia* pasture grazed by cattle.

4. The rotation of several years of rice and *Stylosanthes guyanensis* is prac-tised in some West-African projects and may become important in tropical semi-humid lowlands.

5.3. General characteristics of regulated ley systems

5.3.1. *Principles of crop and ley management*

Cropping becomes clearer and more uniform as farming is carried out within clearly demarcated fields. The various distinct patterns in the spatial organization of cropping that we find in shifting and fallow systems are rarely found in proper ley systems. Fields are more uniformly cultivated, and planted predominatly in pure stands. Pseudo-rotations, with overlapping harvests and planting dates, are replaced by clear-cut crop rotations. Nutrient concentration around the homestead is uncommon because of the available transport facilities (carts, tractors, field roads) on the one hand, and because of the preference for night grazing on the other.

Tropical leys differ in the way they are established:

1. Often a sign of extensive ley farming is a natural grass regeneration. This, however, may not be true where grass growth is prolific. In the high-lands of Kenya naturally regenerated grass is found even in comparatively intensively farmed holdings. Smallholders rely almost entirely on natural grass regeneration.

2. Ley establishment by the sowing of grass seed is in general restricted to large holdings.

3. A number of especially productive grasses, eg. Kikuyu grass and elephant grass, are always established by the planting of vegetation material, not seeds. Planting is also the usual practice where (e.g. in the highlands of Kenya) the

change is made in smallholdings from naturally to artificially established swards. In any case, planting depends on the availability of ample labour.
4. Ley systems achieve the highest productivity where it is possible to sow legumes (*Stylosanthes*) or legume–grass mixtures.

Leys can be distinguished on the basis of management according to:
1. whether the leys are fenced;
2. whether rotational grazing is practised; and
3. the intensity of labour input in grassland management.

The usual way in which leys are used is for grazing. In some cases grass is also cut to make hay or silage. In the tropical highlands, however, hay-making suffers because of the heavy dew, or because it coincides with the rainy season. Thus, it is usually more advantageous to meet the dry-season fodder requirement with roughage and silage.

5.3.2. *Characteristics of the fertilizer economy*

The continuity of cropping in ley systems is usually built on the soil-conserving and fertility-regenerating capacity of grass leys. Regulated systems fulfil these functions better than unregulated systems for a number of reasons.
1. The fencing of leys and the systematic rotation of cropping and grass reduce the danger of over-grazing and the resulting erosion damage.
2. The systematic establishment, management, and use of the leys bring about a better soil coverage and allow the development of a dense root system.
3. Regulated systems are occasionally combined with folding systems, in which there is more manure because of the more intensive stocking.
4. Where ley farming is practised with a legume–grass mixture, not only is there a significant increase in the quantity and quality of fodder, but an increase in nitrogen fixation is also brought about.

It should be stressed, however, that the soil-preserving characteristics of leys as grown in temperate climates or at high altitudes cannot be simply repeated in medium-altitude and low-altitude tropics. Webster and Wilson (1967, p. 318) write in this connection:

> The grass ley probably has a modest, direct, beneficial effect on the nutrient and physical status of the surface soil, and, compared with continuous cultivation, a rotation including a grass break is likely to be advantageous in reducing erosion and decreasing the incidence of weeds, pests and diseases in the arable crops. However, unfertilized grass leys are no more effective in maintaining fertility than are fallows of natural regeneration.

A large number of legume–grass mixtures have recently been developed for the subtropics and some tropical high-altitude areas. A suitable legume for ley farming in the humid and semi-humid tropics is *Stylosanthes guyanesis*, and new systems based on this crop are in the process of being developed.

5.3.3. *Characteristics of animal husbandry*

The obstacles preventing rationalization of the animal economy that are found in unregulated ley systems with communal grazing exist in regulated ley systems only where communal grazing is not replaced by definite grazing rights on each holding. In so far as this has occurred, there is a tendency for more intensive forms of animal husbandry to be employed, whereby the production of beef and heifers tends to be of primary importance on large holdings, and milk production on smallholdings.

Regulated ley systems frequently show advanced husbandry practices. On settlement schemes and large farms, we usually find improved or grade cattle and planned breeding programmes. Whereas fallow cultivators keep their animals overnight in a kraal with fencing, in regulated ley systems night grazing is possible, and the need to walk long distances from the kraal to the grazing areas and water is thus obviated. Furthermore, fencing makes it possible to graze calves, heifers, and cows separately. Another characteristic of regulated ley systems is the more intensive fodder economy: rotational grazing and arable fodder production are standard procedures. However, the use of concentrates, which is important in tropical systems with permanent arable farming, plays a lesser role in ley systems.

The tropical highlands with sufficient rainfall afford especially favourable conditions for low-cost animal production in regulated ley systems. It is much easier and cheaper to keep grade animals healthy in the cool mountain areas than in hotter climates. When the rainfall is evenly distributed, the cattle may graze almost all the year. Arable fodder might provide some additional roughage in a dry spell. A good example of low-cost animal production in regulated ley systems occurs in the large-farm economy of some parts of the Kenya Highlands.

5.3.4. *Characteristics of the labour economy*

Whereas unregulated ley systems are mainly hoe economies, in regulated systems we find plough cultivation, largely because of the traction power required to plough the leys. Where the holdings are too small to keep draught animals or tractors, ploughing is usually undertaken by a contractor, as for example is widely practised in Kenya.

Ley systems as a whole may require a low labour input per unit of output in relation to other farming systems. The weeds that grow on arable land are suppressed during the grass years. In fact, the labour input in weeding is very often less than in systems with permanent farming. Whereas the clearance work in unregulated systems has still to be done largely with the hoe, in regulated systems the sward is broken with the plough. In most areas the use of tractors in ley systems becomes a financially viable proposition. A further important point is that work in animal busbandry involves the relatively regular employment of labour.

Low labour requirements per hectare and relatively high labour productivities are typical for smallholder ley systems in Kenya. Labour requirements per cow and year are around 200 hours (300–400 hours ha^{-1}) which is significantly less than in arable cropping. Labour productivity clearly increases with the change from Zebu to grade dairy cattle, as it is indicated by a comparison of the data in Tables 4.3 and 5.1. However, with increasing land shortage, farmers tend to change from grazing to partial- and zero-grazing systems which produce much higher returns per hectare, but significantly lower ones per hour of work. Swoboda (1978, personal communication) established that two dairy cows require 912 hours of work per year for milking, transporting of milk, dipping, and watering. Including the labour requirements of fodder production, total labour input per cow and year are around 1000 hours (see Table 5.1).

TABLE 5.1

Input–output relations of smallholder milk production in Kenya (1976)

Technique	One grade cow on permanent Kikuyu grass pasture	One grade cow with zero-grazing, feeding with Napier grass (*Pennisetum purpureum*)
Method	Models derived from survey data	
Land requirements (grazing or fodder) in hectares per cow	0·61	0·20
Labour requirements (man-hours) including fodder production	200	1000
Output per cow ($)		
Milk (2500 kg)	301	301
Calf	12	12
25 per cent of cull of cow	24	24
Manure	—	72
Total	337	409
Material inputs per cow ($)		
Concentrates	48	18
Veterinary and dipping	7	7
Replacement	24	24
Risk (5 per cent mortality)	12	12
Depreciation of structures	24	48
Fodder production	—	36
Miscellaneous	2	16
Income before interest and overheads ($)	220	248
Income ($ ha^{-1})	361	1240
Income ($ man-hour^{-1})	1·10	0·25

Source: Stotz (1977*b*).

The low employment content of ley farming is also reflected by the data in Table 5.2. Both shifting and fallow systems in Africa involve more labour inputs than the medium-sized ley farming system of the Nandi which offers high returns per hour of work, but which can only be maintained under the condition of low population pressure.

TABLE 5.2

Use of time in various African farming systems

System	Shifting[a]		Fallow[b]		Ley[c]	
Type of farming	Food crops and cacao		Tobacco–millet		Maize–milk	
Country	Congo		Uganda		Kenya	
Location	Boutazab		Kigezi		Nandi	
Year	1972–3		1971–2		1976	
Method	n.a.[d]		Daily questioning		Daily observation[e]	
Use of time as percentage of daytime, 84 hours per week[f]	Males	Females	Males	Females	Males	Females
Agriculture	23	20	24	25	21	20
Hunting and fishing	5	2	—	—	—	—
Household	10[g]	27	1	15	1	41
Social activities ⎱ Meals and leisure ⎰	44	35	16 / 26	21 / 17	5 / 39	2 / 28
Illness ⎱ Visits ⎰	10	16	3 / 15	4 / 8	1 / 12	1 / 4
Off-farm work	8	—	15	10	21	4
Total	100	100	100	100	100	100
Employment (hours per year)						
Agricultural work	1223	961	1048	1094	920	876
Off-farm work	349	—	657	438	920	175
Household work	437	1179	44	657	44	1796[h]
Total work	2009	2140	1749	2189	1884	2847
Total work in per cent of total day time	46	49	40	50	43	65

[a] A recently commercialized system in a humid tropical lowland under the condition of ample land. [b] A low-income small-farm community with land shortage (see Table 4.7). [c] A high-income medium farm-size community under very favourable natural conditions. [d] The data are for a 4-month period which is roughly representative for the year. The men's load is slightly underestimated. [e] The data rely on 6 months of observation. [f] The original data are on a 13-hour day in Kigezi and a 12·5-hour day in Nandi. They have been brought on a 12-hour basis by deductions from leisure time. [g] Household work such as hut-building, repair of tools, etc. is registered by Guillot under 'other work' which also includes off-farm work. It is assumed that 56 per cent of the 'other work' is related to the household. [h] Including crop processing. *Sources:* Congo: Guillot (1978, personal communication, Table 25); Uganda: Kyeyune-Sentongo (1973a); Kenya: Oboler (1977).

5.4. Examples

Ley farms in tropical highlands tend to show characteristics similar to ley systems in temperate climates, and this is shown by the examples from Kenya.

5.4.1. *Large-scale wheat-dairy farms in Kenya* (Table 5.3)

Regulated ley farming was introduced and adapted to the conditions of the Kenya highlands by British farmers, and the example from Njoro shows the land-use pattern which is still practised on large group-owned farms. The emphasis is on wheat, and there is a large dairy herd with leys as the fodder basis. The dairy economy is relatively extensive with long calving intervals, low inputs of concentrates, and relatively low yields of milk per cow and year.

5.4.2. *Small-scale ley farms with dairy cattle in Kenya* (Table 5.3)

In the 1960s and 1970s grade dairy production on the basis of ley systems was increasingly adopted by smallholders. The dairy ranching farms at South Baringo and the dairy–maize holdings at Kericho developed within a few decades out of sedentary pastoralism. A land tenure reform transferred traditional land rights into private ownership of land, introduced fencing, and thus allowed the development of an important smallholder dairy industry. Smallholders prefer maize to wheat, as is shown in the Kericho example. Maize rotates on the grassland; 2–3 years of maize are followed by 4–5 years of grass. The land is usually ploughed by hired tractors or hired ox ploughs, planted in rows with hybrid seeds, fertilized and weeded, and harvested with hand tools. The change from hoe cultivation to ox-ploughing reduces the labour input for maize by about half (627 compared to 1157 hours ha^{-1}, Frey 1976, p. 6). Owing to the cool climate and the short days of an equatorial area, the maize grows slowly and is harvested about 8–9 months after planting. Maize-growing is closely related to the livestock economy. The stover is grazed, and after a few weeks the harvested field usually shows a dense vegetation mainly of Kikuyu and Star grass. No artificial ley need be established in order to obtain a productive pasture.

Tea, pyrethrum, and potatoes have been introduced as cash crops. Zebu cattle are substituted by European milk breeds. Dairying has become the main activity, relying on fenced paddocks, weekly dipping, and artificial insemination or the use of grade bulls. Fencing is expensive and fenced areas are rarely used for growing maize. The greater the fenced part of the farm, the shorter the number of grass years in the maize–ley rotation. The tendency is towards permanent arable cropping with maize and potatoes on some part of the holding, maintained by mineral fertilizer, while the rest tends towards permanent pastures.

The Baringo example shows the expansion of grade dairy production into drier areas. Here the traditionally unfenced permanent pasture has gradually

TABLE 5.3

Farm-management data of ley farms and other dairy farms in Kenya, Brazil, Costa Rica, and Colombia

Country Location	Kenya Njoro	Kenya Southern Baringo	Kenya Kericho[a]	Brazil San José do Rio Padre, Sao Paulo	Costa Rica Sta. Cruz, Turrialba	Colombia Bogota	Kenya Kiambu
Rainfall (mm)	900	750	1800	1300	2800	1060	1200
Altitude	2000	1500	1700	600	1200	2600	1700
Year	1977	1977	1977	1974–5	1977	1977	1978
Type of farming	Large farm Wheat–milk	Smallholdings Milk–maize	Milk–maize	Small entrepreneur Coffee–milk	Milk	Milk	Coffee–milk
Type of fodder	Leys–silage	Ranching–leys	Maize–leys pasture	Permanent pasture–silage	Permanent pasture	Irrigated pasture	Zero-grazing
Method of data	Case studies of typical farms somewhat above the average			Accounts Case studies	Case study	Case study	Case study
Persons per household	n.a.	12·00	8·50	n.a.	5·00	n.a.	10·00
Labour force (ME)[b]	34·00	5·60	3·20	6·23	1·50	9·00	3·00
Size of holding (ha)[c]	312	40·00	7·00	69·11	9·00	55·00	3·00
Wheat (ha)	178	—	—	—	—	—	—
Maize (ha)	26[d]	6·80	1·37	—	—	—	0·50[1]
Other food crops (ha)	—	2·00	0·61	3·03	—	—	0·60
Coffee or tea (ha)[e]	—	—	0·61	8·40	—	—	1·00
Arable fodder crops (ha)	30	—	—	9·68	—	—	0·90
Leys (ha)	78	17·60	2·41	—	—	—	—
Permanent pasture (ha)	—	13·60	2·00	48·00	9·00[f]	55·00	—
Livestock (numbers)							
Dairy cows	103	12	2·5	30	17	100	3
Dairy followers	54	11	2·9	51	—	108	3
Beef cattle (zebus)	—	4	4·0	—	—	—	—
Sheep and goats	—	—	2·0	—	—	—	3
Performance indicators							
Maize (t ha⁻¹)	2·9	2·7	3·2	—	—	—	3·4
Wheat (t ha⁻¹)	1·9	—	—	—	—	—	—
Milk (kg per dairy cow and year)	1920	1980	2000	1287	2500	2500	—

Stocking rate (ha LU^{-1})[g]	0·7	0·7	0·52	0·83	0·53	0·32	0·23
Concentrates (kg cow^{-1})[h]	210	320	280	347	508	—	228
Milk (kg ha^{-1} fodder)[i]	1156	553	1224	665	4700	6470	6667
Labour (hours cow^{-1})	300	240	n.a.	217	265	192	642
Economic returns ($ ha^{-1})							
Gross return							
Crops	180	53	112	49	—	—	966
Milk	82	62	105	86	1385	1240	242
Other animal products		18	70	32	48	339	69
Total	262	133	287	167	1433	1579	1277
Purchased inputs							
Seeds	n.a.	2	2	1	20	10	24
Chemicals	13	—	9	30	133	191	86
Mechanization	55	18	13	10	10	99[j]	1
Concentrates	16	3	47	17	165	25	10
Minerals		6		3	188[k]	29	7
Other livestock expenses	27	2	6	6	31	11	29
Other				6			9
Income	151	102	210	100	886	1214	1111
Wages for hired labour	42	14	13	81	78	402	48
Family farm income		88	197	—	808	—	1063
Net return before taxes, interest and land rent	109	—	—	19	—	812	—
Labour productivity							
Gross return ($ ME^{-1})	2404	950	628	1852	8597	9649	1277
Income ($ ME^{-1})	1385	728	459	1109	5318	7418	1111
Labour input (man-hours)	70 000	5400	3600	12 552	4200	21 600	4840
Income per man-hour ($)	0·67	0·76	0·41	0·55	1·90	3·09	0·69
MJ per man-hour	77	70	n.a.	n.a.	39	61	n.a.

[a] 1973 data on 1977 price base. [b] Available for farming. [c] Without waste land, bush, etc. [d] For silage. [e] Tea at Kericho and coffee at Kiambu and Sao Paulo. Half the coffee in the San Paulo case is not yet producing. [f] Half is artificial pasture (Cynodon and Setaria spp). [g] On fodder hectarage only. [h] Including farm produced concentrates. [i] Without milk produced with concentrates. Per kg of concentrates, 2 kg of milk have been deducted. [j] Mainly irrigation. [k] The difference between heifer purchases and cull cow sales is included in 'other livestock expenses'. [l] Two crops per year.

Sources: Kenya: (Njoro, Southern Baringo, and Kiambu): Stotz (1977a); Kenya: (Kericho): Stroebel (1975); Brazil: Bemelmans (1978, personal communication); Costa Rica: Avila (1978, personal communication); Colombia: Seré (1978, personal communication).

been replaced by fenced Rhodes grass leys. Land-use intensity is significantly less than in higher rainfall dairy areas. The dry-season feed supply remains an unsolved problem, because smallholdings are not suited for silage production. Labour productivity, however, is high. The South Baringo type of land use provides an idea of the potential of semi-arid tropical highlands for dairying.

5.4.3. *Dairy–coffee farms in Sao Paulo, Brazil* (Table 5.3)

Dairy farmers in Sao Paulo produce under climatic conditions of production and on soils which are rather similar to the examples in Kenya, and prices for milk were relatively favourable in 1974–5 ($0·17/kg). Land, however, is ample and this leads to much more extensive types of land use. Coffee production is highly risky, and smaller farmers in the area tend to emphasize milk. Dairying is not based on leys, even though fertilizers are relatively expensive and motorization common. Farmers prefer to produce milk by combining the use of extensive, unimproved pastures with silage and sugar-cane production, mainly for the dry season. The farm fodder base is supplemented by substantial purchases of concentrates. The example is a below-average farm. Most dairy herds produce around 2000 kg of milk per hectare with a rather similar pattern of land use.

5.4.4. *Dairy farms in Costa Rica and Colombia* (Table 5.3)

Milk production in Latin America is a substantial industry, and this indicates the suitability of tropical highland areas for dairying. The example from Costa Rica is taken from a highland area with high and well-distributed rainfall and fertile soils of volcanic origin. Most of the milk in the area is still produced on integrated coffee–milk–arable crops farms, but the tendency is towards specialized dairy farms on artificial pastures, maintained with intensive grazing techniques, such as small paddocks, high fertilizer inputs, and bush control with herbicides. The division between breeding farms and specialized dairy farms is facilitated by the absence of major cattle diseases such as are to be found in African highlands. In the example all calves are sold after birth, all heifers are purchased, and the cows are kept for 5 to 6 lactations on average. The high degree of specialization allows the operation of the farm with low labour inputs per cow.

Milk production in the example from Colombia is even more intensive than in the example from the Costa Rican highlands. The farm operates close to Bogota and specializes in dairy operations. No arable fodder is grown and no concentrates are bought. Milk is produced with intensive grassland use only. About half of the land is under artificial temperate climate pastures, such as *Dactylis glomerata*, *Lolium multiflorum*, and *Trifolium repens*, and the other half is under Kikuyu grass (*Pennisetum clandestinum*). The farm area is subdivided into 27 paddocks, each paddock is grazed for

5 to 7 days with half-daily portions allotted to the cow herd and rested for 40 days. After each grazing period the pasture is fertilized and sprinkler-irrigated, provided rainfall is insufficient. The farm presented in Table 5.3 is better managed than the average dairy operation in the area. It indicates, however, that the tendency of milk production under tropical highland conditions is not towards ley farming or mixed farming. Expensive fodder conservation activities are avoided. They tend to be substituted by high fertilizer inputs in combination with supplementary irrigation. This then produces roughage of high quality throughout the year so that the purchase of concentrates can largely be avoided.

5.4.5. *Zero-grazing dairy in Kenya* (Table 5.3)

In medium population density highlands with rising wage levels, the tendency in dairy production is towards very intensive grazing systems as shown by the examples from Colombia and Costa Rica. With very high population densities and relatively low wage levels the tendency seems to be towards integrated holdings with minimum or zero-grazing dairy operations, as shown by the example from Kiambu, Kenya. In several areas of Kenya smallholders gradually changed from grazing to zero-grazing systems which produce less value added per hour of work and more per hectare (see Table 5.1). Cash cropping with coffee, subsistence food production, and grade dairy cattle compete for the limited land. The alternation of crop and grass years is abandoned. Coffee production, food crop production, and livestock production become distinct enterprises on different pieces of land. Most of the roughage is provided by Napier grass, grown on slopes and terraces. Farm-grown roughage is supplemented by purchases of concentrates. The amount of milk produced per hectare of fodder is substantial. Manure is mainly used for potatoes and grass. The country-wide tendency towards zero-grazing dairy production indicates that ley systems have to be considered as an intermediate stage in land-use intensification. With increasing land scarcity they tend to be replaced by more intensive modes of fodder production which seem to be particularly suited for smallholder farming. The examples suggest that smallholders perform better in returns per hectare and about as well in return per hour of work as the large dairy farms.

5.5. Problems of the system

We can see from the examples in Kenya and Latin America that the tropical highlands, where climate is temperate and grazing is provided throughout the year and where there is organized marketing, provide favourable conditions for dairy production and ley systems are often the initial step in a process which tends to lead towards more intensive systems. As soon as we leave the tropical altitudes and enter a warmer climate, however, ley systems and dairy systems generally meet with an increasing number of obstacles:

1. The productivity of tropical leys can be very high when measured in tonnes of dry matter per year per hectare, but it proves to be a difficult farm-management problem under tropical lowland conditions to produce an even supply of high-quality grazing or roughage.

2. The obstacles to the keeping of grade cattle in effective production multiply. Problems arise both in keeping the animals healthy and in feeding them. The alternation of wet and dry seasons means that fodder for the dry season has to be produced and conserved.

3. Comparatively high yields of maize, manioc, sorghum, and other crops may be set against the increasing difficulties of ley cultivation and dairy cattle husbandry. In other words, arable farming has a high relative advantage compared with ley farming and dairying.

The warmer the climate becomes, the greater is the technological jump required by a regulated ley system. In the highlands of Kenya, at an altitude of 2000 m, and with a rainfall of more than 1000 mm, we find many examples of a gradual change, even among the smallholders. But as the altitude decreases, or as the rainfall becomes less and is limited to the rainy seasons, we find scarcely any ley systems. A number of reasons account for this situation.

5.5.1. *The low fertility effect of tropical grass leys*

Tropical cultivated pastures are generally of grass alone, and frequently of only one species of grass. Their effect on the yield of arable crops is by no means unequivocally better than that of natural regeneration. Webster and Wilson (1967, p. 318) have concluded:

Hence it is unlikely to be worth-while expending labour and money on planting grass unless an appreciably better return can be obtained from animal production on the ley than from natural regeneration. Furthermore, although increased crop yields are normally obtained from a grass ley, these increases do not usually compensate for the loss of cropping during the ley period, so that the maintenance of fertility is, in part, achieved at the expense of a reduction in the total amount of crop products removed from the land. Clearly, alternate husbandry will only be satisfactory if crop yields during the arable break and animal production from the ley can be raised to levels that together more than compensate for the loss of cropping during the grass break. It seems unlikely that this can be achieved without a use of fertilizers which most tropical farmers cannot at present afford, or without a much greater ability in husbandry than the majority at present possess.

5.5.2. *Unfavourable capital–output ratio*

Regulated ley systems, in contrast with unregulated ley systems, require investments in (1) complete land clearance, for the removal of trees, bush, and roots, (2) fencing, and (3) sufficient traction power—tractors or oxen—and the necessary implements for mechanical cultivation, because regulated ley farming and plough cultivation are highly complementary to each other. In addition, it is frequently necessary to invest in a number of improvements, including: (4) seeding or planting of leys, (5) purchase of improved cattle, (6) cattle pens, dips, etc., (7) water supplies, (8) carts for the transport of manure, (9) roads, (10) drainage and the levelling of termite hills, (11) tsetse-fly control, and (12) land consolidation and the demarcation of fields.

Some of these improvements can be carried out by the available labour force in the slack months, but as a rule they involve considerable financial outlays: ploughing the leys requires a good deal of traction power, and to use this traction power economically requires a size of holding that is rarely found in smallholder agriculture in the tropics. Alkali (1967) quotes, for example, for Northern Nigeria minimum sizes of 8–16 ha for ley farms with ox-plough cultivation. In this context, Webster and Wilson (1967, p. 297) write, 'Land tenure systems under which the land is communally owned, or held by individuals on short-term, insecure leases, may make farmers un-interested in establishing leys. Fragmentation of holdings, resulting in small, scattered and unmanageable plots of land, has a like effect.'

5.5.3. *Dependence on improved animal husbandry*

The return from animal husbandry determines to a large extent the econo-mic practicability of grass leys. Consequently, regulated ley systems remain more of less uneconomic wherever animal husbandry is organized on tradi-tional lines, because under these conditions the farmer does not receive a sufficient return to justify any appreciable expenditure on the establishment and improvement of leys. The introduction of improved animal husbandry, however, depends not only on better feeding and better care of the animals, but above all on the provision of better communications, abattoirs, organized markets, and prices conducive to the production of quality meat, milk, or other livestock products.

5.6. Development paths of regulated ley systems and dairy systems

A ley system must show itself to be better than tumbledown grassland or other fallow systems in (1) having a better fertility-restoring capacity, (2) supporting more livestock production, and (3) allowing a more efficient use of the farm labour; and the combined effect of these advantages must be higher by quite a substantial margin than the costs of establishing a full ley-farming system. These conditions rarely exist.

The fact that regulated ley systems have spread and been able to hold their own in the tropics only in exceptional cases indicates that the expansion path of tropical rain-fed farming usually leads directly from fallow systems to permanent farming. Only occasionally are ley systems being intensified. A more noticeable tendency is for the area of unregulated leys in each holding to be progressively reduced, until finally permanent farming is practised. Usually, it is economically more profitable in smallholdings to expand the cropping of high-yielding cash crops, and if necessary to practice arable fodder cropping, than to intensify the ley system. Alkali (1967), for example, with reference to Northern Nigeria, describes the policy of agricultural advisers as follows: 'The system of alternate husbandry was scrapped mainly because the current design of ox-drawn implements was incapable of ploughing up leys, and also there was no proof that a system of leys was better than permanent pasture or permanent arable cropping.'

The tendency for farmers to prefer permanent cropping systems to ley systems stems not least from basic technical advances. New varieties of maize, wheat, sorghum, cotton, groundnuts, etc. have made arable farming more profitable. With the increased availability and use of mineral fertilizers, the farmers are less dependent than before on the fertility-restoring properties of leys. Thanks to herbicides and mechanical weeders, they no longer need leys to save labour in weed control. The tendency to expand arable farming at the expense of the leys is likely to lead to holdings with permanent arable cropping on the land that is more suitable for crop production and to permanent pastures on the remaining land. Also there is the tendency towards intensive grazing systems for dairy production on improved permanent pastures with arable cropping for dry-season roughage production only. A trend towards ley systems is unlikely to occur in the semi-arid and semi-humid tropics, except under conditions in which both fertilizer prices become much higher as a consequence of a general energy shortage, and much more effective nitrogen-fixing legumes for the tropics become available. Entirely new ley systems may, however, prove to be interesting propositions in humid tropical lowlands, such as the rice–*stylosanthes* combination which is experimented with in West Africa, and the manioc–grass combination (*Bracharia* grazed by steers), which is being planned in Amazonas, Brazil.

Also in tropical highlands the tendency is not towards ley systems, but either towards intensive grazing systems, as shown by the examples from Costa Rica and Colombia, or towards mixed farms with minimum or zero-grazing fodder regimes, as shown by the example from Kenya. Planning proposals for livestock development and balanced farming in the tropics again and again suggest that ley farms would be ideally suited and should be promoted. The evidence is, however, to the contrary. Other systems of land utilization, specialized grazing systems, or systems with permanent grassland and permanent crop land, are usually more profitable.

Generally the humid tropical highlands seem to be ideally suited for dairying, and better so than temperate climates. Almost no capital is required for buildings and fodder conservation, and labour is much less expensive. Intensive dairy systems in tropical highlands depend, however, on the use of much fertilizer which is essential for the provision of high-quality grazing throughout the year.

6. Systems with permanent upland cultivation†

6.1. Definition

A CONTINUAL expansion of arable farming at the expense of the fallow or ley leads to systems with permanent cultivation. This class includes those farming systems in which fallows or leys are only rarely, and for a short term, interpolated between the cultivation of arable crops, with the result that the R value exceeds 70. In contrast with fallow or ley systems, farming systems of this kind are normally characterized by (1) a permanent division within the holding between arable land and grassland, which is seldom or never cultivated; (2) clearly demarcated fields; and (3) a predominance of annual and biennial crops.

Often, permanently cultivated upland plots are a supplementary activity in holdings that are predominantly devoted to irrigation farming or tree crops. In particular, the growing of wet rice in valley bottoms is often combined with the cultivation of permanently cropped upland plots on the slopes. The same applies where the planting of fruit trees has led to stationary housing and where the shortage of land in the vicinity of the tree-crop areas has led to permanent cropping. The following accounts are restricted, however, to those farming systems in which permanent upland cropping is the main activity.

6.2. Types of permanent upland system and their geographical distribution

In densely populated regions with temperate climates the evolution from fallow systems to permanent cropping characterized the development of agriculture. In the tropics, and more particularly in the humid tropics, this has certainly not yet been a general or typical avenue for development. The warmer and more humid the climate, the greater are the soil-fertility problems

† In the first edition of this book the term 'permanent cultivation on rain-fed land' was used, which may be considered as the most appropriate term in tropical Africa and Latin America. In Asia, however, unirrigated wet rice is called 'rain-fed' rice (see § 7.1.1). Therefore the term 'upland cultivation' has been introduced.

and the more pronounced has been the tendency to prefer other ways of obtaining food and cash for growing populations.

1. Irrigation farming is spreading; in particular, wet rice proves to be a simple and highly productive permanent use of land in tropical climates. With the problem of conservation of soil fertility in upland farming, and as the pressure of population increases, there is an obvious tendency to concentrate production on what is topographically the lowest land, because that is where the most fertile soils occur, and often irrigation is possible in these valley bottoms.

2. The planting of perennial crops, tree crops especially, is increasing. By this means upland cultivation, which is a much greater disturbance of a tropical eco-system than a temperate one, is replaced by crops whose effect on the soil is similar to that of forest or bush vegetation.

The possibilities for irrigation farming and the growing of perennial crops are, however, limited by water availability. Moreover, they require high investments, and the return to labour may not be high. Because of this, permanent upland cropping has expanded rapidly in recent years.

6.2.1. *Permanent cultivation in humid climates*

The permanent cultivation of upland in a hot, humid climate presents some of the most troublesome problems of tropical agriculture. The traditional solution is gardening† especially where the population density is very high, where labour is cheap, and where profitable markets are at hand. Permanent gardens can already be found in shifting systems where the hut change is infrequent. In permanent farming, they develop into an important branch of the holding, and one which can become dominant where there is a town in the vicinity or where there is a marked shortage of land. Thus, in densely populated Java, one-fifth of the land used for agriculture is used in plots that are more like gardens than fields.

Examples of permanent cropping based on vegetables, and which preserve the soil, and are economically successful, can be found in Chinese agriculture, e.g. in Taiwan. Here the permanence of cropping is based on an intensive fertilizing with organic and mineral fertilizers, supplemented by measures for erosion control, and intensive ploughing. In other densely populated areas, e.g. in Java, the West Indies, and Eastern Nigeria, root crop systems (manioc, sweet potatoes, yams) have principally developed and are associated with

---

† Garden cropping is distinguished from arable cropping by the following features, which are usually, but by no means in all cases, found simultaneously: (1) production of small amounts of produce from a great number of different food crops, (2) small plots, (3) proximity to the house, (4) fencing, (5) mixed or dense planting of a great number of annual, biennial, and perennial crops, (6) a high intensity of land use, (7) land cultivation several times a year, (8) permanence of cultivation, and (9) cultivation with hand implements.

various tree crops. Farming tends towards a division of labour between outer upland fields and homesteads with a multi-storey physiognomy. However, some systems exist, produce comparatively well, and seem to be stable without manure or fertilizer, as, for instance, the soya–maize–rice holdings on soils of volcanic origin in the coastal plains of Ecuador (see Table 6.9).

There are as yet few examples of economically viable systems of permanent upland farming in the humid tropics, and the few systems that exist are to be found on very fertile alluvial or volcanic soils. Research stations provide us with an increasing number of rotations that would maintain soil fertility, but these rotations usually depend on tractors and much mineral fertilizer. Mostly they include several green-manure or fodder crops, which have a low economic return. Upland cultivation in the humid tropics frequently lacks a lucrative cash crop. In short, there may be technically feasible solutions to the problems of permanent cultivation, but their economic returns are as yet still marginal.

6.2.2. *Permanent cultivation in semi-humid climates*

Extended areas of the semi-humid tropics are already under permanent cultivation. Much of India is farmed in this way. Farming is usually mixed, often in combination with irrigated plots. Fodder production, manuring, terracing, and other soil-preserving measures are applied, but they are normally insufficient to maintain soil fertility at its original level. After the clearance of virgin land a rapid decline in yields usually takes place within the first 2 or 3 years, but after one or two decades yields tend to stabilize. Comparatively high yields may be obtained permanently on soils with a high natural fertility, as it is shown by permanent upland farming in much of Java. In most of southern Asia, however, yields have stabilized but at a very low level (Geertz 1963, p. 29).[†] Erosion is the remaining threat to the stability of the land-use system, and if it is controlled by careful terracing, production at a low level of output may be carried on indefinitely.

[†] Continuous upland farming for more than 20 years on fertile and stable grumosols in Thailand showed no decline from the original 2 t ha^{-1} yield level (compare 6.4.4). In northern India, for instance, yields of unfertilized wheat in a wheat (*kharif*)—fallow (*rabi*) rotation stabilized under experimental conditions at about 900 kg ha^{-1} (Singh 1972, p. 273), while in mid-western Nigeria, on sandy soils in a humid tropical lowlands, maize yields which started with 2 t ha^{-1} after bush clearance, levelled off at 0·3–0·4 t ha^{-1} after 5 years (Forbes 1975, p. 61). No systematic research results comparing yields on newly cleared land and at the low-level steady state seem to be available. The difference in yields clearly depends on climate and soils. My hypothesis would be, that, without mineral or organic manuring, yields in temperate climates (tropical highland), in semi-arid climates, and on alluvial or volcanic soils in semi-humid tropical climates would stabilize at about two-thirds or half of the yield level on newly cleared land. Mostly it is around 0·8–1·0 t grains ha^{-1}. In a semi-humid or humid climate and under unfavourable edaphic conditions the low-level steady state yield might be expected to vary between a third and zero of the original level (see e.g. Bourke 1974; IRAT 1972; p. 67; Angladette and Deschamps 1974, pp. 166 and 262).

Most of the arable cropping in semi-humid Africa is still in shifting or fallow systems, but those small areas which already had permanent cultivation some time ago are expanding like *taches d'huiles*, and new ones evolve yearly out of intensified fallow systems. Table 6.1 provides information about areas in Africa where permanent cultivation is traditional and where improved techniques are applied. Very few are in the humid tropics but there are many more in semi-humid climates.

Where the process of ensuring the food supply for a slowly expanding population has led to the gradual spread of permanent cropping, as for example in Ukara, Lake Victoria (see § 6.4.3), often sophisticated methods of manuring have spread. The traditional methods of maintaining fertility, however, require a high labour input. The holdings are small and their yield in relation to the labour input is low. By contrast, in those areas where permanent cropping has recently been introduced due to cash cropping or a rapidly growing subsistence demand, incomes are higher. But the soils are being mined with little regard for future returns. Substantial fertilizer inputs and tractor use for the establishment of an effective structure of contours to control soil erosion, can give stability to permanent cropping on most soils in the semi-humid tropics, as is shown by some large-scale farming in Brazil (see Table 6.10). Such modes of farming, however, rarely spread until the possibilities of finding new land for low-input cropping and the related soil-mining process are exhausted.

6.2.3. *Permanent cultivation in semi-arid climates*

Most permanent cultivation in the tropics is found in semi-arid areas where leaching problems are less pronounced. A great part of India and a rapidly growing part of semi-arid Africa is cropped annually by rain-fed agriculture. Almost all of it is production at a low level of output with high drought risks and a great deal of erosion. In areas with a bimodal rainfall pattern (Kenya, southern India) much land is planted in both rainy seasons. Crop failures are frequent, but cropping in both seasons increases the chances that at least one crop per annum is sufficient.

6.2.4. *Permanent cultivation in tropical highlands*

The evolution path of arable land use in densely populated tropical highlands is similar to that of temperate climates: shifting cultivation — unregulated ley systems — permanent arable cropping. The decline in soil fertility that accompanies more intensive land use without fertilization is less pronounced than in a hot and humid climate. Yields usually stabilize at a level sufficiently high to make permanent cultivation advantageous, even without heavy inputs of fertilizer.

Permanent cultivation with maize and beans is widely practised in the Mexican highlands and in several smallholder areas of the Andes. Potatoes,

TABLE 6.1

Permanent farming systems with soil-conserving practices in the African tropics

Tribe	Country	Average altitude (m)	Average rainfall (mm)	Population density, (inhabitants per km²)	a	b	c	d	Main crops
Malinke	Senegal, Guinea	620–1200	991	10			×		Millet, rice, maize
Baule	Ivory Coast	620	1194–1397	10–20			×	×	Yams, bananas, taro
Kita	Mali	620–1280	991	10					Millet
Dogon	Benin	750	787	20–50	×	×			Millet, yams, banana
Bobo	Benin	620	787	20–50		×			Millet, yams, banana
Gurensi	Benin	620–1280	787	20–50			×		Millet, yams, banana
Nunuma	Benin	620–1280	787	20–50			×		Millet, yams, banana
Mamprusi	Ghana	620–1280	787	30–50			×		Millet, yams, banana
Losso	Togo	620–1010	991–2007	50–100			×		Millet, groundnuts, yams
Kabre	Togo	1010–1280	1499	219	×		×	×	Millet, yams, rice
Mandara	Nigeria	620	787	50–70	×	×	×		Millet, beans
Kamuku, Kanuri, Chamba	Nigeria	620–1280	787	50–100			×		Millet, yams, banana
Bauchi, Berron	Nigeria	620–1280	787	50–100	×		×		Millet
Sokoto, Kano	Nigeria	620–1280	787–991	120–250	×	×	×		Millet, groundnuts, manioc
Batta, Mundang, Mandji, Bamum, Dama, Musgu	Camerouns	1020–1950	787–991	50–100	×		×		Millet, yams, banana
Bana, Adamawa	Camerouns	1950–2570	787–1499	100–150	×	×	×		Millet, beans
Kuru, Bari	Sudan	620–1280	991–1397	30–50			×	×	Millet
Konso	Ethiopia	1950	991–1194	190	×	×	×	×	Millet, cotton, maize
Tigre	Ethiopia	1950–2570	610–991	100–150	×	×	×		Maize
Kipsigi, Kikuyu, Nandi, Suk, Keyu, Taita	Kenya	1950–2570	1397–1803	50–150	×	×	×		Maize, banana, sweet potatoes
Rundi	Burundi	1950–2570	991–1397	100–150	×	×	×		Banana, millet, sweet potatoes
Ruanda	Ruanda	1950–2570	991–1397	150–200	×	×	×		Maize, banana, beans
Kiga	Uganda	1950–2570	787–1499	50–100	×	×	×	×	Maize, millet, manioc
Matengo, Makonde	Tanzania	1280–1950	991–1194	30–100	×		×		Maize, millet
Kinga	Tanzania	620–1950	991–1394	20–100	×				Millet, maize, beans
Sandawe, Iraque, Fipa, Turu, Gogo	Tanzania	1010–1950	787–1194	10–100	×	×			Millet, maize, beans
Mbugu, Shambala, Pare, Meru	Tanzania	1950–2570	787–1194	50–100	×	×			Millet, maize, beans
Wakara	Tanzania	1560	1600	209	×		×	×	Millet, manioc, rice

ᵃ Terracing. ᵇ Irrigation. ᶜ Manuring. ᵈ Stabling. *Source:* Ludwig (1968, pp. 92–3).

vegetables, and millets are the main crops of permanent cultivators in the Nilgiris Mountains of southern India. In parts of Ethiopia, ley systems have been replaced by permanent cropping with wheat, teff, millet, and numerous other crops. Maize, beans, manioc, and millets are the predominant crops grown by permanent cultivators in the highland areas of East Africa.

6.3. General characteristics of permanent upland systems

Systems with permanent upland cultivation show many of the characteristics of intensive fallow systems, but new elements occur because of the higher land-use intensity and others modify in importance.

6.3.1. *Spatial organization of cropping*

A basic difference between permanent cultivation systems and fallow or ley systems is the fact that the cultivator in a permanent system can no longer select a plot that is suited to the particular crop; instead, he must decide which crop will grow best on the given plot. We commonly find permanent land-use systems with distinct spatial differentiation of cropping, and with rotations including a great number of crops. There are three basic types of spatial organization.

(a) *Organization of cropping according to different soils within a catena.* Where the types of soil vary within a catena, the differentiation of cropping mainly arises from the planting of each type of soil with the most suitable crop or mixture of crops. A case in point is land use in a valley on Ukara, Lake Victoria (Fig. 6.1). Rice, sweet potatoes, vegetables, and sorghum predominate in the valley bottoms, manioc (cassava) and Leguminosae are the 'pediment crops', and the slopes of the Inselberg serve as grazing. Farmers prefer to have their plots distributed over the catena instead of having a larger consolidated field in one location. The dominating characteristic of a great number of holdings—particularly in India—is the combination of 'red' and 'black' soils. The former produce early crops, but their water-holding capacity

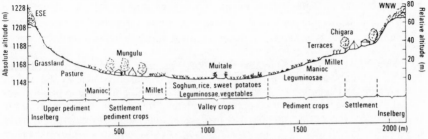

Fig. 6.1. Profile of a valley on Ukara Island, Lake Victoria, showing land use within a catena (from Ludwig 1967, p. 100).

is low and the risk of drought is high. The latter produce late and more reliable crops.

(b) *Organization of cropping in concentric belts.* Fallow farming often shows the 'ring' cultivation pattern with dungland close to the settlement, unregulated ley farming next to it, and shifting cultivation proper at some distance (see Fig. 4.2). In much of West African permanent farming a small circle of dungland is surrounded by large areas of low-yielding, permanently cropped land. The older the settlement and the greater the difficulties of transporting manure, the more markedly does cropping change as the distance from the farmstead increases. In the farmstead or village we almost always find fruit trees, papayas, bananas, etc. Usually there is an adjoining belt of heavily fertilized garden. As the distance increases, the manure input declines and farming occurs at a low-level equilibrium level.

(c) *Organization of cropping according to the micro-climate.* The differences in the micro-climate (the amount of rain, sunlight, land gradient, and danger of wind damage or flooding) help to diversify cropping. In the valley bottoms and flooded plains we frequently find some rice, occasionally surrounded by embankments, indicating the tendency to supplement permanent upland farming with wet rice (see § 7.1).

Such first steps towards a differentiation of cropping within the holding may be found in shifting systems (see § 3.3.1) and are common in fallow systems (see § 4.3.1). In permanent systems, however, they are considerably more pronounced because of the stationary nature of farming and higher population densities.

6.3.2. *Cropping principles*

A further feature of permanent cultivation systems is the variety of forms of mixed cropping, of relay-planting, and of rotations that occur.

(a) *Mixed cropping and relay-planting.* Provided land is scarce and labour ample, mixed cropping and relay-planting, as yield increasing, moisture-saving, risk-reducing, fertilizer- and power-saving, but labour-demanding techniques, are as common in permanent systems as in fallow systems (see § 3.3.2 (a)). Relay-planting is the usual practice in intensively cropped areas with extended rainy seasons, or where several rainy seasons overlap.

> This is exemplified in Kenya by the Kikuyu practice of cropping maize, beans, sweet potatoes, and pigeon peas. When the rainy season begins, at the end of March or the beginning of April, the women plant maize and pigeon peas broadcast in the plot, which has been prepared with hoes by the men. When the seed has germinated, the women go over the field again and plant two different sorts of bean, an early-maturing and a late-maturing variety, among the growing maize and pigeon peas. When the bean seed has sprouted, a third planting takes place. The field, which has produced a great deal of weed in the meantime, is weeded. At

the same time, sweet-potato seedlings are planted in the growing stands of maize, peas, and beans. The early-maturing beans produce the first yield from the middle of June. Maize and late-maturing beans are harvested from the middle of August. Sweet potatoes and pigeon peas are harvested from the plot as required from September onwards (Leaky 1934).

In regions with several overlapping rainy seasons, relay-planting can become continuous cropping, i.e. when one crop is harvested the plot is already being used by the next interplanted crop.

Some general principles of mixed cropping and relay-planting can be taken from Table 6.2:

1. A crop combination of millet and sorghum yields significantly more than a sole stand, but the labour productivity is less if the mixture includes more than two crops.

2. The adding of groundnuts and cowpeas to the mixture increases the value of the output and the net return. Labour productivity, however, is reduced. The reduction in labour productivity is more pronounced in terms of energy than in monetary terms (see Table 6.2).

Mixed cropping and relay-planting are therefore more common in areas with high population densities than elsewhere. In Asian areas with a long history of land shortage, as for instance, on Java, mixed cropping has been replaced by intercropping, i.e. two or more crops, sometimes up to six crops, are grown on different but proximate rows. This allows a better spacing of plants than intermingled mixed cropping and each row and plant can be individually weeded, manured, and fertilized (McIntosh 1975, p. 5).

(b) *Crop rotations*. Generally in tropical farming less emphasis is placed on crop rotations than in temperate climates, partly because the level of land-use intensity is not yet sufficiently high to show the yield-increasing effects of rotations; partly because crops prevail which lend themselves to mono-culture, such as maize; and partly because the fallow vegetation which develops during the lengthy seasonal fallow (often 8 months), has rotational effects. Existing crop-rotation patterns vary widely with rainfall and population densities as the major determining factors.

1. In regions with a distinct and short rainy season, there is only one planting each year and two-crop rotations prevail, as for instance in Senegal, where a groundnut–millet sequence is practised.

2. In semi-arid regions with somewhat more rainfall we find a tendency to grow a crop with a longer vegetation cycle, in particular maize instead of millet. Under high population pressure the tendency is often for a mono-culture of maize (Kenya Highlands, Malawi, Mexico). Elsewhere the tendency is towards crop associations. The humid period is usually too short for multiple cropping. Farmers prefer to plant early- and slow-maturing crops together, since no one crop can efficiently utilize the whole

TABLE 6.2

Productivity of the major sole crops and crop combinations in permanent cultivation systems at Zaria, Northern Nigeria

Major crops or crop mixture	Survey area (ha)[a]	Output per year (kg ha⁻¹)	Gross return in $			Net[b] return per ha ($)	MJ[c] per man-hour
			per ha	per man-hour			
				Total	June–July[d]		
Upland							
Sorghum	30	782 kg sorghum	63	0·25	0·58	38	45·4
Millet-sorghum	93	368 kg millet + 765 kg sorghum	92	0·32	0·70	48	58·3
Millet-sorghum–groundnuts	18	325 kg millet + 405 kg sorghum + 388 kg groundnuts	100	0·16	0·48	54	26·1
Millet-sorghum–cowpeas	14	400 kg millet + 711 kg sorghum + 165 kg cowpeas	107	0·18	0·74	52	31·8
Millet-sorghum–groundnuts-cowpeas	19	365 kg millet + 372 kg sorghum + 427 kg groundnuts + 138 kg cowpeas	118	0·15	0·56	64	24·3
Lowland							
Sugar-cane	23	13 778 kg sugar-cane	298	0·32	2·64	383	35·9

[a] Total crop area in the three survey villages is 397 ha. The difference is accounted for by a large number of less important crops and mixtures. [b] All inputs, including family labour, valued at the market price. [c] This is a supplement to Norman's calculation. [d] Total return divided by the man-hours in June–July (peak season in labour demand). Source: Norman (1972, pp. 79–99, Table A1).

wet season (Baker 1974, p. 5). A case in point is the millet–pigeon-pea–sesame combination around Hyderabad, India. There the millet matures first, and during the second rainy season (rabi) the field is occupied by the interplanted crop.

Overlapping cropping systems are practised in areas with bimodal pattern of rainfall, as for instance in Eastern Province, Kenya: maize and pigeon-peas are planted in the short rains. The maize is harvested in the in-between dry season. The pigeon-peas survive and develop fully in the long rains. Another 'two-season' crop is cotton which is planted in the short rains, survives the dry season and develops fully in the long rains.

3. In the semi-humid tropics, more possibilities are open, and we often find either a diversity of plant associations which change during the rainy season, or simply the preference for a crop with a longer vegetation cycle, like manioc. In regions with a single extended rainy season or with two distinct rainy seasons the tendency is towards two crops per year on at least some part of the land, as is shown, for example, by the upland rice–maize rotation in central Luzon, Philippines.

4. Continuous cropping, where the soil is continuously under various short- and medium-term annuals, requires sufficient moisture all the year round and thus a humid climate with even rainfall distribution. It is a rare type of land use, but cases in point are the examples from Colombia (Table 6.4) and from Ecuador (Table 6.9). Highly intensive examples are the continuous sweet-potato-growing in some parts of New Guinea (Kimber 1974; Waddel 1972) or continuous manioc on uplands in Java.

In looking at rotation patterns it should always be remembered that in most tropical systems crop rotations are plot-specific, and most cultivators aim at having various plots on different soil-types, each having its specific rotation.

(c) *Adapting cropping to the risk of drought.* Cultivators in all rain-fed systems produce under the risk of drought, and centuries of experience have taught them how to adjust. In the semi-arid zones, where most of the permanent cultivation systems are to be found, these adjustments are more obvious than elsewhere. There are three types of rainfall risk: (1) the onset of the rainy season may be delayed; (2) there may be lengthy gaps between the rains, which is usually the most dangerous kind of drought—it is not unusual for the rainy season to have alternations of waterlogging and minor droughts; and (3) there may be an early end to the rains.

The risk of drought is an important obstacle to realizing the production potential:

1. Farmers tend to plant late, waiting until they know that the rains have come to stay. A significant proportion of the nutrients which have become available due to the mineralization of organic matter during the dry season are consequently leached (Krantz *et al.* 1974, p. 26).

2. Farmers prefer drought-resistant crops and varieties which are often low-yielding. Drought-resistant crops are planted at the end of the rainy season as catch-crops.

3. Farmers do not plant at times which would maximize yields, but prefer phased planting to distribute risks.

4. Farmers shy away from high input of labour and fertilizer.

The poorer the farmer and the more dangerous the risk of drought, the stronger the tendency not to use intensive techniques.

(d) *Combining upland annuals with perennials and wet-valley-bottom cropping.* In some cases permanent upland cropping is carried out under a canopy of trees. In the West African lowlands some fallow systems have been intensified to such a degree that cropping with maize and manioc under a cover of oil-palms has become permanent. Here the relation between trees and annuals is usually competitive. Maize yields significantly less under shade than without shade and the difference in yield is the more pronounced the higher the fertilizer input (Flinn 1974, p. 4). The complementary relation between trees and annuals is most conspicuous with the Sérèr of Senegal, who crop millet under *Acacia albida* and who argue that 'five acacias fill a store with millet' (Lericollais 1972, p. 29). Their proverb has been confirmed by research: under the acacia more nutrients and more water are available, and yields are much higher. One tree covers 100–300 m², and a typical Sérèr field may have 12 acacias, which would mean that about 2200 m² ha^{-1} are improved. In addition the acacia provides dry-season fodder amounting to 2743 kg starch equivalents per hectare (Charreau, cited by Lericollais (1972, p. 29)).†

Similar complementary and competitive relations hold between upland and wet-land farming. In traditional Asian farming these relations are highly developed, in particular with wet rice in the valley bottom and on terraces. In Africa there is a continent-wide tendency to develop valley bottoms and to reap the benefits of the interactions between upland and wet-land farming.

(e) *Fallows.* In the dry season the crop land lies fallow. Crop residues, stubble, and weeds serve as pasture. In almost all traditional systems cultivation has to wait for the first rains to soften the soil. Ox-ploughing and hoe-cultivation are not usually feasible in the dry season. The seasonal fallow supplies some fodder, but it consumes soil moisture, and when the rains start run-off and erosion are substantial.

The impact of the remaining fallow years on soil fertility is slight (Charreau 1974, p. 5). If fallowing is to be effective under tropical conditions it has to be

† Other fodder trees with similar effects are *Prosopis cineraria* (Rajastahn) and *Parkia clappertonia* (Northern Nigeria). Both seem to promote food-crop production even though they do not shed their leaves during the rainy season (Willing 1979, personal communication).

a long-term or medium-term fallow. Most of the fallows that still occur in permanent systems can be explained by shortages in the labour economy (mainly insufficient traction power) and the need for grazing.

6.3.3. *Characteristics of the fertilizer economy*

The methods of fertilizing in tropical permanent upland farming are insufficient, and we usually find soil mining and stagnation at a low yield per year. Generally speaking, however, permanent cultivators practise some methods of fertilization, and the more so where the original fertility was low and the longer ago the change occurred from fallow systems to permanent cropping. The principal fertilizing methods in fallow or unregulated ley systems are scarcely practicable in these cases, since permanent cropping requires stationary housing. The moving of the hut sites, and the consequent creation of fertile gardens, can take place only within a restricted area, and is hardly equivalent to the 'human folding' which is so common among fallow cultivators. The cattle numbers are usually insufficient for folding systems, unless extensive grazing areas are available. Also, permanent cropping has to forgo the erosion protection that ley systems provide in the years between cultivation. Permanent cultivators who do practise fertilization and erosion control depend on other methods, with a higher labour input.

Various traditional methods of erosion control have been employed, from ridge and mound cultivation to terracing and the use of baskets to carry back the soil that has been washed away. In particular, stone terraces in traditional farming systems are an indication of a long-term, acute land shortage, which has helped to make erosion control part of the cultural heritage of the local people. Although these practices are traditionally known, they are rarely employed in farming systems where cash cropping has been introduced and where land shortage is a recent phenomenon. In many of these situations, particularly in the drier savannas, gullies increase rapidly in number and size, soil conservation usually being neglected as cash cropping and incomes per head increase, mainly because of the unfavourable short-term input–output relationship of the labour invested. The way out of this undesirable situation probably does not lie in a return to traditional agricultural methods, but in additional cash cropping, which, by changing the economic setting, can make soil conservation economically worthwhile. The tractor-drawn plough becomes very important in this connection, because it enables the farmer to carry out soil-conservation measures like contour ploughing, mechanized ridging, and mechanized terracing, which are difficult to perform with an ox plough.

Of special interest is the development of the fertilizer economy in permanent farming. The primary task is the replacement of nutrients and organic matter. From the standpoint of the evolution of farming systems, a number of stages may be distinguished.

1. Manuring begins as a rule with the collection and transport of household refuse. Permanent cultivators in the Mandara Mountains, northern Cameroun, prefer to have their huts on a high point in their fields so that the household refuse is washed down by the rains to the crops (Hallaire 1971, p. 36).

2. Next comes the application of animal manure, which is sometimes increased by stabling and the provision of bedding.

3. Cultivation of green-manure crops is an indication of a more advanced fertilizer economy.

4. Processing of household refuse, harvest residues, and farmstead earth in compost is characteristic of a higher level of intensity.

5. The use of night soil is traditional with Chinese cultivators. It is rare in Africa and only practised by some tribes with a long-term land shortage (in Ethiopia (Nowak 1954) and Mali (Dumont 1966, p. 90)).

Increase in the intensity of cropping entails the import of nutrients and organic matter. Traditionally, this could take one of the following forms.

1. Areas outside the holding are grazed, and animal manure collects in the farmyard, mainly at night. In this way nutrients are concentrated on the permanently cropped land. This is a principle that is practised generally.

2. Nutrients are imported in fuel. The search for fuel has stripped extensive bush or forest land near densely populated areas (needles, leaves, twigs). The ashes serve as fertilizer, either directly or as a component of compost.

3. An indication of a special lack of nutrients is the collection of leaves and branches as green manure, which is applied directly to the land (see § 6.4.3).

4. In traditional Chinese agriculture, oil-cake, fish, and other organic materials were bought as manures.

A general feature of traditional forms of intensive fertilization is the high labour input; this is the case both in preparation of compost and in collection of animal manure, and lies especially in the difficulties of transport that are part and parcel of most forms of organic manuring. Whereas in shifting and fallow systems the fertilizer is produced where it is used, i.e. it scarcely needs to be transported, the permanent cultivator with stationary housing must transport the material. Consequently, we find the tendency to concentrate the fertilizer on the gardens near the farmyard, particularly where the lack of transport is acute and there are no carts, draught animals, or field tracks. Bottlenecks in transport are the reason why old Chinese towns are often surrounded by a belt of black soils, which are enriched with nutrients and humus. Elsewhere, the saying that 'those fields are most fertile which can hear the cock crowing' reflects the same situation.

A further feature of the fertilizer economy in densely populated areas is the burning of straw and cow manure. Since there is no fallow vegetation to provide fuel, as is the case in shifting and fallow systems, straw and cow manure are frequently the only available and economic types of traditional fuel.

Technical progress has scarcely altered the production conditions in any branch of agriculture so much as in the fertilizer economy of holdings with permanent cropping.

1. Mineral fertilizer is of primary importance; compared with organic fertilizers it proves to be labour-saving because of its high nutrient concentration and the relative ease with which it can be transported.

2. New crops and cultivation methods for green-manure crops are available. The use of tractors can provide more scope for increased green-manuring, since the work is carried out more quickly.

3. Where intensive livestock-keeping is practised, the fertilizer cycle is extended by the import of nutrients as purchased concentrates.

4. Progress in transport vehicles like ox carts, tractors with trailers, or lorries and the development towards a comprehensive road network facilitate the transport of manure, night soil, etc.

Thus we may say that agricultural innovations have drastically changed the fertility problem in permanent upland farming in the tropics. Technical solutions are available. A major problem is that small amounts of fertilizer seem to be relatively ineffective. Permanent arable cultivators apparently have to choose either between production at a low-level equilibrium or with rather heavy chemical inputs. 'In-between' situations do not seem to work. Unfortunately, however, the price relations are often too unfavourable to make intensive fertilizing economic.

6.3.4. *Characteristics of animal husbandry*

Traditional farming on permanently cropped tropical upland is usually combined with small herds per family, but high cattle densities per 100 ha. The role of livestock in these conditions differs from that of stock in fallow systems or ley farming. The aim is not only to produce milk and meat, but increasingly to provide traction power and manure. The hoarding of the cattle is not important, because the available grazing is limited, and also because permanent cropping is usually coupled with private land rights, and the acquisition of land is an alternative way of gaining status and security.

The fodder basis primarily determines which forms of animal husbandry predominate in permanent cropping systems, and to what extent they are carried on. In this connection we can distinguish two principal forms.

(a) *Animal husbandry in systems with extensive grazing areas.* Where extensive areas are still available, cattle husbandry is mostly organized on lines similar to those in fallow systems. The animals are communally grazed and in the charge of herdsmen by day, and they return to the farm at night. Emphasis is put on cows, female calves, and heifers. The husbandry practices

are more or less the same as those of fallow farmers, but there are funda-
mental differences in the fodder economy, which is based increasingly on the
productivity of arable farming, e.g. on the seasonal distribution of grazing
between the arable land and the more distant grazing areas.

A typical example of animal husbandry in an African farming system which
passed recently from fallow to permanent farming is supplied by Guillard (1965)
in his monograph about the village of Golonpoui in northern Cameroun (see Fig.
6.2). At the beginning of the rainy period in June or July, the animals graze on the
arable land that is not yet cultivated or is lying fallow. As cropping progresses, the
animals are concentrated on the fallows, which are 2–3 km from the village. In
October, millet straw from the permanent gardens provides additional fodder. At

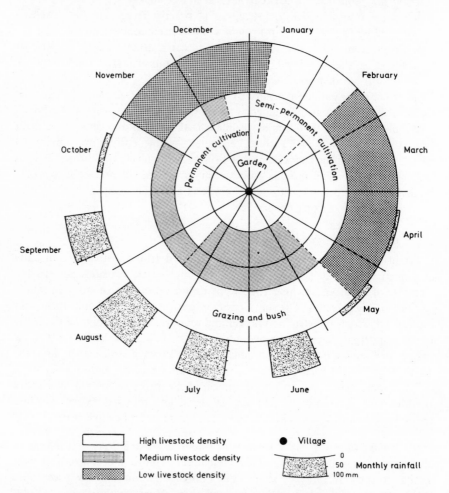

FIG. 6.2. Distribution of cattle on the village land during the different seasons in Golonpoui,
northern Cameroun (from Guillard 1965, p. 305).

the beginning of the dry season, the animals are driven to bush grazing further afield. This is interrupted in January and February, when harvest residues are eaten. The animals spend the rest of the dry season, from February to May, in the more distant grazing areas. In addition, the leaves and fruits of odd acacias (*Acacia albida*) scattered in the arable land are used.

(b) *Animal husbandry in systems without extensive grazing areas.* As arable cultivation spreads at the expense of the grazing area, there is a tendency for farmers to reduce the number of animals kept, or to hire out the cattle to nomads who are hardly ever near the village. Roadside grazing, seasonal fallows, and harvest residues constitute the remaining fodder basis.

Bottlenecks in the fodder economy change. Systems with extensive grazing areas primarily face a dry season feeding shortage while those where almost all the land is annually cropped face a wet season feeding shortage. Fodder reserves have to be established to feed the cattle until crop residues become available after the harvest. The lower limit to the number of cattle kept is determined by the number of traction animals required. Hoe-cultivators may come to have no cattle at all. With the reduction of the grazing areas, the rearing of goats, pigs, and fowls increases in relative importance. In requiring less grazing land and providing smaller units for sale or consumption, this kind of stock is better suited to the economic condition of smallholdings.

The reduction in the importance of livestock that occurs as land becomes increasingly scarce can be halted only by the introduction of fodder crops, which are practically unknown in fallow systems. The need for traction animals is usually the first cause for a step in this direction, since the systematic feeding of traction animals can scarcely be avoided. Where sub-sections of the holding can be irrigated, they are in many cases used to supply the traction animals with fodder. Beyond this, traditional farming systems on permanently cropped land are rarely in a position to support a substantial volume of fodder production. In most cases, the livestock economy, although important in numbers, stagnates at a low level of productivity (see also § 6.6.1).

(c) *Animal husbandry in systems with permanent pastures.* Where permanent arable cultivation has been introduced under large-farm conditions, as for instance in Latin America, we find a most pronounced decline of the inter-actions between crop and animal production. Each becomes a rather distinct activity, the major interaction being that dry season feed for dairy cattle is provided by sugar-cane and silage-crops grown on arable land. The tendency is towards either specialized dairy farms or beef farms (ranches) or crop-producing farms. The soil-fertility maintenance in the latter group tends to rely on fertilizers instead of leys or manures.

6.3.5. *Characteristics of the labour economy*

African permanent cultivators still rely mostly on hoe-cultivation, although forest and bush fallows, which are the compelling technical reasons for hoe farming, no longer exist. Hoe-cultivation usually remains because farm size in relation to family size is so small that there is no room for draught animals. This is true for densely populated highlands like the Kisii and Kikuyu districts in Kenya or for semi-arid farming in densely populated areas of West Africa, e.g. in much of Upper Volta. In Ethiopia, Asia, and Latin America, however, we find a different and varied tool system, which involves plough-cultivation, carts, and animal traction. Ox-plough cultivation rapidly spreads in semi-arid East and West Africa. The plough makes it possible to cultivate larger areas and to improve the timing of cultivation (Haswell, 1975). On the other hand, disadvantages are (1) the fodder requirement of the draught animals, (2) the poorer quality of animal ploughing compared with hoe-cultivation, and (3) the increased danger of erosion. Communal work, which we often find in shifting systems and still sometimes in fallow systems, scarcely plays any part at all, and this is a further feature of permanent cultivation systems. Similarly, the tendency disappears for the land belonging to a household to be divided up into smaller units cultivated by members of

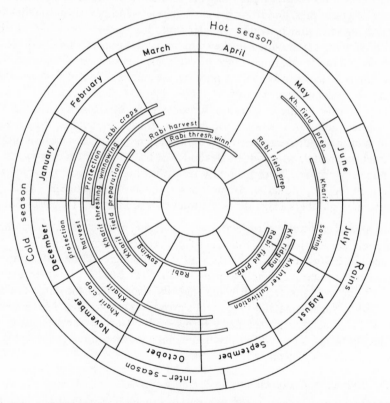

FIG. 6.3. Annual agricultural labour cycle in a system of permanent farming in the West Nimar District, central India (from Leshnik 1966, p. 61).

the family for themselves. In plough-culture, we usually see centrally organized holdings, where the farmer directs the family labour and employs hired labour if need be.

(a) *Seasonality of labour demand*. With the transition to permanent cultivation, the seasonal concentration of labour tends to become even more acute than in fallow systems. The work is concentrated in the rainy season, not least because draught animals and hoe-cultivators are unable to cultivate the soil in the dry season. A fairly regular use of the labour force is found only in a few favourable regions where double cropping can be practised.

> The situation in the West Nimar district in India illustrates this (see Fig. 6.3). Cotton, wheat, and sorghum are the chief crops, and wheat is usually grown in winter (*rabi* season) on a plot which is then left fallow. The main summer crops (*kharif* season) are sorghum and cotton. The distribution of cropping over two seasons brings about a relatively balanced use of labour. The time when there is little labour demand is limited to February–May.

In most cases with permanent upland cropping, however, we find only one pronounced rainy season, with little scope for multiple cropping. Consequently, labour demand is decidedly seasonal. High population densities and a high degree of underemployment on the one hand, and insufficient labour at the beginning of the rainy season on the other are probably not as marked in any other land-use system as in permanent upland systems.

> Fig. 6.4 illustrates the situation in an intensive arable farming area in northern Cameroun. The total available time is recorded under several headings. The distribution of field work is highly seasonal for both the man and his wife. Most of the time between the high labour demands for field work is filled with various kinds of work around the hut. Sickness takes only a small percentage of the total time. It should not be overlooked, however, that on two occasions both were sick together, and once this took place during a time of high labour demand (October 1957). It may appear from Fig. 6.4 that field and homestead work together lead to the full employment of man and wife. It is, however, difficult to present working time in terms of days. The length of the working day differs (work here was measured in half-daily units), and the intensity of work is much higher during peaks of labour demand than in slack seasons.

It is not only the workers who are underemployed. Where plough cultivation exists, the traction animals too are only seasonally employed (80–100 days a year). Yet the seasonal nature of work requires a large number of draught animals, since every cultivator wants to plough his land at the right time and sets value on having his own animals. Lack of draught power is one important reason for late and inefficient cultivation of fields. Consequently, a considerable part of the arable land is used to provide draught animals with fodder, while they work at full capacity for short periods only. This is particularly true in India.

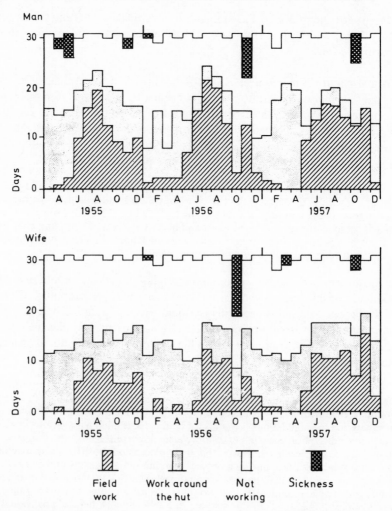

FIG. 6.4. Seasonal distribution of work of a hoe-cultivation holding with permanent upland farming at Golonpoui, northern Cameroun (from Guillard 1965, p. 343).

(b) *Low productivity of labour*. Table 6.3 provides some information about typical labour input per hectare of crop in permanently cropped areas. It is, as a rule, lower than in fallow systems because of the reduced input in land preparation. Particularly where plough cultivation is practised, labour requirements per hectare are significantly lower. The available information is insufficient to relate labour input to yields and to derive general conclusions about variation of labour productivity in terms of grain-equivalents per hour of work with increasing permanency of cropping. However, several case studies indicate that, with increasing permanency of cropping, output per

TABLE 6.3

Some examples of input–output relations in permanent upland farming (per hectare and crop)

Crop / Country / Location	Pearl millet[a] Senegal Bambey	Groundnut Senegal Bambey	Groundnut India Coimbatore	Sorghum India Coimbatore	Teff[b] Ethiopia Modjo	Wheat Ethiopia Holetta	Maize Thailand[c] Ampur Takfar	Potatoes Kenya[c] Meru	Manioc Brazil Minas Gerais	Soya Brazil Minas Gerais	Upland rice Brazil Parana
Year	1971	1971	1973	1973	1973	1973	1970	1976	1977	1977	1977
Technique	Hoe-farming			Ox-plough farming			Partial mechanization		Full mechanization		
Method	Sample	Sample	Sample	Sample	Sample	Sample	Sample	Sample	Accounts	Accounts	Accounts
Technical data											
Labour input (man-hours)											
Land preparation	39[d]	55[d]	38	40	153	99	68	93	6	8	5
Planting	28	38	34	20	—	—	1	194	32	3	3
Fertilizing and manuring	—	—	6	10	75	101	—	46	64	27	—
Weeding	298	152	172	48	—	—	111	587	n.a.	11	4
Plant protection	—	—	3	—	—	—	—	40	—	1	—
Harvesting	138	95	355	128	111	91	164	499	280	1	3
Threshing, sorting	n.a.	136	—	—	85	59	43	—	—	8	—
Total	503	476	608	246	424	350	387	1459	382	61	15
Ox-pair hours	—	—	25	38	119	80	n.a.	—	—	—	—
Yield (t)	0·47	0·72	0·59	0·44	0·99	0·72	3·17	9·40	15·00	1·80	1·50
Economic analysis ($)											
Gross return	32	65	99	32	127	69	150	509	360	375	233
Material inputs											
Seed	1	3	25	3	6	10	2	102	28	26	8
Fertilizer and pesticides	—	—	2	1	22	—	—	91	22	155	116
Mechanization	1	1	13	15	12	12	18	31	43	85	40
Income before interest and overheads	30	61	59	13	87	47	130	285	267	109	69
Income per hour	0·06	0·13	0·10	0·05	0·21	0·13	0·34	0·20	0·70	1·79	4·60

[a] *Pennisetum gambicum.* [b] *Eragrostis abyssinica.* [c] Ploughing is carried out by contracted tractors. [d] Planting without major soil cultivation. *Sources:* Senegal: Lericollais (1972, p. 72); India: Rajagopalan and Balasubramanian (1976, p. 94); Ethiopia: Friedrich, Slangen, and Bellete (1973, pp. 55, 89); Thailand: Grimble (1973, p. 59); Kenya: Dürr (1977); Brazil: Estado São Paulo (1977, pp. 34, 36, 52).

man-equivalent available is likely to decrease (Boserup 1965; von Rotenhan 1966; Ludwig 1967). The average return per hour of work actually performed is usually higher with permanent upland cultivation than with fallow systems. The problem is that marginal returns to labour tend to be low because there is no unused land left to absorb labour productively and the marginal returns to higher labour inputs per hectare are usually low without irrigation.

Systems in Kenya with potato production illustrate that much labour can be productively absorbed where the climate is favourable and markets allow root-crop production. Generally, however, the production function of labour in semi-arid, traditional, permanent, upland farming shows rapidly declining marginal returns. It lends itself easily to mechanization which drastically reduces labour input and increases labour productivity, as shown by the examples from Thailand and Brazil. This contrasts with irrigation farming, in particular wet-rice farming where average returns of labour may often be low, but where marginal returns tend to be high. The poorer the soils in permanent upland cultivation and the drier the area, the lower will be the marginal returns to increasing labour inputs. This is particularly true of systems where subsistence food requirements absorb most of the land, and little land is left for cash cropping. Where this applies, the number of working days per available labourer and the hours worked per day tend to be particularly low (Cleave 1974, p. 32). Only dire necessity will then bring about increasing labour inputs per hectare. It is for this reason that the husbandry patterns of permanent cultivation in Africa are described by Gourou (1966) as 'siege methods', by which he implies that they came into existence because the tribes were surrounded by hostile neighbours, which prevented the extension of farming in fallow systems into expansion areas. The African peasants—the Wakara (see § 6.4.3) being a case in point—are apparently quite aware of the fact that under their system of permanent cultivation they produce yields that are higher per hectare than those obtained in more extensive systems but are lower per man-hour. Wherever conditions permit them to leave their crowded home area, because new land is opened up for settlement, they immediately turn to fallow systems.

The low marginal return per hour of work probably explains much of the high degree of underemployment in low-income countries (Heyer 1966, p. 3) and the problem is particularly pronounced in permanent cultivation systems. Usually much additional work with low but positive returns can be performed —e.g. terracing and other forms of erosion-control—but the utility of the expected increase in income is normally below the utility of leisure.

6.4. Examples

The following examples give an idea of the various forms taken by permanent upland systems. The example from the highlands of Ecuador shows the conditions of production at the high altitude limit. The data from holdings in Ethiopia and Colombia indicate that farming in the temperate climate of a tropical highland tends towards a land-use pattern similar to that found in temperate climates generally, and they show the comparatively high return per hour of work which can be attained by the use of animal power. Land use in Senegal shows the impact of increasing cash cropping on hoe-farms and the tendency towards modern inputs which is made possible by commercialization. The Zaria examples show permanent cultivation in a subsistence system where farmers succeed by the skilful choice of a mixed-cropping pattern to secure their food demands with comparatively few hours of field work. For the rest of the time, however, they remain largely unemployed. The case of the Wakara on an island in Lake Victoria shows the great efforts which are required to sustain an increasing population under traditional permanent upland farming, and the Thailand example shows the comparatively high incomes which can be obtained on medium-sized farms with some mechanization as long as the soil fertility and crop yields remain high enough to carry the cost of tractor ploughing. The examples from coastal Ecuador and Brazil indicate that very high productivities can be achieved provided full mechanization is combined with very high fertilizer inputs.

6.4.1. *Holdings in the tropical highlands: examples from Ecuador, Ethiopia, and Colombia* (Table 6.4)

Tropical highlands with low population densities are usually exploited through grazing systems. With growing population pressure and urban food demand these tend to change within a few decades into systems with permanent arable farming.

(a) The example from Ecuador indicates the returns under decidedly marginal conditions (Bernard 1978). Most of the land is not arable, and is grazed by sheep. The food requirements used to be covered by a fallow system which experienced a gradual reduction of fallowing. Three crop years with mainly barley and some potatoes and beans are followed by only two fallow years. The land is cultivated with hoes, because draught animals cannot be maintained in the area. Llamas provide within-village transport. Yields are very low. With barley, farmers expect 3–4 kg of crop for each kg of seed, and with potatoes not more than 1.6 t ha^{-1} are harvested. Potatoes require 9 months from planting to maturity and four weedings are needed during the lengthy vegetation period. Labour requirements per hectare are consequently high, and returns per hour of work exceedingly low. The potato and bean crops are frequently destroyed by frost. Energy production per hour of work is—next to the case of the pastoral Gabra (Table 9.2) and of the Wakara (Table 6.9)—lower than in any other example cited in this book, and

TABLE 6.4

Farm-management data of holdings with permanent cultivation in tropical highlands

Country	Ecuador	Ethiopia	Ethiopia	Colombia
Location	Guangaje	Holetta	Modjo	Antioquia
Rainfall (mm)	1800	1085	862	2000
Altitude (m)	3500–4000	2400	2200	2100
Year	1974	1972–3	1972–3	1978
Method	Sample	Sample[a]	Sample[a]	Model[b]
Number of holdings	14	12	12	1
Persons per household	4·4	5·33	6·73	9·00
Labour force (ME)[c]	3·0	2·72	2·94	4·50
Size of holding (ha)	4·0	5·51	10·23	6·00
Teff (ha)	—	1·00	5·05	—
Wheat and barley (ha)	0·7	1·81	2·17	—
Maize (ha)	—	—	—	3·50
Beans and peas (ha)	0·7	0·84	1·93	3·50
Potatoes (ha)	0·6	0·10	—	2·00
Other crops (ha)	0·1	0·10	0·68	—
Total crop area (ha)	2·1	3·85	9·83	9·00
Fallow (ha)	1·1	1·24 ⎫	0·40	—
Grazing (ha)	7·5[d]	0·42 ⎭		1·50[e]
Cropping index[f]	67	76	96	200
Livestock (number)				
Draught oxen	1[g]	2	4	—
Cattle	—	4	3	3
Sheep and goats	22	3	5	—
Yields				
Teff (t ha^{-1})	—	0·50	0·91	—
Wheat or barley (t ha^{-1})	0·6	0·74	0·86	—
Maize (t ha^{-1})	—	—	—	1·50
Beans (t ha^{-1})	0·4	0·69	0·91	1·00
Potatoes (t ha^{-1})	1·6	—	—	15·00
Milk (kg cow^{-1} year^{-1})[h]	—	400	500	1800
Economic return ($ per holding)				
Gross return	483	201[i]	847[i]	8142
Purchased inputs				
Seeds[j]	68	27	59 ⎫	
Fertilizer	35	4	105 ⎬	1824
Mechanization	1	4[k]	33[k] ⎭	
Income	379	166	650	6318
Wages for hired labour	—	5	50	1004
Rent and taxes	n.a.	55	136	n.a.
Family farm income	379[l]	106	464	5314
Productivity				
Gross return ($ ha^{-1} total land)	121	36	83	1357
Gross return ($ ME^{-1})	161	74	288	1809
Income ($ ha^{-1} total land)	95	30	64	1053
Income ($ ME^{-1})	125	61	221	1404
Labour input (man-hours)	2826	1413	3278	7392
Income per man-hour ($)	0·13	0·12	0·20	0·85
Man-hours per ME	942	519	1115	1643
Hours of work of ox-pairs	—	529	1250	—
MJ per man-hour	6·1	25	36	34

[a] Random sample of contract farmers in the extension programme. [b] Built on 7 interviews. [c] Including permanent hired labour and excluding off-farm work. [d] 0·9 ha individual pasture and 6·6 ha share in 'paramo' (mountain pasture) open to everybody. [e] Fenced Kikuyu grass pastures. [f] Arable land only. [g] Llama for transport. [h] After suckling of calf. [i] Without returns from gardens and livestock. [j] Including farm-produced seeds. [k] Purchased cost items for ox-power. [l] Plus 255 $ of non-farm income. *Sources:* Ecuador: Bernard (1978); Ethiopia: Friedrich, Slangen, and Bellete (1973, pp. 7–114); Colombia: personal enquiries.

energy production per farm is insufficient to cover energy requirements. 70 per cent of all food is purchased. Families increasingly rely on non-farm income from migrating members. Intensification of the system is hardly feasible. Out-migration to urban areas seems to be the only solution of the poverty problem, which is greater than in any other system described in this book.

(b) Production in Ethiopia occurs at a much lower altitude and provides much more scope for land-use intensification. Teff (*Eragrostis abyssinica*) is the main crop, supplemented by maize in the intermediate altitudes and wheat, barley, and pulses in the higher altitudes. Yield levels are low, not only because soil fertility has been consumed by preceding generations, but also because of the heavy cloud cover during most of the rainy season. Teff is the preferred crop, even though returns per hectare are lower than with other cereals and labour input tends to be higher. Consumers pay a higher price; farm families have a preference for teff consumption; teff is a reliable crop and less sensitive to waterlogging than the other cereals; and the straw is a valuable roughage, low in lignin and much more easily digestible than wheat or barley straw. Storage losses with teff are low, and teff is appreciated in the rotation as a crop before wheat and barley, because the high weeding effort produces a comparatively 'clean' field (Westphal 1975, p. 83).

1. At Holetta the climate is relatively cool, soils are poor, and yields are low. Wheat and teff production predominate. Permanent cropping with cereals leads to a serious weed problem, and this contributes to the large area with pulses which are appreciated as a rotation crop. Expansion of cropping at the expense of fallowing has led to a gradual reduction of the cattle herd, and farmers are compelled to buy grazing rights in distant valleys during the crop season. There is much over-grazing, and erosion is increasing. An initial step to try and overcome the land-scarcity problem is the growing use of mineral fertilizers.

2. At Modjo all the arable land is permanently cropped. There are highly commercialized farms, producing a high-quality teff for the urban market. The livestock density is even lower than in Holetta, and most of the cattle migrate during the crop season to low-lying pastoral areas, returning after the harvest to utilize the straw of teff, wheat, and barley. Yield levels are comparatively high, and most farmers use substantial amounts of mineral fertilizers.

The change towards permanent cropping and the use of mineral fertilizer may be considered as the initial steps in a sequence of changes which lead to more intensive land use. Teff yields hardly surpass 1 t ha^{-1}, while wheat and barley could produce 3 t. With increasing fertilizer use and population density, teff is likely to become less competitive within the system. The valley bottoms in the area, which are extensively grazed, could be developed by drainage works. In the long run, dairying based on cut-and-carry systems would become an interesting proposition, at least close to consumption centres.

(c) The example from Colombia demonstrates the production possibilities of a tropical highland with a temperate summer climate throughout the year, relatively fertile soils of volcanic origin, and high and evenly distributed rainfall. The land is continuously cropped. Most of it carries an association of maize and beans. The maize is planted in the beginning of the year and close to maturity it is interplanted (mid-year) with climbing beans which use the dry maize stalks as support. Both crops are thus closely related to each other. The beans grown in the area have a preferred market ($1 kg^{-1} in 1978) and are the major cash crop providing a reliable income. Several years of the maize–beans association are followed by two potato crops per year. Potatoes yield less per hour of work, but more per hectare. The risk of potato production is high because of widely fluctuating prices. Potatoes are therefore the speculative cash crop of the system. Most farmers keep one or two dairy cows on land which is not suitable for arable cropping for home consumption of milk. The system provides relatively high returns per family and per hour of work, even though the land is cultivated with hoes. The mechanical input is limited to hand tools, while the chemical input is high and tends to grow with the use of more fertilizer and the introduction of herbicides.

6.4.2. *Holdings in semi-arid climates: examples from Senegal, India, Mexico, and Nigeria* (Tables 6.5–6.8)

Agriculture in Senegal shows the development path of a farming system in a semi-arid area where land shortage developed mainly because of the expansion of cash cropping. Where the problem of ensuring the subsistence food supply of a slowly expanding population has entailed a gradual extension of permanent upland farming, as, for example, in Ukara (see § 6.4.3), or on a large scale in China, more-or-less intensive methods of traditional fertilizing have spread, and this guarantees permanent cropping. However, these methods do require a high labour input, and the yield is low in relation to the amount of work done. Where permanent upland farming has developed on the basis of an increase in market production by the local people, the situation is usually quite different. The more rapidly production is commercialized, the less often do the farmers worry about conserving the soil.

An example of this occurs in the groundnut holdings of Senegal. The commercialization of the area began about 1850, and it has accelerated in the last 40 years. The traditional fallow system has been replaced by a permanent cropping system, with groundnuts as the dominating crop, which is unsurpassed in the area in terms of returns per hectare and per hour of work. The guiding principle of land use is a simple one. There have to be sufficient cereals for food (about 0·25 ha per person yielding 200 kg at a yield expectation of 0·8 t ha⁻¹). 90-day pearl millet supplies early food, and 130-day sorghum supplies food later in the season. The rest is planted with groundnuts. Some rice is grown in valley bottoms. Farmers now tend to replace millet with early-maturing maize, which yields much more, produces green maize as food even earlier than millet, requires less threshing work, and no bird-scaring, and is simpler to store. Cropping is traditionally organized in concentric circles, which are still recognizable in most cases:

1. Several small plots close to the village and the houses are permanently cropped with millet by the head of the unit. Yields are comparatively high because this is dungland, manured by a folding system. Kraal sites are changed every 2–3 days.

2. The next circle consists of the main fields, where groundnuts alternate with millets or sorghum. Traditionally the land had a 1- or 2-year fallow after 3 or 4 years of cropping, but most of it is now permanently cropped. Much of the land is not yet fully cleared of bush. Interspersed *Acacia albida* give the countryside a park-like appearance.

3. In low-population areas the main fields are supplemented by outlying fields which are cropped in a groundnut–bush-fallow rotation.

Most farmers own cattle which are taken care of by paid herdsmen. The livestock lives on waste land, fallows, and crop residues in the rainy season. The leaves of *Acacia albida* are an important source of feed in the dry season. The livestock provides draught power and manure for the dungland, in addition to some meat and milk.

The system gives full employment of the available labour force during 5–7 months. 8–9 hours per day are worked in a 2-month peak season, 4–5 hours in the rest of the cropping season, and almost no field work is done in the off-season. There is thus much underemployment. Total field work per man is, however, higher than in most shifting and fallow systems, and much higher than in permanent cultivation systems without a dominating cash crop (see Table 6.5). The most important cost items in the system, except labour, are the draught animals. In the case depicted in Table 6.6 two oxen, one donkey, and one horse are kept but they remain largely underemployed because each type of animal is used for specific operations only. The great number of draught animals seems wasteful. It has to be remembered, however, that the animals live on grazing and crop residues: feeds with low opportunity costs. Some farmers buy immatures, use them for cultivation, and then sell them as mature animals for slaughter.

The farm, housing, and consumption units in this part of Senegal are comparatively large: 17·36 ha, and 12 people belong to the unit shown in Table 6.5, which is typical of the part of the country occupied by the Wolof of Sine-Saloum. This unit is not yet a centralized farm operation. The labour is organized in somewhat independent subunits (see Fig. 6.5).

TABLE 6.5

Farm-management data of a traditional groundnut–sorghum holding of the Wolof at Sine-Saloum, Senegal

	Plots of the leader of the unit	Plots of the leader of the household	Plots of the women and children	Plots of the adult male relations	Total farm unit
Location	Farm Ousmane Loum, Nioro-du-rip				
Rainfall (mm)	914				
Year	1971				
Method	Case study, daily observation				
Type of sub-unit[a]					
Number of persons	1	1	7	3	12
Labour force (ME)	1	1	3·5	3	8·5
Size of sub-unit (ha)	9·86	2·50	2·50	2·50	17·36
Groundnuts	3·80	1·70	2·50	2·50	10·50
Millet[b]	1·60	0·80	—	—	2·40
Sorghum[b]	3·25	—	—	—	3·25
Cotton	0·25	—	—	—	0·25
Fallow	0·96	—	—	—	0·96
Cropping index (per cent)	90	100	100	100	94
Groundnut economy					
Labour input (hours ha⁻¹)	556	976	776	781	730
Animal work (hours ha⁻¹)	74	72	51	33	58
Fertilizers ($ ha⁻¹)	7	6	3	4	5
Yields (t ha⁻¹)	1868	1094	1274	1728	1568
Gross returns ($ ha⁻¹)	172	101	117	159	144
Material inputs[c] ($ ha⁻¹)	32	31	24	23	28
Income ($ ha⁻¹)	140	70	93	136	116
Income ($ per hour)	0·25	0·07	0·12	0·17	0·16
Economic return ($ per unit)					
Gross return	992	209	293	397	1891
Material inputs	328[d]	49	31	29	437
Income	664	160	262	368	1454
Cash surplus[e]	319	160	262	368	—
Productivity					
Gross return ($ ha⁻¹)	101	84	117	159	109
Gross return ($ ME⁻¹)	—	—	—	—	222
Income ($ ha⁻¹)	—	—	—	—	84
Income ($ ME⁻¹)	—	—	—	—	171
Labour input on the land of each subsystem (hours)	4093	2046	1941	1953	10 033
Income per hour of work ($)	0·16	0·08	0·14	0·19	0·14
Labour input per person (hours)	1225	1342	517	1514	—
Labour performed by each group (hours)	1225	1342	3618	4542	—
MJ per hour	16	13	10	14	14

[a] The relations between the two leaders and the two other classes are shown in Fig. 6.5. [b] Average millet yield is 1310 t and sorghum yield is 0·742 t ha⁻¹. [c] 63 per cent of total material inputs are the costs of draught animals and equipment estimated according to local standards. These costs include farm feeds with low opportunity costs. Fertilizer amounts to 19 per cent. [d] $71 of which are payments to other members of the unit in exchange for labour. [e] This is not precisely the cash surplus. The item shows sales minus material inputs and provides an idea of how much remains for the leader of the unit after he has supplied food for the common kitchen. *Source:* Monnier, Diagne, Sow, and Sow (1974, pp. 28–56).

1. The leader of the unit distributes the land, commands the draught animals, does most of the buying and selling, pays taxes, supplies most of the food to the kitchen, and receives labour services from others. He is the dominating power in the unit. His yields are higher, because he has priority access to land, draught animals, and fertilizer, but his gross return per hectare remains comparatively low, because he has to grow the food crops

TABLE 6.6

Use of draught power in a groundnut–sorghum holding, Senegal

Capacity	2 oxen	1 donkey	1 horse
Working hours per year	440	203	591
Working days per year	98	34	113
Average working hours per working day	3·50–6·12	6·00	5·15

Use of each animal as percentage of total working hours

Land preparation at the beginning of the season	41	3	1
Planting	—	47	17
Weeding	20	50	29
Harvesting	23	—	7
Land preparation at the end of the season	16	—	—
Transport	—	—	46
Total	100	100	100

Source: Monnier, Diagne, Sow, and Sow (1974, p. 42).

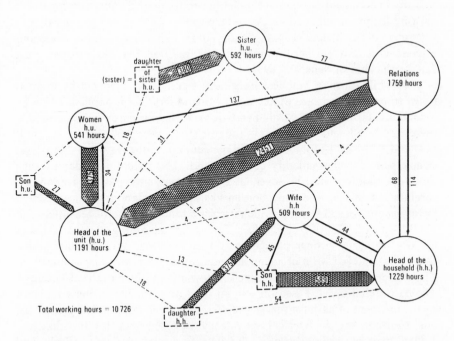

FIG. 6.5. Labour exchange in a groundnut–sorghum holding in Senegal (From Monnier *et al.* 1974, p. 37). The arrows illustrate the flow of labour hours from each type of household member to the fields managed by other household members.

for the common kitchen which, in monetary terms, yield much less than groundnuts.

2. The leader of the household is related to the leader of the unit and is often a son. He has his own kitchen and grows not only groundnuts but also food crops. He supplies hardly any labour to the leader of the unit.

3. Women and children cultivate their own personal plots with groundnuts and supply labour to the fields of the two leaders.

4. Most units house permanent labourers and/or male adult and unmarried relations who cultivate their own groundnut fields but who are supplied from the common kitchen and are obliged to supply much labour to the leader's field.

Fig. 6.5 shows the labour exchange between the various subsystems. The circles show the labour hours of each person on his or her field and the arrows show the services rendered to others.

Because of the relatively large cash-crop areas per worker, farmers in the groundnut areas attain a comparatively high income, but most of it is still attained at the expense of continual impoverishment of soils. Areas whose fertility has been exhausted by some decades of cultivation have been abandoned and new ones cleared. Meanwhile land has become a scarce resource. Permanent cropping of millet has introduced *Striga senegalensis*—a parasite living on millet roots. A country-wide decline in soil fertility due to the spread of permanent cropping made Tourte Gaudefroy-Demombynes and Fauché write that 'further farming with the traditional land-use system must lead to the ruin of Senegal' (1954).

Meanwhile farming practices have been significantly changed by innovations. Improved seeds are widely used. Mineral fertilizer is applied, although in amounts far below the recommendations. Animal traction has successfully been introduced, and yield levels in the case study, with 1.57 t ha^{-1} of groundnuts and 1.31 t ha^{-1} of millets, are 50 per cent above the traditional low-level steady state for permanently cropped land. The potential under average farming conditions lies, however, at 2 t of groundnuts and 3–5 t of millet, sorghum, or maize. The main bottlenecks are in the labour economy. High yields depend on soil fertility, which itself is a function of fertilizer inputs and a full crop with a lot of crop residues. The keys to obtaining effective fertilizer use are primarily proper ploughing, timely planting, and weeding, which cannot be achieved without adequate and centrally organized mechanization. The tendency of the system is therefore towards more centralized holdings. The patriarchal order, based on mutual obligations, is increasingly replaced by the payment of cash for services rendered by hired labour. The greater the land shortage, the stronger is the tendency for changes in this direction.

Data from groundnut–sorghum holdings in a semi-arid area of India and from sorghum–sesame holdings in Mexico contrast with those of the Senegal holdings (Table 6.7). Returns per hectare and per worker in the Indian plough-

TABLE 6.7

Farm-management data of groundnut–sorghum holdings in India and sorghum–sesame holdings in Mexico

	India		Mexico	
Country	India		Mexico	
Location	Cuddapah District Andhra Pradesh		Matamoros	
Rainfall (mm)	688		400–600	
Year	1969–70		1971	
Type of farming	Groundnut–sorghum		Sorghum–sesame	
Method	Sample		Model	
Number of holdings	60[a]		1	
Persons per household	6·18		n.a.	
Labour force (ME)[b]	2·30		3·80	
Size of holding (ha)	6·60		100	
Groundnuts		2·31[c]	—	
Sorghum		1·85[c]	40	
Sesame		—	40	
Other crops[d]		1·71	10	
Total crop area	5·87		90	
Cropping index (per cent)	89		90	
Livestock (number)				
Draught animals	1·45		—	
Cattle	4·30		n.a.	
Yields				
Groundnuts (t ha^{-1})	0·69[e]		—	
Groundnuts ($ ha^{-1})	94		—	
Sorghum (t ha^{-1})	0·72[e]		3·50	
Sorghum ($ ha^{-1})	49		175	
Sesame (t ha^{-1})	—		0·70	
Sesame ($ ha^{-1})	—		140	
Milk (kg per cow and year)	68–203[f]		—	
Economic return ($ per holding)				
Gross return	488[g]		15 000	
Purchased inputs				
Seeds	52		200	
Fertilizers, pesticides, etc.	9		2517	
Mechanization	9		3993	
Other costs	—		1970	
Income	418		6320	
Wages for hired labour	36		2016	
Taxes and rents	13		792	
Family farm income	369		3512	
Productivity				
Gross return ($ ha^{-1})	74		150	
Gross return ($ ME^{-1})	212		3947	
Income ($ ha^{-1})	63		63	
Income ($ ME^{-1})	182		1663	
Labour input (man-hours)	1904		7600	
Income per man-hour ($)	0·22		0·83	
Man-hours ME^{-1}	828		2000	
MJ per man-hour	35		388	

[a] Holdings without irrigated land. [b] Including hired workers. [c] Almost all crops are grown in the *kharif* (summer) season. The figure includes small plots with low yields in the *rabi* (winter) season. [d] Millets and various minor crops in the Indian case, beans in the Mexico case. [e] Kharif season. [f] 68 kg per cow per year, 203 kg per buffalo cow per year. [g] The source does not give the return from unirrigated holdings. The figures have been derived from the data for unirrigated crops. Farm produced seeds are listed as returns and costs. By-products of crops, manure, etc., are not listed as returns and farm bullock labour not as costs. *Source:* India: Narayana (1974); Mexico: personal information.

systems are about the same as in the Senegal hoe systems, but returns per hour of work are much better. The Mexican case provides an idea of the increase in returns per hectare and per worker which can be achieved by the combination of yield-increasing techniques and mechanization in a semi-arid setting.

Rather distinct from those systems with permanent upland cropping, where land shortage is primarily a consequence of cash cropping (as shown in the Senegal example), are those where the gradual increase in population and subsistence demand has led to the change from fallow systems to permanent cultivation. Sorghum–millet holdings at Zaria, carefully analysed by Norman (1972), are an example of this type of change in the land-use system (Table 6.8). The production process relies almost entirely on land and labour, with very little capital involved. Yield-increasing inputs were at the time of enquiry (1966–7) not yet sufficiently known and/or profitable. Farmers react to the challenge of increasing land scarcity in a traditional way.

1. Fallowing is reduced. The land is permanently cropped.

2. More emphasis is given to farming in the valley bottoms (lowlands), where sugar-cane and vegetables are grown. The valley bottoms show much higher marginal returns to labour than the upland and receive, therefore, 1235 man-hours ha^{-1} compared with 494 man-hours on the upland. Lowland farming is, however, not yet irrigation farming.

3. Output per hectare is kept at a relatively high level by growing crop mixtures, and in all three villages, irrespective of land scarcity, crop mixtures predominate. Twenty-five crops are found in the area and are grown in a total of 200 different crop combinations. Seven crops, however —sorghum, millet, groundnuts, cowpeas, sweet potatoes, manioc, and sugar-cane—account for 84 per cent of the total area. More labour and more complex crop mixtures are to be found where land is particularly scarce (see Table 6.8).

The impressive amount of evidence collected by Norman indicates that crop mixtures are economic. Gross and net returns are higher with crop mixtures, and working time in the peak season is more effectively used (see Table 6.2). Crop mixtures allow the production of grain legumes, which are essential for the diet, at low opportunity costs. Also the yields tend to be more reliable. Two or more crops are produced with only one major cultivation. The labour input, at 541 man-hours ha^{-1} for sole crops and 585 man-hours for crop mixtures, remains low (compare Table 6.8).

The above practices increase output per hectare with increasing land scarcity, but do not prevent a decrease in the return per hour of work from $0·17 to $0·15 and $0·12. There is much underemployment of labour in the system. The average adult man performs only 710 hours of field work. Eighty-nine per cent of the work involves male adults, which contrasts with other African systems where much work is performed by adult females and children. Expressing labour capacity in man-equivalents, which include adult females and child labour, is a particularly dubious concept under these conditions, but it is nevertheless interesting to note that only 276–349 man-hours of field work are performed per man-equivalent. Crop mixtures allow the families to produce the required food with comparatively

TABLE 6.8

Farm-management data of sorghum–millet–groundnut holdings in Northern Nigeria

Location[a]	Dan Mahawayi	Doka	Hanwa
Rainfall (mm)	1118	1118	1118
Year	1966–7	1966–7	1966–7
Method	Sample	Sample	Sample
Number of farms	42	44	38[b]
Persons per household[c]	6·8	8·0	10·9
Labour force (ME)[d]	3·9	4·4	6·3
Size of holding[e]	4·82	3·99	2·94
Upland (ha)	4·41	3·47	2·77
of which fallow (per cent)	19	28	2
Lowland (ha)	0·41	0·52	0·17
of which fallow (per cent)	42	12	1
Total crop land (ha)	3·80	2·92	2·87
Cropping index (per cent)	79	73	98
Percentage of crops in[f]			
sole stands	24	29	14
two-crop mixtures	39	33	47
three-crop mixtures	25	18	22
four-crop mixtures	6	14	15
five-crop mixtures	4	3	2
six-crop mixtures	2	3	—
Livestock per holding (number)			
Cattle	—	—	10
Sheep, goats	3	3	6
Poultry	5	7	5
Farm capital ($ per holding)			
Livestock	16	23	426
Buildings, etc.	22	31	31
Tools	6	8	5
Economic return ($ per holding)			
Gross return[g]	279	266	302
Purchased inputs[g]	28	25	21
Income	251	241	281
Wages for hired labour	32	9	30
Family farm income	219	232	251
Off-farm income	68	33	65
Total family income	287	265	316
Productivity			
Gross return ($ ha^{-1} total land)	58	67	103
Gross return ($ ha^{-1} crop land)	73	91	105
Gross return ($ ME^{-1})	71	60	48
Income ($ ha^{-1} total land)	52	60	96
Income ($ ME^{-1})	64	55	45
Labour input (man-hours)	1516	1634	2249
of which hired labour (per cent)	29	6	21
Income per man-hour ($)[h]	0·17	0·15	0·12
Man-hours of field work per family ME	276	349	282

[a] Dan Mahawayi is 32 km from Zaria in an area where only about 21 per cent of the land is farmed and 79 per cent is bush. Doka is 40 km from Zaria and in a densely settled area with only about 26 per cent of the total area being bush. Hanwa borders on the Zaria township and about 20 per cent of the total area is bush. [b] Of which 18 farms had cattle. The other households in the survey area had no cattle. [c] Most households consist of two or more adults with their families (*gandu*). The 'normal' household (one man and wife and children) is becoming more popular (*iyali*). [d] Household members only. Eighty-nine per cent of the field work is done by adult males. The man-equivalent figure includes adult females and therefore overestimates the man-equivalent in terms of local habits. [e] In an area with much bush land it is difficult to establish the fallow land. The size of holding at Dan Mahawayi is perhaps underestimated. [f] Yields: see table 6.2. [g] Including value of farm-produced inputs (seeds and manure). [h] MJ per man-hour are shown in Table 6.2. *Source:* Norman (1972).

low labour input. Most of the field work is carried out in May, June, and July, with 16–17 days of 5 hours per male adult each month. This compares with 2 working days of 2–3 hours each in March. A 5-hour day is certainly not a full utilization of the labour capacity according to international norms, but it is subjectively felt as such in the area. Labour shortage in the peak season and much seasonal underemployment are two essential characteristics of the system.

6.4.3. *Holdings in semi-humid and humid climates: examples from Tanzania, Costa Rica, and Ecuador* (Table 6.9)

Land shortage at Zaria is still comparatively recent. The Wakara by contrast have lived for more than a century tightly packed (500 persons per km²) on a little island in Lake Victoria, 1000 m above sea level. The soils are derived from weathered granite, with little natural fertility. The need to guarantee the food for a dense population on these soils has caused the Wakara to develop and apply highly refined practices of a soil-preserving agriculture, although no influence from immigrant people is recognizable (Thornton and Rounce 1963).

The average Wakara family has only a hectare of arable land at its disposal, and this is cultivated by hoe. The land is cultivated chiefly according to a 3-year rotation:

(1) summer season   Bulrush millet, with manure and interplanted legumes for green manure (*Crotalaria striata*).
     winter season   Growth of the green manure.
(2) summer season   Bulrush millet, with green manure dug in; groundnuts planted among the ripening millet.
     winter season   Growth of groundnuts.
(3) summer season   Bulrush millet, with manure.
     winter season   Sorghum or manioc (in the case of manioc, a fourth year of manioc follows).

In addition, small patches of irrigated rice are cultivated in the valley bottoms, and hill land in the interior of the island is grazed.

The crucial problem in the Wakara holdings is that of acquiring manure, which can be done only by the production of fodder. The two or three cattle and the one or two goats and sheep that are kept on the average holding are fed with harvest residues, leaves of growing sweet potatoes and manioc, weeds collected while hoeing, fodder grasses planted and watered on the island shores, and ratoon rice, which grows on rice plots at the beginning of the dry season. There are also some grazing areas. Sometimes the animals are tethered, in order to economize on the available grazing. The animals occupy one-half of their hut and stand in deep litter. Bedding consists of the remains from harvesting and fodder, leaves and branches of trees and bushes, old roofing grass, and any other suitable material that becomes available. The

TABLE 6.9

Farm-management data of holdings with permanent arable cultivation in semi-humid and humid climates: examples from Tanzania, Costa Rica, and Ecuador

	Tanzania	Costa Rica	Ecuador
Country	Tanzania	Costa Rica	Ecuador
Location	Ukara	Pérez Zeledon	Quevedo
Rainfall (mm)	1600	2944	2200
Year	1964	1976–7	1977–8
Type of farming	Millet–bambara nuts	Beans–rice–maize	Soya–maize–rice
Technique	Traditional hoe-farming	Traditional hoe-farming	Modern mechanized
Method	Sample	Case study	Case study
Number of holdings	25	1	1
Persons per household	10·9	16·0	—
Labour force (ME)[a]	4·5	5·0	13
Size of holding (ha)	2·83[b]	12·94	122
Homesteads, roads, bush	0·10	0·20	2
Pasture	0·95[c]	4·1	—
Crop land	1·78	8·57	120
1. Season			
Maize	—	3·72[f]	70
Beans, soya, bambara nuts	0·32[d]		—
Rice	0·43[e]	—	50
2. Season			
Millets	0·78	—	—
Maize	—	1·80	—
Beans	—	2·10	120[g]
Rice	—	1·70	—
Other crops	0·30	—	—
Both seasons manioc	0·35	0·20	—
Total crop area	2·18	9·52	240
Cropping index (per cent)	122	111	200
Livestock (number)			
Cattle	3	9	—
Sheep, goats, and pigs	4	8	—
Yields			
Millet (t ha⁻¹)	1·0	—	—
Maize (t ha⁻¹)	—	0·6	3·2
Beans (t ha⁻¹)	—	0·5	1·7
Rice (t ha⁻¹)	—	0·5	3·3
Economic return ($ per holding)			
Gross return	182	3343	135 544
Purchased inputs	3	312	84 619
Income	179	3031	50 925
Wages for hired labour	—	179	31 067
Family farm income	179	2852	—
Net return before taxes, land rent, and interest	n.a.	n.a.	19 858
Productivity			
Gross return ($ ha⁻¹ total land)	64	258	1111
Gross return ($ ME⁻¹)	40	669	10 426
Income ($ ha⁻¹ total land)	63	234	417
Income ($ ME⁻¹)	40	606	3917
Labour input (man-hours)	6300[h]	3967	31 140
Income per man-hour ($)	0·03	0·76	1·63
Man-hours per ME	1400	793	2400
MJ per man-hour	6	27	306

[a] Family and hired labour. [b] Sample of farmers with children in the local Middle School who are expected to have above average farm sizes. [c] Including communal grazing. [d] Bambara nuts. [e] 0·1 ha irrigated rice and 0·3 ha other crops. [f] Mixture. [g] Soya. [h] Based on 1400 h ME⁻¹. Ludwig reports that family members usually perform 10 hours of work per day, which would imply a much higher labour input than given here. *Sources:* Tanzania: Ludwig (1967, pp. 177, 206); Costa Rica: Navarro (1977); Ecuador: personal information.

household ash is applied to the nursery beds for rice. As well as using green manure, the Wakara use leaf manure; leaves from trees that are not suitable as fodder are carried to the field and dug into the soil.

Even though the agriculture of the Wakara may be remarkable, it is largely a relic of a poorer past. Their careful cultivation is not motivated so much by the desire to increase their income as to guarantee the livelihood of a growing number of people. The Wakara obviously work harder than the neighbouring farmers, but they are nevertheless poorer. Energy production is just sufficient to cover energy requirements. When they migrate to the mainland, they change to the fallow farming of the Wasukuma, which, considering the labour input, provides a disproportionately better income. The Wakara's land use is an isolated example. The population in the surrounding areas regard staying on the island, where people are obliged to work from morning till evening merely to cover their needs, as stupid, and the methods that they are obliged to use to manure their land as 'dirty'.

The Wakara's example shows that, in traditional conditions on moderately fertile tropical soils, there must be a special emergency situation before smallholders will adopt the labour-demanding techniques that are essential features of permanent upland farming with soil conservation.

The marginal situation of the Wakara and their desperate efforts to obtain a living on small plots of highly leached soils contrasts with the example from Costa Rica. Here the family commands a large piece of relatively fertile soil (of volcanic origin) at about the same altitude as the Island Ukara in a climate with high and evenly distributed rainfall (10 humid months). The resource endowment is a most favourable one. Both are pre-machine systems, and are operated with hand tools only, almost without purchased inputs. But clearly the Costa Rican farmer is much better off, with a far higher income obtained with far less labour. The system is, however, very wasteful in the use of resources. Land use is extensive with about one crop per hectare and year (cropping index 111), even though two or three crops would be feasible. Mostly beans, intercropped with some rows of maize, are grown and some upland rice. Land preparation for planting implies weeding without further cultivation. Most crops are weeded once only. Yields are very low mainly because of weed competition. Much of the cropping occurs on sloping land, and erosion is serious. Cattle are kept on unimproved pastures, and pigs are fed with purchased concentrates. There is need for crop–livestock interactions as in the case of the Wakara. The example shows that permanent arable crop production in the humid tropics is extremely difficult with pre-machine techniques. Permanent crops, such as coffee and plantains, would be a much better choice.

The example from Ecuador shows the characteristics of production of one of the rare cases where large-scale continuous arable production is carried out under the conditions of a tropical, humid lowland. The production occurs on soils derived from volcanic ash, rich in nutrients and with a high water-holding capacity. They drain easily and the fact that most of the rain falls at night allows effective mechanized operations throughout the year. The production relies on heavy fertilizer inputs (100 kg N, 100 kg P_2O_5, and 65 kg K_2O), the use of herbicides, and several sprayings of pesticides for each crop. Upland rice production is fully mechanized, but maize and soya are still harvested by seasonal workers which explains the relatively large labour input in the otherwise mechanized production process. The important characteristic of the system is the high gross return per hectare and per worker, while the income per hectare and crop is not much higher than in many smallholder systems. The high chemical and mechanical inputs consume a great part of the value that is produced. Also it has to be considered that the system relies on very cheap petrol prices which prevail in Ecuador, and heavily subsidised credit (interest rates are significantly below rates of inflation). It is unlikely that the system would be economic under prices which prevail on the world market. The example indicates that modern permanent upland farming in a humid climate may produce high outputs but requires very high inputs, much more so than in semi-arid or more temperate climates.

6.4.4. *Holdings with partial and full mechanization: examples from Thailand and Brazil* (Table 6.10)

These two examples provide an idea of the modes of production which are feasible in semi-arid (Thailand) and semi-humid (Brazil) areas which lend themselves to mechanization.

The example from Thailand illustrates the organization of production in settlement areas with maize as the major cash crop. Here we find medium-sized partially mechanized holdings. The area per farm worker is large, and on the grumosols in the Pra Buddhabat settlement farmers can normally produce about 2·5 t ha^{-1} of maize (1974, with 1·75 t ha^{-1} was an exceptionally poor year), and this for 30 years without any visible decline in yields (Boonma and Klempin 1975). The farm consists of (1) a homestead with about 50 useful trees, some banana clumps, and some vegetables; (2) some wet-rice if there is a valley bottom; and (3) a large open field. The fairly even distribution of rainfall from April–May to September–October allows two arable crops per year. Maize, the main crop, is followed by a catch crop and maize–beans, maize–sorghum, or maize–cotton are typical rotations in the area. The growing of sorghum and mung beans as catch-crop are low-input activities. The seed is broadcast in the maize stover. One buffalo-ploughing follows. The catch crop is not weeded, and yields are low and unreliable because moisture may be lacking at the end of the season. Contractors perform the disc-ploughing in the dry season. Buffaloes, maintained at low opportunity costs, perform the remaining traction work. Actually the farmers buy hardly anything for their operations other than some seeds and tractor services.

TABLE 6.10

Farm-management data of mechanized holdings with permanent cultivation

	Thailand	Brazil
Country	Thailand	Brazil
Location	Pra Buddhabat	Casa Branca, Sao Paulo
Rainfall (mm)	1200	1300
Year	1974	1977–8
Type of farming	Maize–sorghum–beans	Soya–maize–rice
	Partially mechanized	Fully mechanized
Method	Sample	Model[a]
Number of holdings	38	1
Persons per household	6·01	n.a.
Labour force (ME)	3·66	18
Size of holding (ha)	9·20	740
Homestead, roads, etc.	0·23	20
Crop land	8·97	720
Season 1:		
Maize	7·85	180
Soya beans	—	360
Upland rice	—	180
Season 2:		
Maize and sorghum	3·84	—
Beans	1·92	—
Other crops	1·43	—
Total crop area	15·04	720
Cropping index (per cent)	167	100
Livestock (numbers)	—	—
Yields		
Sorghum (t ha^{-1})	0·73	—
Sorghum ($ ha^{-1})	73	—
Maize (t ha^{-1})	1·75	4·0
Maize ($ ha^{-1})	197	542
Beans (t ha^{-1})	0·27	2·1[b]
Beans ($ ha^{-1})	48	514
Rice (t ha^{-1})	—	2·0
Rice ($ ha^{-1})	—	500
Economic return ($ per holding)		
Gross return	2084	372 870
Purchased inputs	414[b]	165 230[c]
Income	1670	207 640
Wages and salaries	168	48 975
Family farm income	1502	—
Net return before taxes, land rent, and interest	—	158 665
Sales as percentage of gross return	90	100
Productivity		
Gross return ($ ha^{-1} total land)	226	504
Gross return ($ ME^{-1})	569	20 715
Income ($ ha^{-1} total land)	182	281
Income ($ ME^{-1})	456	11 536
Labour input (man-hour)	4005	43 200
Income per man-hour ($)	0·41	4·81
Man-hours per ME	1094	2400
MJ per man-hour	65	728

[a] The actual farm includes flower production as major activity. The model has been derived from the data on food crop production only. [b] Mainly tractor hire services. [c] In per cent: machines 30, seeds 9, fertilizers 36, pesticides 5, herbicides 2, and other purchased inputs 18. *Sources:* Thailand: Boonma and Klempin (1975); Brazil: Schoenmaker and Mansholt (1978, personal communication).

About half of the field work is carried out by seasonal hired labour. Farm families receive comparatively high incomes owing to the fact that large areas are cultivated within an effective system of partial mechanization. The 1974 example shows the high gain due to recent changes in maize prices in relation to input prices. There is scope for intensification of the system. Some settlers in higher-rainfall corners of the settlement area tend to plant more and more fruit trees. Mineral fertilizer is not effective on the grumosols and is too expensive. The main avenue for farm development is the change to a more productive second crop. Earlier planting of maize (in a dry seed bed at the end of the dry season) and relay-planting of beans or cotton into maturing maize are techniques which are already applied by a number of farmers at Pra Buddhabat. In particular, the introduction of cotton as a second crop would mean a major change in the system.

The example from Brazil provides an idea of the land and labour productivities that can be achieved with large-scale mechanized upland farming under the condition of a semi-humid, almost sub-tropical, climate on poor 'cerrado' soils. Without fertilizer, yield levels are around 0·5–1·0 t of grains ha^{-1}. High and regular inputs of fertilizers are used, including lime. Chemical inputs amount to 43 per cent of all purchased inputs and greatly surpass the machine costs of the fully mechanized production process. Soil erosion is controlled by careful maintenance of contours. The example seems to indicate that under semi-humid conditions high levels of labour productivity are tied to high chemical inputs, while in semi-arid farming high labour productivities may be achieved with mechanical inputs mainly.

6.5. Problems of the system

The scope for increasing production within traditional agricultural methods diminishes as land reserves are exhausted and yields per hectare stagnate at a low level. Permanent farming carried out on impoverished soils may well be considered as a final stage in the land-use development process that begins with shifting cultivation, leads to fallow farming yielding somewhat higher incomes per labour, and ends with a low-level steady state in permanent upland farming.

The growing population's increasing subsistence demands and the extension of cash cropping, to cover increasing cash demands, are the driving forces behind changes in farming systems. Three overlapping phases may be distinguished in respect of these processes.

1. The first step consists usually of extending a given farming system by establishing new smallholdings in expansion areas, wherever they are available.

2. When there are no more expansion areas available, or when the distance between the expansion areas and the original settlement area becomes too large, intra-farm land reclamation acquires importance, and cultivation is extended at the expense of the fallow. The transition from shifting cultivation to fallow and permanent farming occurs slowly but surely.

3. Cropping patterns change at the same time. As land use becomes more permanent, the relative suitability of some crops diminishes, while that of

others increases. Thus a change in the type of crop ensues. The cultivation of such crops as manioc and sweet potatoes gains ground, since they yield high and certain calorie returns per hectare and per hour of work, and therefore release land and labour for the cultivation of cash crops. Millets and maize diminish in relative importance.†

As long as these three possibilities still exist we can count on rates of production growth that are at least as high as those of population growth. However, as soon as production reserves of this type are exhausted, the law of diminishing returns applies. Up to this point farmers may have obtained constant or even increasing returns to increasing labour inputs based on the exploitation and consumption of soil fertility. From this point on, farming moves into another stage in land use which, within traditional techniques, means decreasing soil fertility, decreasing yields, and therefore decreasing returns to labour. There are certainly wide differences in the speed of the process. The lower the rainfall and the less favourable the edaphic conditions, the more rapid the fall in labour productivity. The higher the rainfall, the greater usually are the chances of intensification by multiple-cropping techniques and of a more gradual decline in labour productivity. The general tendency, however, is obvious: the impact of diminishing returns is much more pronounced in upland systems than in irrigation systems.

Following the nomenclature of Geertz (1963) agricultural 'evolution', with constant returns to labour, turns with increasing land-use intensity into 'involution', with decreasing returns to labour, and will tend over time towards a maximum population living at a low-level steady state. Increasing subsistence demand leads to the reduction of the cash-crop area, and low-yielding though high-quality foodstuffs, such as millet and maize, are replaced by high-yielding but low-quality food crops, like manioc and sweet potatoes. The greater the population density and the smaller the farms, the higher is the percentage of crops like manioc and sweet potatoes and the lower the percentage of maize in the subsistence food. The final stage of this type of involution consists of very small holdings, without room for cash cropping, which grow root-crops almost exclusively. The families living by this type of farming are particularly poorly nourished, diseases are more widespread than elsewhere, and the extent of underemployment is particularly great.

The system of land use on Ukara in Lake Victoria is a remarkable attempt to overcome this process of pauperization. But even when all imaginable measures open to pre-chemical men are applied to retain the fertility of the soil, as they actually are on Ukara, it is apparently not possible to avoid decreasing returns per hour of work when land use is intensified on tropical

† If we assume the growing periods of millet, maize, and sweet potatoes to be 4 months and that of manioc to be 18 months, the monthly return (in Giga Joules ha^{-1}) are: millet 2·15; maize 4·48; sweet potatoes 8·08; and manioc 7·87. Root crops produce two to four times more energy ha^{-1} than grain crops under similar conditions.

soils within the traditional state of agriculture. The average returns per hour of work are particularly low on Ukara.

A similar involution can be observed in the livestock economy. In unregulated ley systems with sufficient grazing, the herds per family are large and the animals well fed. A spatial increase in arable farming reduces the grazing area until almost all of the land is annually cropped. The production of fodder crops is neither customary nor competitive. This usually gives rise to a situation with high numbers of poorly fed cattle. In some cases, cattle are replaced by goats and sheep. Only in exceptional cases is the cattle economy integrated into some type of mixed farming. The cattle numbers per family decrease, therefore, with increasing permanency of cultivation, although livestock densities per 100 ha may be high. The final phase of this involution is reached when the livestock economy is reduced to small herds of animals per family—mainly goats—which live on crop residues and on areas unfit for cultivation.

Neither of the methods to increase output that are practised so skilfully by smallholders—extending the cropping area and changing the cropping pattern—is sufficient to secure a rising income for a growing population. There is a great danger of stagnation at a somewhat higher level of income, and of a decline in incomes, once these possibilities are exhausted.

Increasing land shortages reduce not only the marginal returns of labour but often also those of land. The smaller the holding, the lower is in some cases the average gross return and the marginal gross return per hectare, simply because the smaller the holding the smaller is the scope for cash production with crops yielding high returns per hectare (Ruthenberg 1968, p. 330).

The general tendency is revealed, for example, in three neighbouring districts in Sukumaland. In the thinly populated Shinyanga area the family income rises rapidly and constantly as the labour capacity increases. In the more densely populated Kwimba area labour productivity is already considerably lower. Finally, on the island of Ukerewe, Lake Victoria, where permanent cropping is practised, there is scarcely any connection between family income and the family's labour capacity (see Fig. 6.6).

Because of the priority attached to subsistence-food production the smallholdings with permanent upland farming are caught in a situation that might well be called a 'low-level equilibrium trap'. The only way out, except through migration, is to attain high returns per hectare and per animal by introducing yield-increasing innovations.

6.6. Development paths of permanent cultivation systems

6.6.1. *Improved permanent cultivation systems*

The possibilities for intensification and improvement of permanent arable

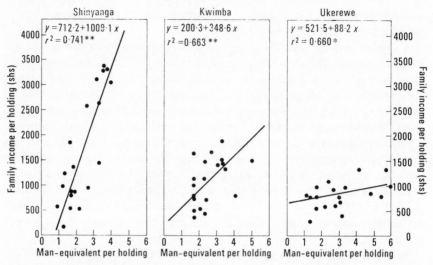

FIG. 6.6. The relation between land availability and labour productivity in three areas of Sukumaland, Tanzania (from von Rotenhan 1968, p. 78). Twelve families out of the total of seventy-five studied have been omitted because they receive non-agricultural incomes.

farming have been drastically changed by technical progress. Intensive cropping with decreasing soil fertility and low yields per hectare used to be the final stage, a blind alley in the development of tropical farming systems for those who had not enough land for fallowing and not enough water for irrigation. Now, we have an almost unlimited number of useful innovations, which are of limited applicability in shifting and fallow systems, but which, from a technical standpoint, are fully applicable in permanent farming. The farm-management aspects of these innovations differ widely and we distinguish therefore between the development tendencies of the major climatic zones.

(a) *Semi-arid regions.* Most permanent cultivation systems of the tropics are in the semi-arid regions. Here the increase of returns to land and labour is closely related to higher 'rainfall efficiency' which is defined as the agricultural production (in kg or the monetary equivalent) in relation to the seasonal precipitation (in mm) (Krantz et al. 1977, p. 2). In most of the existing systems the proportion of annual precipitation used is low, while moisture stress frequently decreases yields. A great number of innovations are readily applicable because field systems already exist, while in semi-humid and humid regions most rain-fed farming is in fallow systems, and a change in system is required before yield-increasing innovations can be applied.

1. The basis of most changes are shorter-maturing and higher-yielding varieties associated with mineral fertilizers. Higher and more reliable yields are obtainable and maize, a high-yielding crop which has hitherto been too much of a drought risk, has become a reliable proposition in extensive

areas. Tourte, for instance, puts his expectations as to the future yield level in southern Senegal at 3–5 t ha^{-1} of grain (1974, p. 931). This would require a fertilizer input per year of 90 kg N, 30 kg P_2O_5, 75 kg K_2O, and 170 kg Ca.

The up-breeding of grain legumes to yield levels comparable to those of the major cereals may become of equal importance. This, and possibly the development of other species with the capacity for nitrogen fixation, would contribute to reducing the high dependence of permanent systems on mineral fertilizers.

2. Shorter-maturing varieties often allow higher cropping intensities even in rather dry areas. The new opportunity does not stem from sequential cropping, because the rainy season is too short for two consecutive crops (Ryan, Sarin, and Pereira 1979). Much more promising are improved types of intercropping, and ratooning, since few crops can efficiently utilize the whole moisture supplied by the rains. The possibility for improvement is well demonstrated in Fig. 6.7, which shows the growth pattern of a long-

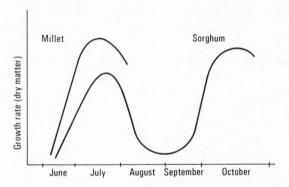

FIG. 6.7. 'Competition gap' in millet–sorghum mixtures, Northern Nigeria (from Baker 1974, p. 11).

season sorghum interplanted in a short-season millet. Dry-matter forma-tion of the sorghum is stunted as long as the millet occupies the field, but takes off in the latter part of the rainy season. Existing varieties thus show a 'competition gap' which implies a loss in output. The general implication is that research should aim to 'tailor' crops and varieties which optimize total output in a crop association (Baker 1974). To fully utilize the water and soil resources it is essential to develop cropping systems which will provide a continuum of productive crop growth from the onset of the rainy season to the post-rainy season as long as soil moisture can sustain plant growth (ICRISAT 1977, p. 41).

3. The drier the area the more important are the interactions between mechanization and improved varieties. Tillage immediately after the harvest, creation of a proper soil structure, reduction of erosion, early

sowing, possibly before the onset of the rains, traction power for speedy resowing in cases where gaps in rainfall kill the young crop, and proper and timely weeding; these are yield-increasing techniques which depend above all on sufficient traction power. The drier the area the more pronounced are the advantages of animal traction compared with hoe-cultivation and of tractors compared with ox ploughs. Mechanization is probably essential for the success of any cultivation system in semi-arid areas. Opinions differ regarding the proper approach. Tourte, relying on a great amount of research in West Africa, considers proper and deep ploughing to be essential (1974, pp. 927, 931). Gibbon, Harvey, and Hubbard point to successful trials with minimum tillage techniques under semi-arid conditions (1974, p. 229). At ICRISAT water- and energy-saving soil husbandry methods have been developed (Kampen *et al.* 1975). All, however, stress the importance of water-conserving cultivation techniques in order to obtain the interaction benefits of new varieties, fertilizers and a greatly improved soil moisture regime. Modernization also tends to produce more reliable yields. Bigot established for Senegal that traditional groundnuts yielding 1 t ha^{-1} have a coefficient of variation in yields of 44 per cent compared to 28 per cent for modernized groundnuts yielding 1·6 t ha^{-1}. The coefficient of variation with traditional pearl millet was 49 per cent compared with 31 per cent with modernized pearl millet (1974, pp. 1.1 and 1.2). The joint action of intensification and mechanization under small farm conditions usually leads to the doubling of output per hectare with roughly unchanged labour inputs. Traditionally the labour demand is highly peaked for weeding operations. Modernized systems tend to a better distribution of labour requirements. Less labour is required for weeding and more for harvesting (Monnier 1976, p. 66).

4. Higher and more reliable yields can be obtained through proper 'drought strategies' (Krishnamoorthy 1974). If the onset of the rains is delayed then other varieties, crops, crop mixtures, seed rates, and fertilizer applications have to be chosen. If gaps in rainfall (during the humid season) occur, then the capacity for re-sowing should be available, weeding has to be more careful, ratooning of millets and sorghum can be practised—the first growth is cut as fodder and the ratoon crop produces the grain—and fertilizer application has to be split. If the rains stop too early, then moisture-demanding plants of the crop mixture ought to be removed; the crop of maize and sorghum may be saved by removal of the lower leaves.

5. In permanent systems the functions of the fallow are taken over by imported inputs, and growing green-manure crops for soil fertility reasons is usually not a profitable proposition in semi-arid regions. Important increases in output may be expected by mixed farming due to the inclusion of leguminous fodder crops in the rotations, the application of manure, and the full utilization of crop residues, grasses grown on waterways, and so on.

Mixed farming on a low level of intensity is widely practised in India and yields low returns, but the change to more productive mixed-farming systems requires heavy inputs in stock, stables, provision of water, fodder conservation, transport equipment, etc., and is likely to be a slow one.

6. Innovations change the potential of areas with bimodal pattern of rainfall. Shorter-maturing varieties reduce the risk of crop failure due to

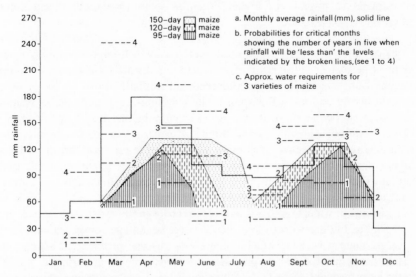

FIG. 6.8. Water requirements and water availability under the bimodal pattern of rainfall in Boro Division, Siaya District, Western Kenya (from Collinson 1977).

the drought. The provision of more traction power in interaction with water-conserving cropping is in itself a risk-reducing device. Collinson (1977) demonstrates the possibility with an example from Kenya (Fig. 6.8). The long rains provide sufficient moisture for a 120-day crop in about three out of five years, and the same can be expected of a 95-day crop in the short rains. Two crops are thus feasible in a semi-arid area where a highly risky 150-day maize is normally grown.

But most of the wide range of output-increasing innovations which would change permanent cultivation in semi-arid regions into a reliable, productive, and balanced way of farming depend on sufficient traction power, and the labour-absorption capacity is far less than in semi-humid or humid regions.

The encroachment of smallholder hoe-farming towards semi-arid land, which takes place in many parts of Africa, is therefore a particularly dangerous tendency. The expansion of smallholder hoe-farming has much to recommend it in areas of high potential, in particular in valley-bottom development. It becomes very dangerous in semi-arid upland farming where yields are unreliable and depend to a high degree on the proper timing of cultivation, planting, and weeding. Smallholder hoe-farming ties people to a situation with very low marginal returns to labour. The famine risk is high. The probabilities of drought are known and affect a great number of hoe-farming families. Where the land is not yet occupied by smallholders, efforts should be made to organize settlement in mechanized medium- or large-scale holdings, while the surplus population of the semi-arid areas should be encouraged to migrate to higher potential semi-humid and humid areas wherever this is possible.

(b) *Semi-humid regions.* The same principles which apply to the development of permanent cultivation in semi-arid regions are valid for semi-humid regions, but there are important differences in emphasis. Most arable farming in semi-humid regions still employs fallow systems, and a change in system has to occur before innovations can effectively be applied. This change in system requires even more of a 'jump' in technologies than is required in semi-arid regions, because it is more difficult to preserve soil fertility. Much higher levels of soil-preserving inputs have to be provided to attain a balanced system. There is, however, much more scope for long-term increases in output and productive employment because of the extended wet season.

1. A much wider range of improved crops and varieties has become available, but grain crops lose their relative advantage compared with root crops mainly because of cloudiness and disease problems. Particularly interesting prospects for producing low-cost starch are opened up by the improvement of several root-crop species in interaction with mineral fertilizers. The improved grain legumes may prove to be of no less importance, in particular cowpeas and pigeon peas.

2. The attainment of the double objective 'more output in a balanced system' means aiming for a continuum of cropping which begins with the onset of the rains and stretches as far as possible into the dry season, whereby much emphasis can be laid on multiple cropping and ratooning, in addition to intercropping and relay-planting (Krantz *et al.* 1974, p. 18).

3. In semi-humid climates the functions of the fallow are much more essential for the maintenance of soil fertility than in semi-arid areas, and the question usually is whether to substitute the fallow by (i) high inputs of mineral fertilizer and/or (ii) by green-manure crops and/or (iii) by mixed farming, i.e. by growing fodder crops for livestock on the farm.

Table 6.11 gives an idea of the differences in performance of multiple-cropping systems in traditional and modernized situations. The Costa Rican and the Philippine examples are both from fertile soils in a humid climate. Here output is highly elastic in its response to increasing chemical and mechanical inputs. Output in terms of edible dry matter is four to ten times higher with the more intensive rotations. Also labour input increases, even

TABLE 6.11

Examples for input–output relations in multiple-cropping rotations on upland in the humid tropics (per hectare and year)

Country	Costa Rica[a]	Philippines	Philippines
Year	1976–7	1972	1972
Type of farming	Smallholdings	Field experiments, IRRI	
Technique	Traditional hoe–cultivation	Modernized smallholder farming with two-wheel tractor	
Rotation per year	Beans–maize + upland rice	Rice–sweet potatoes + maize	Rice + cowpeas + sweet corn/ sweet potatoes + sweet corn/ sweet potatoes
Technical data (total for all crops of the rotation and per year)			
Labour input (man-hours)			
Land preparation }	365	184[b]	144
Planting }		152	240
Weeding	329	480	496
Harvesting }	305	328	1456
Other work }		168	504
Total	999	1312	2840
Yields (t ha^{-1})			
Rice	0·6	2·0[c]	2·0[c]
Beans or cowpeas	0·7	—	4·0
Sweet potatoes	—	9·0	25·0
Sweet corn	—	—	12·0
Maize	0·4	4·0	—
Dry matter, edible	1·4	8·2	14·6
Dry matter in straw, etc.	3·8	9·5	15·9
Total dry matter	5·2	17·7	30·5
Carbohydrates (marketable)	1·0	6·6	12·4
Protein (digestible)	0·23	0·4	0·7
Economic analysis ($)			
Gross return	549	924	3150
Material inputs			
Seeds	15	21	66
Fertilizer	—	73	147
Herbicides	—	4	17
Pesticides	1	40	163
Mechanical operations	1	151[d]	173[d]
Total	17	289	566
Income before interests and overheads	532	635	2584
Income per man-hour	0·53	0·48	0·91

[a] In interpreting this table it has to be considered that experiments show much better results than actual farming, and the difference tends to be the greater the more complex the technique or rotation. Also it has to be considered that prices in 1976–7 in Costa Rica expressed in US-dollars were about twice as high as in 1972 in the Philippines. [b] Unusually high because of heavy rainfall. [c] Unusually low due to grassy stunt disease. [d] Mostly for tractors. *Sources:* Costa Rica: Navarro (1977); Philippines: Banta (1972).

though tractors are used. However, much of the additional output in value terms is consumed by high additional material inputs. The advantage of intensification with modernization lies in the additional output and employment that arise, and not so much in the additional return per hour of work.

More research and a lot of trial and error will be required to find out which types of multiple cropping are suitable for different locations. From the farm-management point of view the following observations can be made:

1. The permanent cropping sequence, without green manuring and fodder crops, is by far the simplest solution. If research continues to confirm that a high level of yields in a balanced system can be maintained with high inputs of mineral fertilizer and crop residues, then this will be economically the more promising approach in most locations, provided price relations between the produce and fertilizers are sufficiently attractive.

2. Green manuring is technically easier, substantially cheaper, and probably maintains soil fertility as well as does mixed farming with livestock. The capital costs of livestock and the costs of transporting fodder and manure are not incurred: the green-manure crop is ploughed into the soil where it grows. Green manuring, however, requires fertilizer inputs and sufficient traction power for ploughing, which often has to be done in the dry season.

3. Mixed farming, the growing of fodder crops in a rotation, and the utilization of the roughage by cattle is probably the most productive approach to permanent cultivation, but also the most demanding in terms of skill, capital, and markets. Technically, the approach has become much more interesting through the provision of new and improved grasses and fodder legumes. One hectare of fodder properly managed and highly fertilized is capable of supplying enough roughage of sufficient quality to feed four or even six cows (Whyte 1967, p. 126). However, there are numerous problems. In some areas it is still extremely difficult to maintain a healthy and productive dairy herd, in particular in the West African lowlands (Ruthenberg 1974, p. 84), and beef-cattle raising is often too extensive an activity for smallholders with permanent cultivation systems. Mixed farming is very complex, and trial and error may well show that specialized fodder holdings, based on artificial pastures, are a more economic approach to beef production than mixed farming.

4. A very important asset for agricultural development are the hydromorphic soils in the valley bottoms of the semi-humid zones. They are usually well suited for wet rice, vegetables, grasses, and root crops, and they offer much scope for productive employment.

The impressive list of possible avenues for the development of permanent cultivation systems in the semi-humid tropics has, however, to be interpreted with much more caution than in semi-arid areas. Permanent arable production on soils of alluvial and volcanic origin can usually be carried out with technical and economic success. On other soil types, arable cropping in semi-humid and in humid tropical low-land climates tends to run into a number of problems which have been well summarized by Flinn, Jellema, and Robinson (1974).

1. Yields suffer because of frequent periods of moisture stress, even during the rainy season, because of the poor water-retention capacity of most upland soils. Droughts of only 5 days may depress yields significantly if they occur during the period of tillering.

2. High soil temperatures, especially at the end of the dry season and the beginning of the wet season, tend to inhibit germination and reduce photosynthesis.

3. Intense rains lead to much leaching and severe erosion even on moderate slopes unless the rainfall is broken up by a canopy of plants or a surface mulch.

4. High night-time temperatures and lower radiation intensity, especially during the latter part of the rainy season when there is a high incidence of cloud cover, reduce the photosynthetic capacity (Kassam and Kowal 1973, pp. 39–49).

5. Weed problems in fallow systems are to some degree reduced through the weed-suppressing capacity of the fallow vegetation. Weed problems in irrigation farming, particularly in wet rice, are reduced by deliberately changing moisture levels in the soil. In permanent cultivation systems weed problems are particularly serious. Most systems involve seasonal fallowing, and the effort to prepare the weedy land for cultivation becomes expensive. The weeding requirement during the growing period often coincides with the height of the rainy season when the field is so wet that mechanical operations are not feasible.

6. Plant-disease problems tend to be more pronounced in semi-humid than in semi-arid zones.

7. Fertilizers tend to be less effective than in drier climates or in irrigation farming because soils are predominantly acid, are low in organic matter and deficient in silt and clay particles. Their cation exchange (i.e. nutrient holding) capacity is low. Nitrogen fixation by legumes is also lower (Kassam and Kowal 1973, p. 49).

The introduction of new varieties, mineral fertilizers, and pesticides which yielded impressive increases in the output of sub-tropical farming and some semi-arid tropical farming situations cannot therefore be expected, as yet, to produce similar effects in permanent cultivation systems in the semi-humid tropics. The problems are particularly pronounced on large farms with large fields. Permanent cultivation in semi-arid areas tends to demand fields which are large enough to allow tractor cultivation, but under the condition of population pressure in a semi-humid climate the smallholding seems to be a more effective size of unit. The smallness of plots and the presence of numerous hedges with useful trees tend to prevent gully formation and weeds are more easily controlled by a combination of much family labour and some herbicides than by mechanized operations.

Most of the problems often encountered in permanent upland systems in a semi-humid climate could be solved by a viable zero-tillage system, i.e. by a system in which the ploughing and harrowing before planting would be replaced by spraying with paraquat 3–5 days before planting. Zero-tillage in a semi-humid climate on suitable soils apparently produces a steady yield over a long period, control of run-off and erosion, soil-moisture storage, reduced soil temperatures, soil structure maintenance, and savings in power and savings in time, because no cultivation is required before seeding (Lal 1975, p. 31). However, it requires fertilizers, herbicides, careful insect control, and crops which produce a great amount of residues or the import of mulching material, and it may be suitable for certain soils only. Whether this approach to permanent cultivation or ploughing systems is better suited to maintain yields and soil fertility in the long run will have to be shown by more research and trial and error in practical farming. The development of effective zero-tillage systems would have far-reaching economic and social implications. Ploughing systems show pronounced economies of scale, and smallholder farming may not be competitive, while a zero-tillage system would imply that agricultural progress could be introduced in small and large farms with about the same effectiveness.

(c) *Humid tropics.* Permanent upland cultivation in the humid tropical climate in a balanced system of high productivity is an even more unsolved problem than in semi-humid areas. Wet rice in the valley bottoms and tree crops on the upland are the proven alternatives which deserve preference. If, however, population densities are such that upland has to be used for food crops, then the most promising alternative to the low-level equilibrium seems to be a high-intensity cropping system (Okigbo 1974*b*) involving:

1. several short- and medium-term crops which cover the land most of the year (particularly promising seem to be manioc, yams, maize, rice, pigeon peas, and cowpeas);†

2. plantains and bananas;

3. a cover of tree crops;

4. valley-bottom developments to produce food, cash crops, and mulch material for the upland;

5. small animals—the goat would appear to be most suitable; pig

† Zandstra and Carangal (1977) describe such a system on Sumatra as follows: 'This pattern starts with the planting of upland rice intercropped with corn (2 m spacing). Cassava is planted into the corn rows 30 days after rice seeding. Corn is harvested at 85 days and rice is harvested at 130 days. The cassava canopy is at the level of the upland rice canopy at the time of the rice harvest. The space vacated by the rice crop is then used for the planting of peanuts followed by rice beans. The cropping pattern minimizes land preparation, keeps *Imperata* down by virtually continuous shading and light tillage and distributes labour and cash flow throughout the year.'

enterprises based on processed manioc could be added; the manure produced by these animals is particularly valuable in these systems which require much humus.

Such a system would tend towards zero-tillage on the upland and would have to be maintained by mineral fertilizer and mulching. Some functions of tillage would have to be taken over by herbicides (weed control) and pesticides (insect control). The success of the approach would depend on the 'tailoring' of varieties in such a way that light is comparatively uniformly distributed amongst the crops which participate in the multi-storey plant system.† A continuous cover of crops, established by intercropping techniques and maintained by much fertilizer, would have to take over the various functions which otherwise are performed by the fallow vegetation or by the many weeds which are traditionally found in the farmers' fields. It is obvious that such a system is highly labour-demanding, and trial and error will have to show whether labour productivity is sufficiently high to make it attractive in terms of the farmers' objectives. The social and institutional implications of creating an economically advantageous intercropping system with zero-tillage or minimum tillage are very important, because it would mean that there would be hardly any economies of scale and smallholder farming with high employment would be preferable to larger mechanized farms.

(d) *Tropical highlands.* In most of these areas the development from fallow systems to permanent cultivation has already occurred, and the principles of sound farming for temperate climates usually apply. Two tendencies deserve particular attention.

1. Ley systems are usually not competitive in a situation of growing population densities. The general tendency is towards permanent fields and permanent pastures.

2. Dairy production based on permanent pastures or even on cut-and-carry systems are particularly interesting propositions (Whyte 1967, p. 69).

(e) *Summary.* It appears from more recent trends in agricultural research and development that permanent upland systems need no longer be the 'dead end' of tropical farming. However, we must not overlook the fact that economic and social changes are required to turn a technically feasible innovation into an applicable one. All improved types of permanent cultivation rely on purchased inputs, and in the tropics this is much more so than in

† Experimental work indicates that under a traditional system of husbandry the mixed crop is more profitable than the sole crop, while under a modern system of husbandry the sole crop is more profitable (Flinn 1974). This seems to apply in situations where only one upland crop per year is feasible, as may be the case in semi-arid climates and some locations with semi-humid climates. The situation is different in humid climates. Modernized arable cropping in a humid climate would tend to favour a continuous use of the land, which may be achieved by mixed cropping (two or more intermingled crops), by intercropping (two or more crops in different but proximate rows), and by sequential cropping (one sole crop after the other).

temperate climates. The way out of the low-level steady-state trap of traditional systems depends on cash production and markets. High-output, balanced systems of permanent cultivation in a tropical climate require favourable price relations between purchased inputs (fertilizer, pesticides, herbicides, machines, and tractors) and the produce. The key to success lies in high yields, and this is well summarized by Singh, who points out '. . . that high production is good for the soil, that minimum tillage promotes soil tilth and conserves soil organic matter, that high fertility promotes high yields and lessens the losses of soil humus, that large amounts of decomposable organic matter in soils is essential for good tilth, and physical soil conditions and cover in the form of living mulch is good protection against erosion' (1972, p. 274). Because of this stable permanent systems may yield high outputs, but they are not 'low-cost' systems. Costs per unit of output are probably higher than with low-output traditional systems. In addition institutional changes are required: new types of land rights, secure land tenure, land consolidation, efficient rural supply and marketing systems, and all the rest. These preconditions do not usually exist. Consequently the gap between actual farming and the optimum solution is particularly great in permanent cultivation systems.

6.6.2. *Change from permanent upland cultivation to other farming systems*

(a) *From upland cultivation to irrigation farming.* Generally speaking, we find among permanent cultivators in the tropics the tendency—wherever there is the opportunity—to develop irrigation at the expense of rain-fed farming, especially where holdings are small. Even small irrigated plots can be important in reducing the risk of harvest failure, in using labour and traction power regularly and to their full capacity, and in producing a regular supply of fodder.

The traditional answer has been the change from upland systems to wet-rice systems which occurred gradually over several centuries in Asia. Wet-rice developments begin, as a rule, by using valley bottoms, and as the land becomes more scarce terracing is developed on the slopes. In Madagascar this process is in full swing, and in East Africa it is in the initial stages. Relatively large areas could be made suitable for wet rice in Africa and Latin America.

Whereas we may expect an accelerated change to wet rice in Africa and Latin America, it seems doubtful whether this tendency will continue in the densely populated regions of south-east Asia. The conversion of marginal hill land to irrigated terraces requires a high input of capital and manual labour. The rice terraces on the slope are mostly narrow, and it is difficult to mechanize the labour operations. With the new means of increasing the yield on the existing rice land, by using new varieties, mineral fertilizer, pesticides, etc., it is in general more profitable to use the existing areas more intensively than to create new rice terraces. Technical progress has, moreover, considerably improved the productivity of upland farming, so that the transition to wet rice no longer represents the only way of increasing the productivity of land and labour. Under the conditions of traditional agriculture, however, it was more or less the only way.

Another approach, which is particularly interesting in semi-arid areas, is the support of permanent upland cropping by 'life-saving' irrigation (Krantz *et al.* 1974). This means storing the water in minor ponds on the watershed and pumping it into the fields in case of a drought during the rainy season. The expansion of 'life-saving' irrigation would imply a change in irrigation pattern. Today's irrigation usually means the application of a lot of water on a few hectares, while 'life-saving' irrigation would mean a few millimetres on larger areas. The improvement in water-use efficiency is obvious. The problems occur in the high cost of irrigation equipment which would be used for only a few days per year.

(b) *From arable to perennial crops*. The change-over to perennial crops, particularly bananas (plantains), for food and to various tree crops for cash, is another obvious tendency as population density rises and commercialization increases. Bananas produce almost as much energy per hectare as tropical root crops, with less labour per unit of starch. Banana groves offer an almost complete cover for the soil and improve its fertility. They are clearly an attractive alternative to permanent arable cropping wherever soil and climate are suitable.

Over and above this, there are many indications that tree or bush crops are gaining in importance as development takes place. On the slopes of suitable areas, small plantations are frequently spreading at the cost of permanent arable farming. Examples are the coffee and the tea plantations of the Kikuyu in Kenya, and the tea and fruit plantations of the smallholders in Taiwan. It can usually be observed that the displacement of permanent arable farming by tree or bush crops depends on the cash-cropping opportunities, especially on the availability of attractive markets. Also there is reason to expect that fodder bushes, such as *Leucaena*, will be introduced into permanent arable cropping systems, as they are well suited for off-season fodder supplies.

A fascinating solution for the great number of problems offered by permanent arable land cultivation in semi-humid and humid tropical lowland climates would be the development of a perennial starch-producing crop which could compete with maize as to output and palatability of the produce (Holliday 1976, p. 144). Such a perennial could establish a full leaf canopy after the onset of the rains. High rates of effectiveness in turning solar energy into edible products could be achieved, and the support energies could remain at relatively low levels, due to the soil-, water-, and nutrient-preserving root system of the perennial.

7. Systems with arable irrigation farming

7.1. Definition and genesis

7.1.1. *Definition*

'IRRIGATION' describes those practices that are adopted to supply water to an area where crops are grown, so as to reduce the length and the frequency of the periods in which a lack of soil moisture is the limiting factor to plant growth. This is done in two main ways. Water that has fallen as rain on another piece of land may be collected and led onto the area that is to be irrigated. This procedure is usually called *artificial irrigation*. Alternatively, water falling on a given area may be impounded and held there, in order to reduce run-off. Farm systems where most of the output originates from fields husbanded with these two techniques are covered by this chapter.

In irrigation farming we therefore include a number of primitive forms of water delivery and water impounding, which do not necessarily ensure a regular and reliable water supply to the plants. In widespread areas of the tropics, rice is first irrigated in the sense of being wet-field rice. The fields are ploughed and planted in the rainiest season of the year and have earth walls round them, within which the rain-water falling on the field itself or coming from neighbouring areas is impounded. Later, at some point before the maturity of the plants or the ripening of the grain, many of these fields dry out, owing to the cessation of the rains and the absence of water supplies. This cannot accurately be called irrigation rice, but neither is it properly termed upland rice. Certainly in terms of agronomy and farm management the features of this transitional form correspond to those of irrigation farming rather than to those of upland farming. I therefore include wet rice, grown on lowland and on terraces which impound rainfall, in this chapter, although it is not properly farmed by irrigation methods. However, most of these rain-fed paddies actually receive water from impounded paddies higher up the slope, and are therefore irrigated.†

† Following Barker and Herdt (1978) we distinguish between: (i) irrigated rice (where water is added to the field from the canals, river diversions, pumps, etc.); (ii) shallow rain-fed rice (with maximum water depth from 5–15 cm); (iii) medium deep rain-fed rice (with maximum water depth from 15–50 cm); (iv) deep-water rice; and (v) upland rice. Rice

7.1.2. *Genesis*

Most irrigation farms are the result of a process of land development which occurs gradually over time.

(a) *From fallow to irrigation systems.* The higher the intensity of fallow systems the more acute become the soil fertility problems, and the stronger is the tendency to develop hydromorphic soils in the valleys as a means of gaining access to fertile land. Cropping moves from the upland down to the valley bottoms, where preference is given to wet rice, vegetables, and bananas. Flood control is a problem of valley-bottom farming, and because of this there is a strong tendency to progress to artificial irrigation proper. With a continued demand for productive land, irrigated fields are extended. Farming moves up the slope again, using terraces supplied with water and usually cropped with rice. This is the genesis of most wet-rice systems.

(b) *From permanent upland farming to irrigation farming.* Irrigation is man's major response to the challenge offered by increasing population densities in degraded systems with permanent farming.

1. Larger irrigation projects imply a 'jump' in technology. Within the perimeter of a project rain-fed farming techniques have to be changed without delay to irrigation farming techniques.

2. Perhaps more important in over-all economic terms is the gradual extension of minor irrigation developments. New sources of water are developed. Motor pumps are substituted for animal power. More water is pumped out of the ground, but, with increasing demand for water, less water is usually applied per hectare, so that the area of irrigation farming is extended.

7.2. Types of irrigation farming and their geographical distribution

The guiding principle of upland farming is the adaptation of cropping to the conditions of soil and climate. Irrigation systems, on the other hand, because they involve increasing control of water and reliability of water supplies, are characterized by a 'control of nature'. Both water supply and soil conditions become increasingly subject to human control. Therefore, we classify the forms that irrigation farming takes according to the water supply system, the cropping system, and the exploitation system.

7.2.1. *Water-supply systems*

Water for irrigation can be obtained from natural flows by diversion,

---

classified under (i) occupies 32 per cent of the land and produces 50 per cent of the crop. Class (ii) has 34 per cent of the land and 33 per cent of the crop. With class (iii) the percentages are 15 and 8. Deep-water rice is grown on 8 per cent of the total rice hectarage while production amounts to 4 per cent. 10 per cent of the rice hectarage is on upland which produces 5 per cent of the total crop (Palacpac 1977).

damming, or pumping. Different ways of damming water are particularly numerous. They range from relatively simple methods like retention of tidal water (rice irrigation on the coast of Java, date irrigation in Schatt al Arab, Iraq), growing of swamp rice in naturally flooded areas of river valleys, and impounding of rainfall in paddy fields, to more sophisticated and artificial methods, especially the damming of water courses, which can involve capital works of enormous size and cost.†

Water is conveyed either on the surface in canals or ditches, above ground through flumes, or underground in pipelines. The main methods of applying water in the fields are by flooding, basin, furrow, and sprinkler irrigation. *Flooding* is the practice of letting water on to a larger plot, the water being distributed either into basins or by furrows. *Basin irrigation* is the practice mostly applied in intensively cropped smallholdings: wet rice is almost always basin irrigated. Here water stands in the basin during all or part of the vegetation cycle. Water distribution with the help of small basins is also usual for irrigated vegetable production. *Furrow irrigation* is usually associated with the use of a tractor and comparatively large, well-levelled fields, as, for instance, fields for cotton-growing in Gezira, Sudan. There are, however, numerous intermediate forms between basin and furrow irrigation. Sprinkler irrigation in the tropics is still in its initial stages and mostly applied to perennial crops like sugar-cane or coffee.

The cropping patterns and farm-management characteristics of holdings with irrigation farming are strongly related to (a) water availability, (b) reliability of water supplies, and (c) water quality.

(a) *Water availability and land use.* Water availability on the holding can be considered from a number of different points of view:

1. The availability of water in relation to area. Holdings on which it is possible for all the land to be irrigated are found principally in the flood plains of large rivers (China) or near canals (Punjab).

Where irrigation is supplied by small streams or springs, usually only a part of the holding area can be irrigated. In these cases the holding contains two distinct types of farming: irrigated production and supplementary upland cropping.

2. The availability of water in relation to time. Holdings whose land is irrigated the whole year round, thanks to water supplies that flow all the year, are in the especially privileged position of being able to be cropped

---

† Particularly ingenious methods for obtaining water are used in highly populated but dry areas. An example is the method of water procurement in the Yemen. Rainfall on rocky surfaces is conveyed into adjoining earth-bound fields. In some areas—for instance at Sunnatayn, 150 km north of Sanaa—I found a systematic alternation between fields and completely eroded rocky surfaces in a ratio of about 1:3, i.e. 3 ha of rocks with a high run-off convey about 50 per cent of the 300 mm of rainfall they receive to 1 ha of arable land.

intensively and continuously. This is true, for example, of oasis agriculture and of farms below large dams or with tube-wells.

Water discharges from rivers generally have peak periods, and irrigation farming has to be adapted to the availability of water during the year. In holdings whose water supply is sufficient for irrigation in only one growing season, the question arises as to whether rain-fed farming is possible for the rest of the year. In the semi-arid zones, the irrigation area lies fallow in the dry season, or serves as pasture. In south-east Asia, the production of monsoon rice in paddies is typically confined to the wet season. Where the rainfall is high enough, however—e.g. in Sri Lanka and the highlands of Madagascar—irrigation farming and rain-fed farming alternate on the same land: in the summer it supports wet rice; in the drier, winter season non-irrigated grain or legumes. The availability of irrigation water in the dry season is of particular value in areas with pronounced and heavy rainy seasons, because during the rains crop development is often stunted owing to waterlogging, cloudiness, and plant disease. Crop yields are usually much higher during the dry season.

3. The availability of water in relation to quantity. Over and above this, we distinguish between: (i) crops that rely fully on irrigation, and could not be grown at all if additional water were not available; (ii) the supplementary irrigation of crops that could also be grown without additional water, but yield higher returns with irrigation; and (iii) 'life-saving' irrigation, which means the application of very small amounts of supplementary water to bridge a gap in rainfall during the wet season.

In the first two cases a further distinction may be made according to whether irrigation holdings have plenty of water; or water is scarce in relation to plant needs, or must be bought at high prices.

Where there is no lack of water—as for example in the Burmese rice holdings—the objectives of farm management lie less in maximizing the value of output per cubic metre of water than in maximizing the value of output per hectare or per worker. In other cases—e.g. in the cotton holdings of Gezira, in the Sudan—a high productivity per hectare is of less importance than a high productivity per unit of water.

Limited availability of water leads necessarily to systems of water distribution, which then influence land use. In most irrigated areas, a common water supply is shared by a number of users. Division of the water in such cases may be based upon established priorities for use, or the users may share equally in the available supply, with division based upon the area irrigated by each farmer. The total flow is sometimes divided among the users so that, according to this entitlement, each obtains his share as a continuous flow throughout the season. A more common practice, however, is to rotate the use of the full stream among users. The length of time during which each user is entitled to make use of the water is usually based upon the area irrigated by him, and is limited so that all users will have an opportunity to share in the use of the water at reasonable intervals. Another method of distributing irrigation water, which is used on the more highly developed irrigation projects, is delivery of water to the users upon request. This requires the maintenance of a flow of water in the entire canal system throughout the irrigation season.

The cost of the water, which is met from water charges, taxes on irrigable land, or betterment levies, is also an important factor in determining the irrigation method used. Where water costs are high, water-saving irrigation methods will be applied, and crops will be grown that yield comparatively high returns per unit of water (Boucher 1967, p. 3).

(b) *Water reliability and land use.* It is the ultimate objective of irrigation works to put the irrigator in such a position that he has water under his control at all times. This ideal arrangement is, however, not usually attained. Most irrigators in the tropics work under the threat of flood or drought. The reliability of water supplies is a decisive factor in the organization of irrigation systems, which can be classified according to this.

1. The water supply is particularly unreliable where it depends on seasonal flooding. Thus, for example, in river-flooded areas on the west coast of Madagascar, in the Niger Delta, in Mali, on the flood plains of the Amazon Basin, or in the Rufiji Delta, Tanzania, rice is planted on land where water is likely to collect during the flood season. The floodwater is normally sufficient to produce a reasonable rice crop.

2. In swamp-rice systems use is made of naturally dammed water. For example, in Sri Lanka and West and East Africa, damp, swampy hollows are planted with rice in the rainy season.

3. Water supplies fluctuate widely where rainfall is impounded in paddies. The farmer has, as a rule, no influence on the water supply for his rice, but he tries to adapt the run-off from the basin to the requirements of the rice crop.

4. Water supplies obtained by diverting the natural flow of streams, or by storing water in tanks and small dams, will normally fluctuate in quantity during the irrigation season, and will vary from year to year. The same often applies in the case of wells, which may become dry after years with poor rainfall. Consequently some of the irrigators will have to reckon with reduced water supplies in some years.

5. The most productive forms of irrigation farming are associated with a controlled and reliable water delivery. This is usually the case where the water is obtained from a large surface reservoir or by pumping from an underground reservoir (Boucher 1967, p. 2).

An irrigation farm usually consists of several fields, some of which have a reliable source of water and some of which are 'semi-irrigated' and receive water only when there is a surplus from fully irrigated fields. A case in point are farms associated with tanks in India whose irrigation capacities depend on the monsoon. In wet-rice farming we often find holdings with (i) plots in the centre of the valley bottom where floods are to be expected, (ii) plots on the border of the valley bottoms and the lower slopes, which are not flooded but receive a regular supply of water, and (iii) plots higher up the slope where the supply of water may be insufficient in dry years.

With a decrease in the control and reliability of the water supply, production risks increase, and these become greater as the intensity of farming increases. Farming systems with uncontrolled and unreliable water supplies are therefore in general more extensively managed and offer reduced scope for introducing agricultural innovations. As the water supply becomes less reliable, crops like sorghum, with a lower productivity but a low susceptibility to partial drought, are given priority over other better, but more vulnerable, crop types.

(c) *Water quality and land use.* Another important factor influencing land use is the quality of water, in particular its content of salt, which may determine irrigation methods and cropping patterns as well as investments in drainage. Where the salt content of the water is high, we find a preference for salt-tolerant crops such as date-palms, cotton, rice, barley, lucerne, and cabbage. Farming may become an intermittent affair, i.e. land is developed for irrigation and farmed, but for a few years only. In the initial phase salt-sensitive crops (wheat) are grown. With increasing salt accumulation, farmers change to salt-tolerant crops (barley, date-palms), and finally the salted land turns again into a salted pasture (El-Hakim 1973). As far as possible, farmers will try to irrigate in those seasons when the salt content of the water is relatively low, and they will prefer irrigation methods like sprinkler irrigation, which reduce the import of water and thus of salt. Under extensive farming conditions, fallowing, shifting of the crop area, and rotational irrigation are used to prevent the loss of land due to salination.

7.2.2. *Cropping systems*

The close relationship between water availability, reliability, and quality, on the one hand, and socio-economic conditions, on the other, has caused various cropping systems to develop. The principal difference is between wet-rice systems and other arable irrigation systems.

In wet-rice systems, the land is flooded and the crop is grown in wet soils from the time of transplanting or seeding until harvest approaches. The methods used are not common to other crops, because the physical, chemical, and biological conditions in flooded rice fields differ substantially both from those met in upland farming and from those that occur in other types of irrigation systems (see § 7.4.1). Wet-rice systems have for centuries been the most important form of land use in south-east Asia, in terms of both area and the economy. In Africa and Latin America, wet-rice cultivation is spreading quickly.

Other major tropical irrigation crops are cotton, sugar-cane, groundnuts, wheat, and vegetables. Large areas of irrigated cotton are grown in India and the Sudan. In India and Pakistan, we find also important farming systems with sugar-cane and wheat as the dominant irrigation crops. Numerous minor irrigation systems, frequently based on various arable crops and on vegetable production, are found in practically all parts of the tropics.

7.2.3. *Exploitation systems*

While we can say that shifting and fallow systems are the province of smallholders, that regulated ley systems tend to be best suited to units that are sufficiently large for animal or tractor ploughing, and that tropical permanent upland farming is again the province of smallholders, we cannot generalize about farm size in tropical irrigation farming. Traditionally, irrigation farming and high population densities are two characteristics of the same setting. But while high population densities tend to favour small farm sizes, there is the tendency to manage all the area irrigated from one source as a single, large unit, and the more expensive the irrigation system was to develop, the more pronounced is this trend.

Where smallholders are independent in their water supply, either because they have springs or tube-wells of their own, or because impounded rain-water is sufficient for wet-rice production, they are equally independent in organizing their land use. In minor irrigation schemes of up to about 300 ha, numerous smallholders usually operate either as tenants, getting their water from the landlord, or as owner–farmers, sharing the water according to traditional, purchased, or co-operatively-agreed entitlements. Under such conditions, land use depends almost entirely on water rights, in other words on the timing and quantity of water distribution. Water use in medium-sized (300–5000 ha) and large (more than 5000 ha) irrigation schemes is usually organized by some kind of scheme management or irrigation administration, and sometimes the scheme management is in a position to enforce very precise patterns of land use.

7.3. General characteristics of irrigation farming

7.3.1. *A comparison with upland farming*

Irrigation farming has distinct advantages compared with upland farming. Because of irrigation, the ample and regular supply of solar energy that is characteristic of the tropics is allied with the ample and regular supply of water created by investments and secondly with the nutrients brought by the water. Moreover, irrigation, particularly in the form of impounding, avoids the stress to plant growth that is connected with very high temperatures on the soil surface. At the same time, the conditions for agricultural production are improved, particularly when the water supply is ample, regular, and reliable.

1. Irrigation farming compared with upland farming under similar conditions produces higher gross returns per hectare. The basic possibilities (the realization depends naturally on the local conditions) are that:

(i) higher yields per hectare of a particular crop may be achieved;

(ii) several harvests a year may be produced;

(iii) crops may be grown that produce comparatively high yields per hectare, some of which may demand a water supply all the year round;

(iv) continuous rice cultivation becomes possible, thereby extending the area of this relatively high-yielding crop. Double- and triple-cropping with rice increases the incidence of disease, but the problems are still manageable. At the International Rice Research Institute (IRRI) in the Philippines, three crops of rice per year have been grown for 10 consecutive years, and a high yield level has still been maintained.

2. Irrigation farming allows permanent land use, provided that salt damage can be avoided. In the case of rice cropping, it means that permanent agriculture can be carried on for centuries with high yields per unit area and without impairing soil fertility. The control of water reduces erosion.

3. Yield fluctuations from year to year are reduced. With reliable water supplies, production is almost independent of the weather. Annual fluctuations of yield are less than those in upland farming. In Indian districts with predominantly upland farming the value of output shows average co-efficients of variation of total output of 22 per cent compared to an average of 17 per cent in irrigated districts (Binswanger, Jodha, and Barah 1979).

4. Irrigation farming leads to a more continuous production process. The chance of a regular household supply of food and cash income improves. The same applies also both to the provision of cattle with fodder and to the employment of the labour force, draught animals, implements, and buildings.

5. Arable irrigation farming is relatively adaptable with regard to both type and intensity of production. Holdings with perennial crops are obliged to adhere to one type of production for a long time. In holdings with upland farming, the rainfall determines the range of possible crops, and the less it rains the narrower is this range. In irrigation farming, such limiting factors are much less compelling. By eliminating the problem of drought, irrigation makes it possible to diversify production: it lends itself much more easily to multiple cropping than does upland farming and production changes in response to changing economic conditions can be made relatively easily.

6. Irrigation farming allows the productive employment of a relatively

large number of workers per hectare. It therefore enables a relatively high income to be earned without the use of expensive equipment, or it reduces the cost of such equipment when measured against the return. In hoe-systems, one family can scarcely cultivate more than 2–3 ha, which in upland farming permits only a low standard of living to be obtained, but in irrigation farming the same area will allow a high standard of living. The increase in production on a small, irrigated plot means that the part of the proceeds spent on traction power is relatively small.

7. Irrigation may permit speculative use of upland that would not otherwise be used. Upland farming on soils with marginal rainfall, or stock-keeping on dry pastures, is in many cases only possible because irrigation production guarantees some reliability of farm income.

These arguments may be summarized by saying that, in comparison with upland farming, irrigation farming shows some of the advantages of manu-facturing. Matters are not left so much to chance; there is a more continuous production process; there is a better control of the production factors; and there is a spatial concentration of production.

However, against the many advantages of irrigation farming must be set definite costs and requirements.

1. An important cost item in projects relying on damming may be the loss of the dam site for cropping. Tank irrigation in southern India requires about 1 ha of tank area for 4 ha irrigated.

2. It necessitates a high level of investment in water supply, delivery, and distribution works, and in land preparation. In the case of large schemes, development costs of $8000–15 000 (1979) per hectare are typical. High levels of investment entail a heavy burden of fixed costs, so that water has to be bought for a high price. It is still often more profitable to practise upland farming and accept its disadvantages than to invest in irrigation. Often, in arid areas, water is considered to be a unique resource, and therefore engineering or agronomic feasibility are considered sufficient conditions to justify investment. This may lead to a misallocation of resources (Carruthers 1968, p. 6).

3. In addition to the high level of the initial investment, irrigation farming demands a continuous high level of labour or machine input to supply the water (drawing, pumping, maintenance of the system), and to prepare and maintain the fields and water-distribution or drainage channels.

4. Irrigation farming, as a rule, requires the co-operation of several farmers, or their willingness to accept instructions from a private or public water lord. Moreover, an irrigation holding requires communal work to maintain and improve the water-delivery system. In the larger irrigation systems, a disciplined schedule and scale of water distribution among the beneficiaries is one of the most important pre-conditions for economic success (Wittfogel 1931).

5. Irrigation farming may offer great production possibilities, but a good deal of knowledge is necessary if they are to be fully exploited, for production is technically complicated. One problem of particular gravity is the spread among the irrigation farmers of diseases like malaria, river-blindness, and bilharziosis, which are encouraged by the water, so that medical costs have to be considered in addition to the costs of irrigation development.

It must also be borne in mind that production techniques and land tenure are interrelated. Whereas in shifting and fallow systems communal land ownership with established rights of usage prevails, we generally find in permanent farming individual land ownership with owner-farmers or tenants. In irrigation farming, landlord–tenant relationships, provided that they have not yet been abolished by land reforms, are even more typical than in upland farming. The prevalence of landlord–tenant relationships is perhaps due to the fact that, for water distribution to several irrigators, institutional arrangements on a larger scale are needed.

7.3.2. *Spatial organization of cropping*

Having outlined some of the generally important farm-management characteristics of irrigation farming, we can now consider the more specific organizational features. Again, we find that the various plots of a holding are, as a rule, not uniformly cropped, but that definite spatial cropping patterns prevail. There is, however, a basic difference in this respect between upland and irrigation farming. In upland farming, the spatial organization of production is based primarily on the varying intensity of manure application, in other words on the distance from the farmstead and the consequent transport problems. In irrigation farming, it is based primarily on the availability of water. Three different forms may be noted in this respect, based on whether only a section of the farm area is irrigable land; or all of the land is irrigable, but water supplies are insufficient to provide all of it with the optimum amount of water; or water reliability varies from plot to plot.

(a) *Spatial organization of partly irrigable holdings.* Where only a part of the holding area can be irrigated, land use is diversified, more often than not in the form of a catena. The cultivator usually concentrates his work on the lowest topographical level, which is generally more fertile and easier to irrigate. The seasonally or continuously irrigated fields which are the centre of the holding, and which are usually close to the water supply—a spring, river, or canal—may be supplemented in different ways:

1. There may be gardens close to the farmstead. The combination of irrigated rice fields and comparatively large gardens, containing various vegetables, tree crops, and ponds, is the basic pattern of rice holdings in large areas of south-east Asia.

2. Irrigation farming close to a spring or small river in the valley bottoms, with upland farming on the surrounding slopes, is typical of many Indian farms and of rice holdings in south-east Asia, Madagascar, and other places.

3. Perennial crops may be grown on upland. Commonly, we find only small plots for crops for home use, but sometimes there are substantial plantations of cash crops like rubber and coconut-palms.

These activities are found in different combinations and proportions, and provide a variety of farming systems, differing widely in complexity and intensity. Two examples are given in Figs. 7.1 and 7.2.

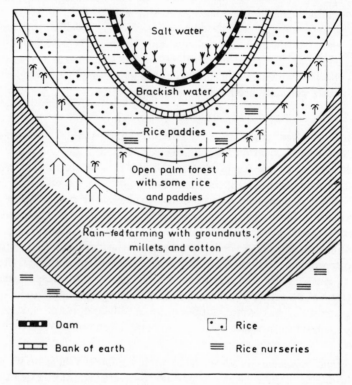

Fig. 7.1. Cropping pattern in the region of the Lower Gambia River (from Mohr 1969, p. 67).

The combination of upland arable crops and lowland rice depicted in Fig. 7.1 is typical of most holdings with wet-rice production in Africa south of the Sahara. One of the major advantages of the combination of wet rice with upland farming is the better distribution of labour requirements.

The spatial diversification of cropping is even more developed in south-east Asia, as shown in Fig. 7.2, which is taken from Singapore.

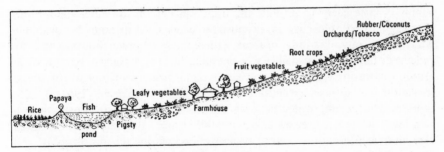

Fig. 7.2. Profile of a catenary cropping sequence in north Singapore (from Ho 1964, Fig. 1).

1. The valley bottoms are used for rice fields.
2. In addition, on the low land there are ponds which produce fish and pig feed (water hyacinth).
3. In the immediate vicinity of the ponds are the pigsties, the waste food from which is put into the ponds to feed the fish.
4. Then come the main vegetable plots with water-demanding crops like cabbage, lettuce, etc. The water for irrigation is taken from the ponds.
5. Vegetable crops like beans, tomatoes, and cucumbers, which require little water, are grown on the higher land.
6. After this come non-irrigated plots with root crops like manioc and sweet potatoes, which are sometimes supplemented by a little maize.
7. On the upper slope, fruit trees are principally grown, but there is also some tobacco.
8. Finally, the high plateau is utilized for tree crops like rubber and coconuts.

The various activities are not organized independently of each other; in particular, arable cropping on irrigated land and on upland is closely related. Where the irrigated plot is small, and contributes only 10–30 per cent of the holding return, it is used essentially as a supplement to upland farming, principally producing fodder in the dry season, and regularly supplying the household and neighbouring markets with perishable food like vegetables. Where staple food crops are grown, the primary intention is to guarantee the household's food supply. Since in most cases the irrigated land is cropped with particular intensity, any organic fertilizer from the holding is applied to the irrigated land. Where there is a relatively large irrigated area, upland farming becomes a supplement to irrigation farming. The marginal return of additional labour on the irrigated land is usually higher than on the upland with the result that the holding has on the one hand carefully and intensively cropped irrigated fields, and on the other hand extensively used upland areas supplying seasonal grazing, fuel, litter, or undemanding food crops like manioc.

(b) *Spatial organization of totally irrigable holdings with limited water supplies (rotational irrigation)*. Where the total holding area is irrigable, but supply is insufficient to provide the optimum amount of water for a given crop, it is advisable only in exceptional cases to distribute the available water uniformly. Instead, systems with rotational irrigation are usually practised, i.e. the water delivery to the field follows a definite pattern from season to season or from year to year. We find primitive forms of rotational irrigation in the oasis agriculture of the Sahara. Nicolaisen (1963, p. 203) reports oasis farming where the irrigated plot is cropped with cereals twice a year for 6 to 7 years, after which it is left fallow and a different plot is irrigated and cultivated. In this shifting system of irrigation, the problems of soil fertility and of salting are solved without effort, because of the fallow period.

In intensive agriculture, systems with rotational irrigation are incomparably more complicated, but also more productive. They are especially highly developed in Taiwan.

In the example shown in Fig. 7.3 only a third of the area can be cropped with wet rice in summer. The other two-thirds support sugar-cane and various short-term arable crops, which can grow properly with the natural rainfall. For the third supporting sugar-cane, which is not irrigated in summer, the little water available for irrigation in winter and spring is enough for initial growth, and as autumn approaches and the crop begins to ripen water is no longer necessary. Since the rice on the first third also needs no more water, sugar-cane can then be planted on the last third and irrigated until May of the second year. The crop rotation for each of the three plots is arranged in such a sequence that the full rotation on the holding's three plots is: sugar-cane (first year until January of the second year)—green manure (February to May of the second year)—irrigated rice (July to October of the second year)—sweet potatoes (November of the second year to April of the third year)—non-irrigated rice, soya beans, or groundnuts (May to September or October of the third year).

We can see from the example in Taiwan that the spatial emphasis in systems with rotational irrigation changes from year to year. The type and intensity of land use is, however, the same on all plots included in rotational irrigation.

(c) *Spatial organization of holdings with uncontrolled water supplies.* Other types of spatial pattern in irrigation farming may be found on holdings with uncertain water supplies and where the degree of uncertainty is different on different plots. Here again, the cultivators tend to adapt themselves to the conditions of the various fields by the selection of crops, varieties, and planting dates that fit the expectations of water supply.

An example of this is the wet-rice cropping in the vicinity of Tananarive, Madagascar. The plots in the valley bottoms, which have a relatively ample and sure supply of water for a long period, are planted first with slow-maturing varieties in April–May, and these produce relatively high yields. The terraces on the slopes are planted with the standard varieties from October to December. Plots on the upper slopes, where the yield is particularly uncertain, are planted up to March, according to the volume of rainfall, and resulting water availability. Similarly, the Bangladesh rice–jute farmer plants different

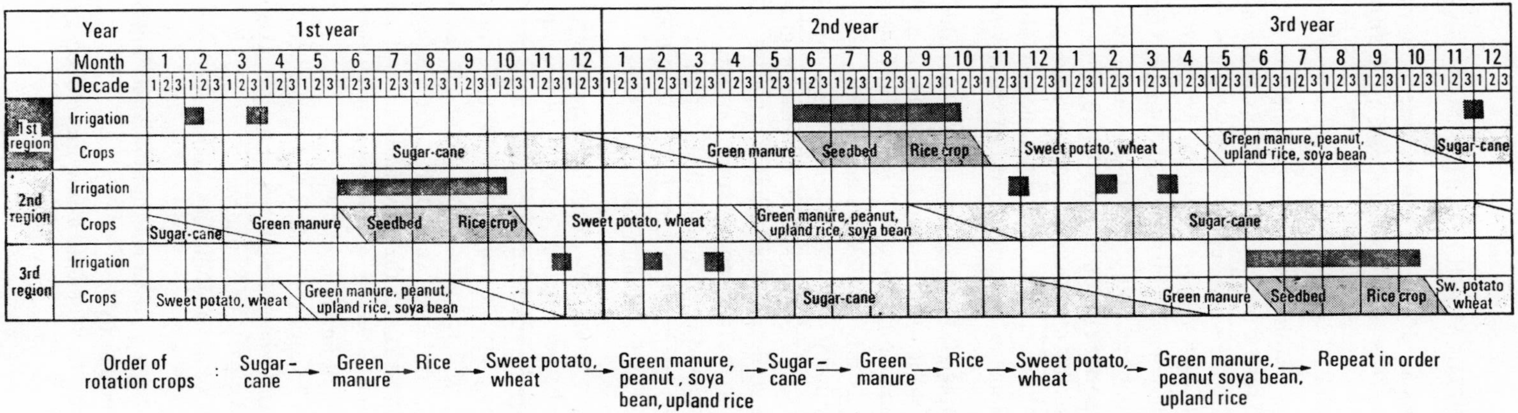
Year
1st year
2nd year
3rd year
Month
Decade
1st region
2nd region
3rd region
Irrigation
Crops
Sugar-cane
Green manure
Seedbed
Rice crop
Sweet potato, wheat
Green manure, peanut, upland rice, soya bean
Sw. potato wheat
Order of rotation crops : Sugar-cane → Green manure → Rice → Sweet potato, wheat → Green manure, peanut, soya bean, upland rice → Sugar-cane → Green manure → Rice → Sweet potato, wheat → Green manure, peanut soya bean, upland rice → Repeat in order

FIG. 7.3. Standard diagram of 3-year rotation irrigation (from the Chia-Nan Irrigation Association, Taiwan).

varieties of rice on fields at various levels according to the danger of flooding and the ability of the variety to withstand it (see § 7.4.1(d)).

7.3.3. *Cropping principles*

In irrigation farming, distinct cropping principles are applied, in addition to definite spatial patterns of organization. Diversification in this respect is the more common where the water supply is ample and constant, the climate is warm, and the population density and local purchasing power are high. This does not mean that irrigation farming and diversified farming necessarily go together. Irrigated monocropping with sugar-cane, cotton, and in particular with rice, is very important in some areas. There is, however, a tendency in irrigation to increase yields per hectare—and in some cases also per man-hour—by following multiple-cropping practices such as ratooning, sequential cropping in crop rotations, and various forms of intercropping.

(a) *Ratooning*. In upland farming, ratooning, in which two or more consecutive crops are taken from a single planting, is a practice generally applied to sugar-cane and sometimes to sorghum, although it is of scarcely any significance as far as other crops are concerned. In irrigation farming, the possibilities for ratooning increase. Ratoon rice is found in low-intensity rice production, where it serves mostly as dry-season grazing and only in exceptional cases is the grain harvested. Highly intensive forms of ratooning have been developed in the Philippines and Taiwan. The sequence: rice (110 days)–sorghum (85 days)–sorghum, first ratoon (80 days)–sorghum, second ratoon (80 days) is mentioned by Bradfield (1968) as a production possibility of tropical irrigation farming within one year.

(b) *Crop rotations*. The crop rotations in irrigation farming may be classified according to: (1) whether seasonal or continuous cropping is practised; (2) whether the rotation is based on one, two, or more crops a year; (3) whether short-term, long-term, annual, or biannual crops are grown; (4) which crops there are; and (5) what input demands the crops make (see § 6.3.2(b)). With rice, for instance, we find the following sequences:

1. One rice crop per year is still the prevalent mode of wet-rice farming in monsoon Asia and indicates the dependence on a distinct rainy season and/or the availability of ample land.

2. Three rice crops in 2 years are to be found where wet rice is not tied to a distinct rainy season and where the land shortage is not yet very pronounced as for instance in some parts of Sumatra.

3. Two rice crops per year usually indicate controlled irrigation and growing land shortage.

4. Five rice crops in 2 years are to be found in areas with proper irrigation and a high population density.

5. Three rice crops per year indicate a fully developed and reliable irriga-

tion system and a high degree of land shortage. They are grown for instance in Tamil Nadu, India, or on Luichow Peninsula, mainland China (Kanwar 1972, p. 62; Kung 1975). Three rice crops plus one wheat crop per year and hectare have been observed at Tai Li Commune, Southern China (IRRI 1978, p. 25).

6. Continuous rice is a production possibility on land with year-round irrigation. In the IRRI 'thrice-weekly rice planting and harvesting system' a 1 ha field is divided into 40 plots of 250 m². A plot is transplanted every other day, and another plot is harvested on alternate days. Four crops of rice are grown per year. During 1977, about 23 t were harvested, and 3 men were fully employed throughout the year (Morooka, Herdt, and Haws, 1979).

These sequences with wet rice may be considered as the backbone of more diversified crop rotations, including catch crops after the rice and dry-land crops between two rice crops. Among the many factors determining the crop rotation, the availability and reliability of water supplies play a decisive role in irrigation farming. In addition, interactions between rainfall and irrigation have to be considered. Multiple cropping is much easier to organize in areas with only one rainy season and irrigation during the rest of the year than in areas with two rainy seasons, where planting and harvesting have to be adapted to the rainfall. Consequently, the most intensive and diversified types of irrigation farming are found in areas with a single rainy season, as is the case in Taiwan. We can see from the rotations practised in Taiwan that are illustrated in Table 7.1 how the crop rotation, depending on the availability of water, varies under otherwise similar production conditions.

1. The most frequent crop rotation on permanently irrigable land consists of two crops of rice and one non-irrigated winter crop, which is most frequently sweet potatoes, maize, or wheat. On poorly drained plots, the usual rotation is rice–rice–fallow or rice–rice–green manure. As the size of holding decreases, vegetable cropping becomes more important. Fig. 7.4 shows that progress in plant-breeding in Taichung District, Taiwan, has reduced the time-demand of wheat in three steps from 170 days to 112 days and then to 107 days while the time-demand for rice was reduced from 120 days to 100 days and then to 91 days. Finally in 1966 it became possible to grow a wheat crop between rice in the autumn and in the spring.

2. On plots with summer irrigation only, and prior to the summer rice, we find as a rule vegetables or sweet potatoes, and the final crop is usually a green-manure plant.

3. In areas with rotational irrigation, the limited availabiiity of water necessitates alternating combinations of the principal crops rice and sugar-cane, so that rotations of 2–3 years with 5–6 crops are combined with the usual 1-year rotations with 2–5 crops which predominate where the plots are permanently irrigated.

TABLE 7.1

Crop rotations on irrigated land in Taiwan

Fields irrigable for the whole year
1. Rice (February/March–July), rice (July/August–November), fallow (November–February)
2. Rice (February–June), rice (July–November), green manure (November–January)
3. Rice (March–July), rice (August–November), wheat (December–February)
4. Rice (February–July), rice (July–November), vegetables (October–January)
5. Rice (February–June), vegetables (June–July), rice (July–November), sweet potatoes (November–February)
6. Rice (February–June), rice (July–October), tobacco (October–February)

Fields irrigable for half a year
7. Rice (March/April–July), sweet potatoes (August/September–January/February)
8. Rice (July–November), vegetables (October–March), green manure (March–June)
9. Rice (July–November), sweet potatoes (October–May), green manure (May–July)
10. Rice (May–September), sweet potatoes (September–April), vegetables (May–December), soya beans (December–May)
11. Rice (March–July), vegetables (July–March)

Fields with rotational irrigation
12. Rice (June–October), sweet potatoes (November–March/May), non-irrigated rice (May/June–September/October), sugar-cane (October–January of the third year), green manure (February–May)
13. Rice (July–November), sugar-cane (October–January of the second year), sweet potatoes (January–June), groundnuts (July–December), groundnuts (January–July)
14. Rice (July–November), sweet potatoes (November–May), groundnuts (July–October), vegetables (October–May), groundnuts (July–October), sweet potatoes (November–May)

Source: Chen (1963, p. 239).

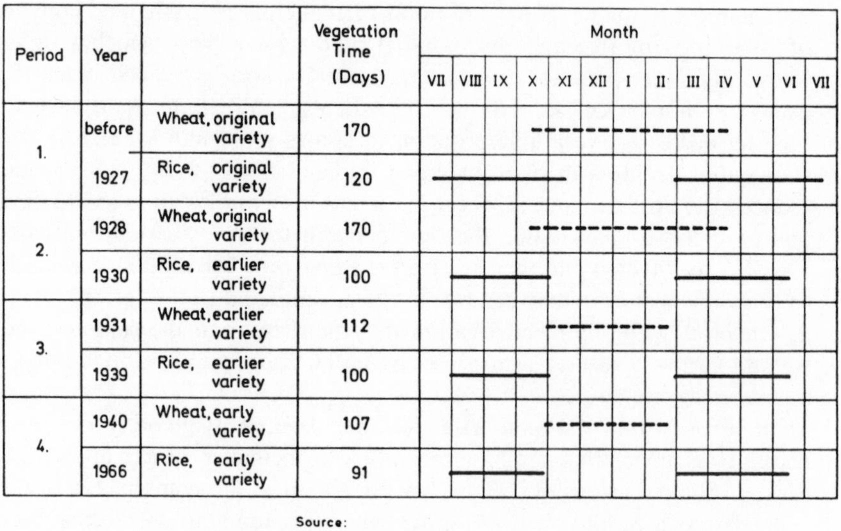

Period	Year		Vegetation Time (Days)	Month												
				VII	VIII	IX	X	XI	XII	I	II	III	IV	V	VI	VII
1.	before	Wheat, original variety	170													
	1927	Rice, original variety	120													
2.	1928	Wheat, original variety	170													
	1930	Rice, earlier variety	100													
3.	1931	Wheat, earlier variety	112													
	1939	Rice, earlier variety	100													
4.	1940	Wheat, early variety	107													
	1966	Rice, early variety	91													

Source:

FIG. 7.4. The impact of plant-breeding on multiple cropping (from Wang 1967).

Plate 22 gives an example of the extraordinarily high levels of land-use intensity that are possible in tropical irrigation farming. The rotation practised in Taichung District, Taiwan, includes 4 or 5 crops in one year, not all of which are short-term crops. Cropping within 1 year on a single piece of land is as follows.

In February the rice nurseries are laid out. In March the rice is planted, and it is harvested in June. Summer rice is not planted until August. In the interim the so-called summer crops can be grown, such as jute, melons, soya beans, sweet potatoes, and various green-manure plants and vegetables. Some of these crops can be planted as early as May in the growing rice stand (relay-planting). Between the second rice harvest and the planting of the spring rice, in the 4 months of the cool season when growth is slow, the so-called winter crops are cultivated, including tobacco, maize, soya beans, peas, seed rape, sweet potatoes, wheat, flax, various green-manure plants, and vegetables. Some of these crops can similarly be planted in the growing summer-rice stand, and sometimes by adopting this method the use of the field can be concentrated so that two winter crops are possible.

Generally, in production conditions like these, there is a strong tendency to grow those crops that produce as high a yield as possible per month of field use. The time for which various crops occupy the field can be shortened considerably by raising the plants in nurseries. Indeed, cultivators try to leave the plants for as long as possible in the nurseries, and crops that lend themselves to this method are preferred. Moreover, it is necessary to prepare the field immediately after the harvest for the plants: a plot which in the morning is yellow with ripening rice should be green in the evening with newly planted rice.

(c) *Intercropping through relay-planting.* Of no lesser importance than sequential cropping are the various types of intercropping (two or more crops are grown in proximate but different rows). Some intercropping systems, for instance the one depicted in Plate 22, depend to a high degree on relay-planting techniques—the rotation crop is planted in the ripening stand of the previous crop. In the case of rice, relay-planting takes place up to 7 weeks before the harvest and, so that the relay crop can get enough light, the panicles of the ripening rice above the row of maturing plants are separated. Some other relay-planting combinations that have been introduced in Indian irrigation farming are wheat in the furrows of potatoes, legume fodder in cotton, grain legumes in maize, groundnuts in cotton, etc. (Kanwar 1972, p. 34).

With some crops, relay-planting is combined with the practice of raising the plants in nurseries.

An example of this is jute-growing in Taichung District, Taiwan. The growing period of jute from sowing to harvesting is 120 days in the summer, but between the spring and summer rice the plot is available for only 40 days. The problem is solved by keeping the jute plants for 40 days in the nursery, and they are then planted in the rice field 40 days before the rice harvest. The pure stand of jute is thus restricted to the 40 days between the harvesting of spring rice and the planting of summer rice. Disturbance to the rice is kept to a minimum as the jute is planted in holes filled with compost. The losses in yield of rice, compared with a pure stand of rice, amount to 5–10 per cent, and of jute to 20 per cent, but these losses are more than outweighed by the additional yield of jute.

The extent of relay-planting depends on the available labour force and on the soils. On soils that tend to be heavy, the land must be prepared after the harvest and before it is used again. In other cases, crops for relay-planting need to be planted in dry land, and drainage facilities may not be sufficient. It must also be borne in mind that the additional water requirements can be met only where the rainfall or the irrigation capacity is adequate.

(d) *Intercropping through interplanting in medium- or long-term arable crops.* In areas with a high degree of land shortage short-term crops are sometimes planted at about the same time as a crop requiring 8 months or a year to mature, e.g. sugar-cane or manioc. The intercrop uses land and light while the main crop is still small. Sugar-cane on Taiwan, for instance, is planted in the autumn and develops slowly in the somewhat cooler winter months. The space between the rows of cane can therefore be used for short-term crops, like sweet potatoes, groundnuts, flax, or vegetables, without making any appreciable difference to the growth of the sugar-cane. When there is a sufficient supply of water, the combination of relay-planting and interplanting has proved its worth. The sugar-cane is interplanted in the ripening rice in the autumn (muddy-in method). After the rice harvest, an intermediate crop is planted in the rows of sugar-cane. Experience has shown that planting should take place in every second row only, so that it is possible to weed and manure each crop individually.

Fig. 7.5 gives examples of combined relay-planting and interplanting. The sugar-cane is planted in September in the ripening rice, so that it can benefit from the relatively warm days. After the rice harvest every second row between the growing cane is planted with sweet potatoes, flax, or vegetables. An even more complex example of intercropping is given by the vegetable planting pattern in a commune of the Canton area, China, shown in Fig. 7.5.

(e) *Intercropping of crops with similar vegetation cycles.* Tall-growing crops may be combined with short-growing, shade-tolerant crops. Maize, for instance, responds to increased row spacing by a higher yield per plant without a corresponding increase in leaf area. Such a 'tall' crop, planted at wider spacing allows an almost fully yielding intercrop, such as sweet potato (Harwood *et al.* 1977, p. 16). We also speak of intercropping if two varieties of a given crop are grown in a given field at the same time in different rows.†

† Kung reports on intercropping techniques with two rice varieties in China (mainland): 'Both varieties are sown in seed beds at almost the same time at about the middle of April. The early variety is transplanted in the middle of May in rows 18 inches apart, and a fortnight later the late rice is interplanted between the rows of the early variety. The early crop is harvested by late July or early August and the second crop, whose growth has been retarded by the shadow of the first, is then able to develop rapidly and is harvested at the end of October or November'. (Kung 1975).

The various multiple-cropping techniques depicted here allow a fuller use of two resources: light and time. Research at the IRRI established that inter-cropping achieved a 30–40 per cent increase in light interceptions (Harwood 1973, p. 9). Farmers need no longer aim at high yields per crop and season but should look for maximum net returns per month of field occupation. These are sometimes higher with lower yields per crop, but more crops per year. Multiple cropping is, however, a technique which requires (1) ample labour, (2) much animal or machine power, (3) all-the-year-round water supplies, (4) warm winters, (5) supplemental nurseries, (6) early-maturing and shade-tolerant varieties, (7) high chemical inputs (fertilizers and pesticides), and (8) a very effective organization of water supplies (Kung 1975). If these conditions prevail, land productivity may become very high, and there is much scope for productive employment. Multiple cropping seems to work, however, under the law of diminishing returns to labour. Average labour productivities may be high, but very intensive forms tend to show a steep drop in marginal returns to labour. A decade of rapid industrialization in Taiwan sufficed to slow down the general trend towards multiple cropping as farm workers found more productive employment elsewhere.

7.3.4. *Characteristics of the fertilizer economy*

The discussion of the various cropping principles indicates the high degree of intensity that irrigation farming can attain. It is obvious that, in order to exploit the production potential of irrigation farming, high yields, multiple cropping, and leaching have to be accompanied by a highly intensive fertilizer economy, and by efforts to prevent salting.

Most of the irrigated land in the tropics, particularly most of the wet-rice areas, is still cultivated year after year without much manuring. Rice-growing over long periods without fertilizer application does not appear to lower soil fertility, for the following reasons.

1. In several large river basins, soil fertility is maintained by the annual deposition of silt.

2. In addition to silt, nutrients are brought in solution with the water. The paddies with wet rice are often on different levels and are slightly inclined, so that the water runs from the high forest land to the sea, draining slowly from one rice field to the next, and thus conveying nutrients from the higher levels to the lower plots.

3. Nitrogen fixation in wet rice is substantial. The role played in this by blue-green algae and several nitrogen-fixing bacteria has been known for some time. Recently it was found that the rice rizosphere could also fix considerable amounts of atmospheric nitrogen under flooded conditions (Dart and Day 1975). Under the reduced conditions of a wet-rice field most forms of iron phosphate release phosphorus and make it available to plants.

Rotations on Taiwan

(1) Crop rotation of the rice and the winter and summer cash crops

(2) Crop rotation of the rice and the sugar-cane, the several crops being interplanted at the beginning of growth of the sugar-cane.

Vegetable cropping patterns for the Canton area
(as provided by the Agriculture and Forestry Association, Kwangtung Branch)

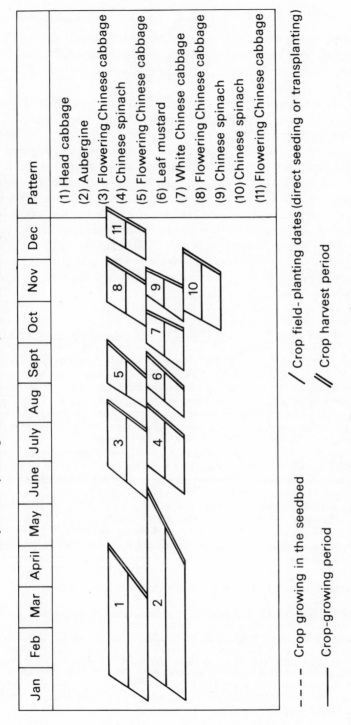

Jan	Feb	Mar	April	May	June	July	Aug	Sept	Oct	Nov	Dec	Pattern

Pattern:
(1) Head cabbage
(2) Aubergine
(3) Flowering Chinese cabbage
(4) Chinese spinach
(5) Flowering Chinese cabbage
(6) Leaf mustard
(7) White Chinese cabbage
(8) Flowering Chinese cabbage
(9) Chinese spinach
(10) Chinese spinach
(11) Flowering Chinese cabbage

---- Crop growing in the seedbed
—— Crop-growing period

/ Crop field-planting dates (direct seeding or transplanting)
// Crop harvest period

Fig. 7.5. Rotations with relay-planting and intercropping in Taiwan and China. *Sources*: Taiwan: Iso (1964); China: Harwood, Plucknett, and Romanowski (1977).

4. Angladette (1966, p. 17)1 remarks on the observation that soils improve their quality for wet-rice production with time because of the impounding of water and its influence on chemical processes in the soil. This process takes 50–100 years.

Intensive irrigation farming, however, requires manuring and fertilizing as well as the beneficial influences connected with water application and water impounding. Traditional types of intensive manuring have developed in those areas where water supplies are sufficient and reliable and where population densities are high. Nowhere else in the world has the fertilizer economy of traditional farming systems reached such a level of intensity in the effort to reduce the loss in nutrients as in Chinese irrigation farming. The father of organic chemistry, von Liebig, in comparing nineteenth-century German agriculture with contemporary Chinese farming (1878, p. 453), viewed the former as the procedure of 'a child compared to that of a mature and experienced man'. Webster and Wilson (1967, pp. 210–11) summarize the intensive manuring practices used in south-east Asian irrigation farming as follows:

All animal manure, household and crop wastes, and plant ashes derived from fuel, are carefully collected and applied to the land, usually in the form of com- posts. Night soil is widely used, commonly being stored in earthenware recep- tacles, diluted with water and thriftily applied to individual plants at appropriate intervals.† Mud, rich in organic matter, is periodically dug out from canals, reservoirs and fish ponds and spread on the land, or used in composts. Green manures may be included in the rotation, sometimes for direct ploughing-in but more often for composting. Hill sides bearing trees and shrubs are regularly cut over to provide material for trampling into the mud of the rice fields, or for the preparation of composts. Great use is made of composts, care being taken to see that they are well rotted, with much of their organic matter broken down, so that they rapidly release nutrients to a growing crop. It should be mentioned, however, that although the fertility of the densely populated alluvial plains of China has been maintained at a relatively high level by deposits of silt from irrigation of floodwater, by application of night soil, and, to a lesser extent, by green manuring with vegetation imported from hilly land, this has been done partly at the expense of other land. Much hilly land in China has been depleted of fertility, and the general level of fertility of lands other than the alluvial plains is only moderate.

It is interesting to observe that the concentric rings of land with decreasing

---

† Night soil is a traditional fertilizer of high productivity. It is, however, a carrier of various intestinal diseases. Buck (1930, p. 265) writes this about traditional China: 'From an economic standpoint it would probably be cheaper to throw away night soil than to incur the losses concurrent with ill-health which result from its use.' The transition to mineral fertilizer results in reduced losses through disease and reduced costs of health care. A remarkable way of fertilizing rice is reported from China and Vietnam, where farmers inoculate wet-rice fields with a small fern, *Azolla*, which contains *Anabaena azollae* in the substomatal chamber; this association is estimated to fix up to 120 kg ha^{-1} of nitrogen per season (Dart and Day 1975, p. 27).

fertility at increasing distance from the farmstead that are frequent in upland systems are rarely found in irrigation systems. With the exception of the gardens near the farmyard, which receive more fertilizing, manures and composts are usually distributed more or less equally over the irrigated land. Most tropical upland farming is still carried on without mineral fertilizers or with only small amounts. In irrigation farming, however, traditional techniques of manuring, which are very labour-demanding, have been widely complemented or even substituted by mineral fertilizers. Scarce amounts of mineral fertilizer are used in preference on the irrigated land, because returns per unit of input are higher and more reliable. Traditionally, the flow of nutrients was from the upland to the valley bottom, because most of the compost and manure collected went to the irrigated land. Nowadays the flow of nutrients tends to go the other way, particularly in wet-rice upland systems (Raison 1970, p. 372): irrigated land receives the mineral fertilizer and the upland receives the manure, which is more effective there because of the humus' capacity to store moisture and nutrients.

Significant changes in the fertilizer economy can be observed in highly commercialized areas close to urban centres of consumption, where larger numbers of pigs and poultry are kept on the basis of mainly imported concentrates. Here mineral fertilizer tends to be substituted and supplemented by nutrients imported with feeds. In these cases the genesis of the fertilizer economy develops from (1) the recycling of nutrients by traditional manuring and composting to (2) the supplementation and expansion of this cycle by nutrients imported as mineral fertilizer and (3) the import of nutrients as feeds, which contribute to the provision of more and better manure.

7.3.5. Characteristics of animal husbandry

The intensification of irrigation farming is closely tied to an intensive fertilizer economy. Cattle activities, however, are not in general intensively organized. In the irrigated farming systems of the subtropics, stock-keeping is of growing significance. In Israel, Japan, and Maghreb the production of milk and meat is increasingly based on irrigated fodder cropping. However, conditions in the tropics are still different. Arable cropping receives priority, as it always has, although it is usually supplemented by various livestock activities.

Great numbers of cattle are kept in irrigation systems in India, Pakistan, and Bangladesh, but their performance is usually very poor. Cattle supply mainly draught power and manure and very small amounts of milk. Table 7.2 provides an idea of inputs and outputs of milk production in irrigation holdings with an advanced dairy economy. Milk yields per cow per year are several times the average, but are nevertheless low when compared with the performance in most subtropical systems. The return per hour of work is negligible or even negative. It should not be overlooked, however, that much of the roughage is a by-product with low opportunity costs, and the value set on manure seems low.

In a few cases, irrigation farmers practise a division of labour with semi-nomadic herdsmen. The fallows of Gezira, in the Sudan, for instance, are grazed in the dry season by migrating herds of cattle, sheep, and goats. As a rule, however, we find stationary animal-keeping with low livestock numbers per holding but rather high livestock densities, rising to two livestock units

TABLE 7.2

Some examples of input–output relations of cattle activities in Indian and Pakistani irrigation farms

Country	India	India	Pakistan[a]
Location	Punjab	Punjab	Sind
Year	1967–70	1967–70	1970–1
Method	Sample	Sample	Sample
Unit	1 Buffalo cow	1 Zebu cow	Herd: 1·9 bullocks, 1 cow, 1·3 young stock
Labour input (man-hours)	360	393	1295[b]
Outputs per year			
Milk (kg)[c]	1040	580	858
Manure (t)	n.a.	n.a.	10·8
Bullock-pair days	—	—	96
Milk ($)	127	92	115
Manure ($)	6	5	22[d]
Bullock days ($)[e]	—	—	102
Gross return ($)	133	97	239
Material inputs ($)			
Roughage[f]	56	54 ⎱	321
Concentrates	35	31 ⎰	
Depreciation	20	7	—[g]
Other inputs	1	1	3
Income before overheads ($)	21	4	–85
Income per hour ($)	0·06	0·01	—

[a] See also Fig. 7.8. [b] 7 hours per day. [c] Average yield for cows-in-milk and dry cows. [d] Value per ton as given in the Punjab Survey. [e] At $1·06 per day and pair. [f] Punjab: value as given in the Punjab survey. Pakistan: $0·06 per kg TDN (total dogestible nutrients). [g] Depreciation = gain in liveweight of the young stock. *Sources:* Kahlon and Migliani (1974, pp. 169–76); McConnel (1972, p. 47).

per hectare in intensive irrigation areas of India and China (Buck 1930, p. 255). The main objective in keeping cattle and buffaloes is the provision of traction power for ploughing and pumping. The provision of manure is a secondary function, and the production of meat and milk is usually of less importance. The irrigation farmer's needs for animal protein is not usually met from cattle and buffalo products, but by small animals, sheep, goats, pigs, poultry, and fish. Pigs are of particular importance in south-east Asia, being fed partly on crop residues and partly on household waste. Ducks are typical

1. Fallow vegetation, burnt by shifting cultivators, Cercle d'Allada, Benin

2. Mixed cropping of groundnuts, sunflower, and maize in a shifting cultivator's field, Urambo, Tanzania

3. Valley bottoms in the savanna zone of the Ivory Coast being transformed into fields with irrigated rice

4. Fallow farming in the derived savanna of Eastern Nigeria. Weeding in this 'crop–*Imperata cylindrica*' complex eliminates the above-the-ground material only. The objective is to subdue the growth of *Imperata* long enough so that the leaf canopy of the crop can establish itself (compare § 4.3.3 (d))

5. Tobacco is a crop well suited for smallholder shifting cultivation in the medium-altitude savannas of Africa. Tobacco production at Urambo, Tanzania

6. Valley bottoms in the rainforest zone of the Ivory Coast being transformed into fields with irrigated rice

7. Cattle feeding on leaves of *Faidherbia albida*, in Sérèr country, Senegal

8. Transporting manure from the stable to the field, Ukara Island, Tanzania

9. Applying stable manure and green manure on Ukara Island, Tanzania. The manure is placed at the base of the ridge

10. Stable of a coffee–milk farm with zero-grazing in Embu District, Kenya

11. A maize–climbing beans combination in the highlands of Colombia. Maize occupies the field in the first half of the year and the stover serves as support for the beans which grow in the second half (compare § 6.4.1)

12. Land preparation for potatoes in the Nilgiris Mts., Southern India

13. Threshing of pidgeon peas produced in a permanent upland cultivation system in a semi-arid climate near Sholapur, India

14. Land preparation for rice in Sri Lanka. The work is being performed by seasonal paid labour

15. Soil preparation for intercropping of sweet potatoes in maturing rice, Taichung District, Taiwan

16. Relay-planting of sugar-cane in growing rice in Taiwan. Note the seed cane in the mud between the rows of rice

17. Rice terraces and upland farming in the Highland of Madagascar

18. Improved nursery for intensive irrigated rice production near Los Baños, the Philippines

19. Two rice fields close to Madras, India. The field to the left shows improved traditional varieties, which tend to lodge. The field to the right shows IR 8, a non-lodging variety which can productively absorb high doses of nitrogen (*see* Fig. 7.11)

20. Preparation of a rice field for planting by trampling with buffaloes, in Sri Lanka. The work is performed after hoe or plough cultivation. In Madagascar the same method is applied and called *piétinage*

21. View into the banana–coffee bean plot of a Bahaya holding, Tanzania. The soil is covered with mulch grass

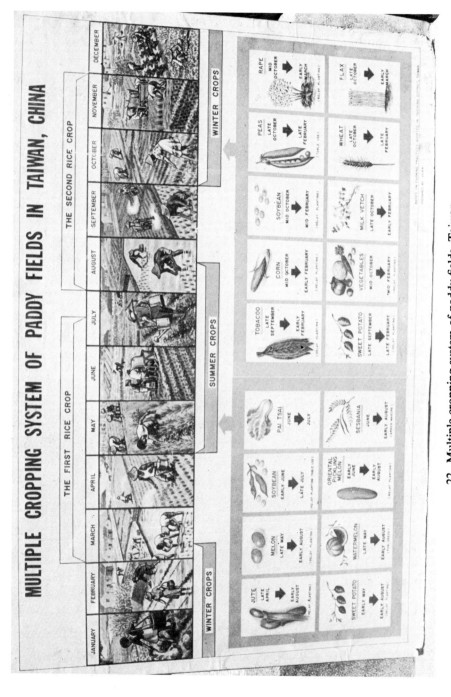

22. Multiple-cropping system of paddy fields, Taiwan

23. A tenant in the Managil Extension of the Gezira Scheme, in the Sudan, irrigating his cotton

24. Sugar-cane harvesting on a smallholding in Taiwan

25. Young coffee densely planted under plantains at Restrepo, Colombia

26. Coffee plantation, Nyeri District, Kenya

27. Badly eroded tea plantation, Sri Lanka, medium-altitude level

28. Young tea plantation established with new clones, planted at high density, surrounded by tea which is about 50 years old, Sri Lanka, high-altitude level

29. Smallholder plot ready for the planting of oil-palms. The woman is planting a cover crop (*Centrosema pubescens*)

30. Weeding of young oil-palms, Sumatra. The plantation was established on *Imperata cylindrica* grassland with herbicides as the major input for land preparation

31. Plantation of mature oil-palms in Benin

32. Root system of two perennial grasses in a dry savanna climate, Kongwa, Tanzania

33. Over-grazed and eroded land in Baringo District, Kenya

of the Chinese type of holding, and chickens are found almost everywhere in the tropics. Part of the animal and human excreta may be used to fertilize fish-ponds, and mud from the pond may be applied to the gardens.†

Attention should also be drawn to forms of combined irrigation farming and fish production. Fish are found almost everywhere in the water distribution system. In many places, the fish-pond by the house is a typical feature of irrigated holdings. Hickling (1961) notes the integration of various forms of fish production into crop rotations.

1. In Travancore, southern India, after the rice harvest the plot is filled with brackish water, and in this way prawns are brought into the field. When the tide recedes, the water passes through nets, and the prawns are collected. The yield amounts to 700–1600 kg of prawns per hectare in the dry season (Hickling 1961, p. 263).

2. In Java, Madagascar, and other places, fish are sometimes found in the growing rice. For this to be possible there must be a sufficient and continuous depth of water. Fish are brought to the plot in the irrigation water, or they can be put in 5 days after planting. Hickling (1961, p. 265) quotes a yield of 28–50 kg of fish per hectare in 100 days.

In Kriak, Perak, Malaysia, it was found in a sample of 732 farmers that double-cropped rice paddies yielded 96 kg ha^{-1} of fish with a farm value of $18 per hectare compared with 203 kg ha^{-1} and $54 per hectare on single-cropped land. (Ta Cheng, Chong Boo Jock, Sier Hooi Koon, and Moulter 1973, p. 47.) In modernized farming rice and fish tend to become separate enterprises.

3. In some cases, land is used alternately for rice cropping and as a fish-pond. The water is let in after the rice harvest. The yield of fish that swim in with the water amounts to about 3 kg ha^{-1} in 5–6 months. By introducing tilapia, up to 300 kg ha^{-1} can be produced in 3 months.

7.3.6. *Characteristics of the labour economy*

As we might expect, the labour input of irrigated holdings shows great variation, depending on the system of water distribution, the intensity of land use, and the degree of mechanization. Where labour is particularly plentiful, hoe- and spade-systems occur, relying entirely on manual labour. In most irrigation systems, farmers employ animal power for ploughing, transport, and pumping. India's farmers usually employ two oxen, while in countries with a Chinese cultural heritage only one draught animal is employed (Delvert 1972, p. 128). The use of the tractor is spreading rapidly for irrigation farming can support higher monetary inputs in cultivation than can upland farming. Despite the varied forms of organization, however, some

---

† Fish-pond farming may yield very high gross returns per hectare. Shang estimates the output in eel farming on Taiwan at 10–14 t of produce per hectare valued at $57 000–82 000 (1973, p. 8).

features and tendencies are found that distinguish the labour economy of irrigation farms from that of upland farming.

(a) *High labour input.* Irrigation farming in the tropics almost always involves a high input per hectare of manual labour. This is because of the extra operations in irrigation, such as maintenance of channels, control of water distribution, and land preparation. The manual labour input is particularly high where water has to be pumped by hand or with the aid of animals. According to Buck (1930, p. 306), farmers in Hopeh Province, China, in 1923 spent on irrigation 58 per cent of the total labour input in wheat and 27 per cent of the total labour input in vegetables. In regions that depend mainly on inundation, the labour demand for irrigation is comparatively low. Much of the human and animal labour once required to procure water has been replaced by motor pumping. However, the provision of controlled water supplies and of careful water delivery is still connected with a high labour input.

Of even greater importance is the labour input resulting from the choice of intensive crops, and the greater amount of labour involved in cultivation, weeding, and harvesting. The methods of producing and applying manure, compost, green manure, etc. require a great deal of labour, especially where multiple cropping, relay-planting, and intercropping play a major part. This kind of production method is not only expensive in terms of field labour; a considerable input is also needed to maintain the irrigation system and nursery beds. Thus the basic difference from upland farming is not simply that more work is involved in cultivating each crop, but also that irrigation often facilitates a higher frequency in cropping.

Tables 7.3 and 7.4 illustrate the labour requirements per hectare for several crops in irrigation farming. The input varies widely according to crop and technique, but as a rule it is much higher than in permanent upland systems (see Table 6.3). Labour inputs in wet rice are obviously a function of several factors:

(i) Technology: Hoe-systems usually show very high labour inputs, in particular if wet rice is a newly introduced technique (compare the example from Cameroun). Ox-plough systems require much less labour. Full mechanization of the production process brings labour requirements down to a fraction of what is traditionally used, in particular if the seeding method is applied (Ghana, Brazil).

(ii) Land scarcity: The choice of technology is clearly related to the degree of land scarcity, and labour inputs are relatively high where farms are small. The ample land availability in most of Thailand is reflected in rather low labour inputs. In Bangladesh land is very scarce and labour input high. Even higher are the inputs in Nepal, where much time is needed to produce compost and to maintain terraces on steeply sloping land.

(iii) Communal farming: The highest labour inputs per hectare seem to occur where land scarcity coincides with communal farming as it is carried out in China. Here the farmer no longer decides whether the additional output is worth the effort. 2400 hours per year have to be worked by each member, irrespective of the marginal return to labour. Wiens reports that typical labour inputs per hectare of rice land in Chinese communes are around 4800 hours (two crops).

(iv) Multiple cropping: Labour inputs go steeply up with multiple cropping compared to single cropping. Wet-rice systems in Taiwan with 3–4 crops per year require 4000–5000 hours ha^{-1}. With multiple vegetable cropping in Singapore 10–15 000 hours ha^{-1} are invested and on Chinese communes they are reported to reach 24 000 hours ha^{-1} (Wiens 1977, p. 38).

(b) *High marginal returns to labour.* The production functions of labour in irrigation systems are, as a rule, different from those in upland farming. The marginal returns to labour in upland farming usually decrease more rapidly with greater employment of labour than in irrigation farming. For a wide range of labour inputs, marginal returns in irrigation farming decrease slowly or remain almost constant. This is indicated by the data in Table 7.3. The average return per hour of work in most wet-rice systems is not higher than in most permanent upland systems. The important point is the greater employment factor which is provided by comparatively high marginal returns. The data also indicate that returns per hour of work tend to decline with increasing land shortage and continued application of traditional techniques, but they rise again with the application of modern inputs.

More often than not, the improvement of water supplies and of farming techniques on irrigated land shows higher marginal returns on labour than the construction of new, but less adequately managed irrigation facilities and irrigation plots. Referring to wet rice, Gertz (1963, p. 35) writes: 'The capacity of most terraces to respond to loving care is amazing.'

A comparison of returns per hour of work with wet rice and other irrigation crops tends to indicate that with wet rice marginal returns of high labour inputs are relatively high and average returns low, while with other irrigation crops the important aspect seems to lie in high average returns but low marginal returns for high labour inputs particularly in modernized and mechanized systems (compare Tables 7.3 and 7.4).

(c) *Less pronounced peaks in labour demand.* Farming systems with seasonal irrigation suffer, like upland farming, from seasonal underemployment of the available capacities of labour and traction power. In large irrigation schemes with one major cash crop, the peak in labour demand and the resulting problems of the employment of seasonal labour (the provision of credit for their payment, the necessity of using expensive machinery,

TABLE 7.3

Some examples of input–output relations in wet-rice production (per hectare and crop)

	Mali	Ghana	Liberia	Cameroun	India	Bangladesh	Nepal	Thailand	Taiwan	Peru	Brazil
Country	Mali	Ghana	Liberia	Cameroun	India	Bangladesh	Nepal	Thailand	Taiwan	Peru	Brazil
Location	Lac Horo	Northern	Toya	Bamunka	Assam	Comilla	Lumle	Chao Phya	Central	Tinachones	Rio Grande
Year	1971	1974	1971	1977	1971	1976	1974	1976	1972	1974	1977
Technique[a]	Flood rice	Flood rice	Swamp rice	Swamp rice	Rain-fed rice	Rain-fed rice	Irrigation rice	Irrigation rice	Irrigation rice	Irrigation rice	Irrigation rice
	Traditional	Modern	Traditional	Traditional	Traditional	Modern	Traditional	Traditional	Modern	Modern	Modern
	Seeding	Seeding	Seeding	Trans-planting	Seeding	Trans-planting	Trans-planting	Trans-planting	Trans-planting	Trans-planting	Trans-planting
Tool	Hoe	Tractor	Hoe	Hoe	Ox-plough	Ox-plough	Hoe	Ox-plough	Tractor	Tractor	Tractor
Method	Sample	Sample	Sample	Sample	Sample	Sample	Sample	Sample	Sample	Accounts	Accounts
Technical data											
Labour input (man-hours)											
Land preparation[b]	60	20	346	714[c]	153	198	1610[d]	125	40	256	98
Nurseries	—	9	—	64	—	181	—	13	52	—[e]	—
Planting	6	9	152	473	146	267	1680	250	88	128	2
Fertilizing	—	—	—	—	—	19	—	—	56	32	6
Plant protection	—	—	—	—	—	18	—	—	80	96	4
Weeding	90	26	247	1393[f]	—	339	35	137	196	240	4
Harvesting	210	28	298	113	206	294	840	212	224	176	88
Threshing	—	—	156	264	63	—	791	88	—	16	—
Other work	—	—	—	280	13	267	252	—	248	16	17
Total	366	92	1199	3301	581	1583	5208	825	984	960	219
Yield (t)	1·00	1·44	1·57	1·57	1·57	3·70	2·20	3·50	3·34	7·00	4·2
Yield (kg day^{-1})	n.a.	n.a.	270	130	165	150	140	170	100	240	n.a.
Field occupation (days)	n.a.	n.a.	5·8	12·1	9·5	24·7	15·7	20·6	33·4	29·2	n.a.
Economic analysis ($)											
Gross return[g]	100	241	173	309	131	530	344	385	613	1371[h]	651
Material inputs											
Seed	5	21	7	24	3	5	4	13	10	147[i]	51
Fertilizer[j]	—	13	—	—	1	44	—	29	80	195	51
Pesticides and herbicides	—	—	—	—	—	11	10	3	38	16	10
Mechanization[k]	1	78	1	1	16	40	1	30	47	112	159
Irrigation	—	—	—	—	—	—	—	—	20	167	n.a.
Other	—	8	—	—	1	—	—	—	13	—	4
Income before interest and overheads	94	121	165	284	110	430	329	310	405	734	372
Income per hour	0·26	1·32	0·14	0·09	0·19	0·27	0·06	0·38	0·41	0·76	1·70

[a] Modern — use of improved varieties and mineral fertilizer. [b] Including maintenance of terraces. [c] Plants are purchased. [d] Including clearance of fallow vegetation. [d] Including maintenance of irrigation works. [e] Exceptionally high 1974 world market prices. [f] Bird-watching, mainly children-hours. [g] Without by-products. [h] Exceptionally high 1974 world market prices. [i] Plants. [j] Including value of manure in Indian systems. [k] Ox-plough labour valued at local prices. *Sources:* Mali: SEDES (1972, Vol. 2, p. 195); Ghana: Winch (1976, pp. 63–76); Liberia: van Santen (1974, p. 31); Cameroun: Fotzo (1977); India (Assam): Goswami and Bora (1977); Bangladesh: Mian (1977); Nepal: Feldman (1977, personal communication); Thailand: Thailand 1975 and 1977 Taiwan: Lih-yuh Shy Tsai and Herdt (1976); Peru: Dietz (1975, personal communication); Brazil: Estado de São Paulo (1977, p. 190).

TABLE 7.4

Some examples of input–output relations in irrigation farming (per hectare and crop)

	Sorghum Yemen Tihama 1972 Uncontrolled	Jute India Assam 1971 Flooding	Sorghum India Mysore 1970 Canal	Pearl millet India Mysore 1970 Canal	Wheat India Punjab 1969–70 Canal and tube-well	Cotton India Coimbatore 1973 Canal	Groundnuts India Coimbatore 1973 Canal	Cotton Sudan Gezira 1973–4 Canal	Sorghum Sudan Gezira 1973–4 Canal	Groundnuts Sudan Gezira 1973–4 Canal
Farming technique	Traditional ox-plough systems		Modernized ox-plough systems					Modern tractor systems		
Method	Sample	Sample	Sample	Sample	Sample	Sample	Sample	Sample	Sample	Sample
Technical data										
Labour input (man-hours)										
Land preparation	56	211	239	202	81	140	141	28	17	10
Planting	60	13	126	44	23	74	71	28	33	77
Irrigation	49	—	138	94	97	151	138	77	58	116
Weeding[a]	—	332	244	160	74	558	238	127	4	236
Harvesting	119	371	158	163	121	423	335	369	222	305
Threshing, processing	107	397	106	101	124	—	—	—	n.a.	—
Other	—	12	—	—	21	—	—	147	7	10
Total	391	1336	1011	764	541	1346	923	776	341	754
Ox-pair input (hours)	n.a.	n.a.	139	98	101	52	64	—	—	—
Yield (t)	1·23	1·55	2·79	2·02	2·51	1·23	1·35	1·84	1·98	1·34
Economic analysis ($)										
Gross return[b]	148	392	230	159	240	358	258	1169	101	184
Material inputs										
Seeds[c]	1	6	2	4	8	3	42	56	13	23
Fertilizer and pesticides[d]	—	2	86	66	20	57	9	—g	—g	—g
Irrigation[e]	—	—	n.a.	n.a.	10	43	19	—g	—g	—g
Mechanization[f]	16	17	18	18	17	14	36	75	4	10
Other	—	1	—	—	2	—	—	—	—	—
Income before interest and overheads	131	366	124	71	183	241	152	1038	84	151
Income per hour	0·33	0·27	0·12	0·09	0·34	0·18	0·16	1·34h	0·25	0·20

[a] Including top-dressing and pesticide application. [b] Main product only. [c] Including farm-produced seeds. [d] Including value of manure. [e] Charges or pumping costs, not including farm labour. [f] Purchases or depreciations, value of farm-produced ox power, hired services. [g] The irrigation charge is paid through a sharecropping system with cotton: 28·5 per cent of the material input is directly charged to the tenant (— $35·90); 71·5 per cent are joint costs, i.e. $42·30. The tenant receives 47 per cent of the gross return; net of joint costs and after deduction of his material inputs he is left with $428·89, i.e. $0·55 per hour. At 1970–1 prices this was about $0·20 per hour. [h] $0·59 at 1970–1 prices. *Sources:* Yemen: Mann (1973, pp. 61, 116); India: (Mysore) Nagbiswas *et al.* (1972, p. XXV), (Assam) Goswami and Bora (1977), (Coimbatore) Rajagopalan and Balasubramanian (1976, p. 94), (Punjab) Kahlon and Migliani (1974, p. 84, pp. 110 et seq.); Sudan: Sudan Gezira Board (1975).

and so on) may be even more pronounced than in more diversified upland systems.

Timing is a particularly serious problem in rain-fed wet-rice systems. About 400 mm of rain are required to moisten the soil before land preparation for transplanted rice can start. Much of the rainy season is thus lost for plant production and thereafter labour capacities are often insufficient to plant and weed in time.

Farmers tend to keep one or two draught animals even on very small holdings, and recently there has been a strong tendency to hire tractor services.

TABLE 7.5

Use of time in a traditional Indian wet-rice system

Location	Nowgang District, Assam		
System	Wet rice		
Year	1970–1		
Method	Weekly questioning		
	Adult male family workers	Adult female family workers	Male servants
Use of time as percentage of day-time (84 hours per week)			
Crops (per cent)	28	4	25
Other agricultural work (per cent)	13	2	17
Non-farm work (per cent)	5	1	3
Household work and social affairs (per cent)	9	24	7
Meals, leisure, illness (per cent)	45	69	48
Agricultural work (hours per year)	1788	272	1863
Non-farm work (hours)	202	42	135
Household and social affairs (hours)	407	1056	297
Total hours	2397	1370	2295

Source: Goswami and Bora (1977).

Rain-fed wet-rice farming often depends more on hired labour for transplanting than irrigated rice, where labour requirements are more evenly distributed over the year. Even on tiny holdings more than half of the total field work is often carried out by hired labour. Nevertheless, an important part of the monsoon wet rice is seeded and poorly weeded or planted late, because the labour capacity is insufficient to plant and weed at times which would maximize yields. Generally the labour force in wet-rice holdings is more fully employed than in holdings with upland farming (compare Tables 7.5 and 7.6 with Tables 6.4 to 6.9 and 5.2). In areas with pronounced peaks in labour demand, however, the degree of underemployment of man-power and draught animals is proverbial. Nakajud, Archavasmit, and Boonyakom established that the average labour input on crop and livestock per working person in wet-rice farming, Thailand, is only 1114 hours per year (less than in some African shifting and fallow systems), and that 54 per cent of the total falls in two peak seasons totalling 84 days (1971, p. 82). The animals work several hours a day for about 40–70 days. During the rest of the time they graze on embankments, waste land, and other such places, but are not required for work.

TABLE 7.6

Energy requirements of irrigated rice, sugar-cane, and wheat in the Punjab, India

Technique	Tractor	Ox-plough
Farm size (average)	16·1	4·6
Number of holdings	14	12
Area irrigated (per cent)	100	100
Rice (yield in t ha⁻¹)	5·60	4·88
(energy in h.p.h.ᵃ ha⁻¹)	3212	1704
(h.p.h. t⁻¹)	573	349
(h.p.h. kcal⁻¹)	0·27	0·17
Wheat (yield in t ha⁻¹)	3·20	2·78
(energy in h.p.h. ha⁻¹)	1540	791
(h.p.h. t⁻¹)	481	284
(h.p.h. kcal⁻¹)	0·14	0·08
Sugar-cane (yield in t ha⁻¹)	36·50	32·40
(energy in h.p.h. ha⁻¹)	6966	1831
(h.p.h. t⁻¹)	191	57
(h.p.h. kcal⁻¹)	0·32	0·10

ᵃ h.p.h. = horse-power hour. 1 h.p.h. in tractors = 1 hour's work of a pair of bullocks = 0·067 man-hours. *Source:* Kahlon (1975, p. 15 *et seq.*).

This is demonstrated in Fig. 7.6. Rice holdings in the central plain of Thailand with but one crop a year show a highly peaked demand in labour for planting and harvesting. Diversified cropping with rice and various other irrigation crops in the northern zone of Thailand shows a better distribution of work, but we still find a pronounced peak in labour demand between November and March. Multiple cropping in the middle rice region of Taiwan is connected with an almost even employment of the family labour. The peaks in labour demand are met by seasonal labour.

Seasonal underemployment in irrigation farming may, however, be reduced or even abolished by an improvement in the water supplies, and wherever continuous cropping is practised employment is found throughout the year. Relay-planting helps to reduce labour peaks at harvest time and when the next planting takes place. Commercial stock-keeping can be another way of creating regular employment for the available labour. It is noticeable, more-over, that irrigated and upland farming combine very well in this respect. The labour requirements of both kinds of holding frequently do not coincide, and the cultivation of upland, especially in holdings with seasonal irrigation, can help to even out the demand for labour.

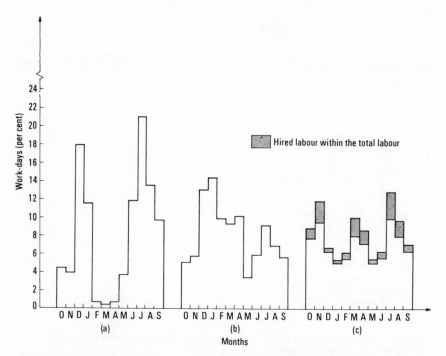

FIG. 7.6. Monthly distribution of working days on rice holdings in Thailand and Taiwan. (a) Thailand central plain; monocropping with rice. Total labour input: 1332 working days at 6 hours per day on 4·7 ha. (b) Thailand northern zone; diversified cropping with rice and other arable crops. Total labour input: n.a. (c) Taiwan middle rice region; multiple cropping with two rice crops and other arable crops. Total labour input: 501 working days at 10 hours per day on 0·96 ha. From Kulthongkham and Ong (1964, p. 69); Taiwan Department of Agriculture (1966, pp. 174–7); Thailand Ministry of Agriculture (1959, tables 35, 39).

(d) *Tendency towards motorization.* Most wet rice is still cultivated by hand or with animal power, but motorization is increasing. Most other types of irrigation farming show an even stronger tendency towards motorization, and farmers often gain sufficiently high incomes to pay for tractor services. Tractor use is clearly an interesting proposition where timing is a problem, as, for instance, in rain-fed wet rice, in multiple-cropping systems, or in large schemes with production under close supervision, where it would be too complex to organize cultivation with animal power. It should not be over-looked, however, that motorization is often not as essential in irrigation farming as in semi-arid upland farming.

Kahlon's estimates show that with several irrigated crops in the Punjab, the energy required per tonne of produce or per kilocalorie is clearly less with ox ploughs than with tractors, and this for the simple reason that yield levels in both systems are about the same, while the tractor system requires more horse-power hours. Motorization, which may be effective from an economic point of view, is not effective in terms of energy. This may be true for irrigation systems with no pronounced peak in labour demand. Different results can be expected from irrigated multiple-cropping systems or from upland farming in semi-arid areas, where tractor use, and thus the proper timing of field opera-tions, is likely to produce much higher yields per hectare than ox-ploughing (see Table 7.6).

(e) *Better planning of the labour economy.* Another advantage of irrigation farming is better labour planning. The labour operations in upland farming depend on rainfall, which fluctuates in quantity and in timing. The irrigator, however, usually knows when he will acquire water. In systems with con-trolled irrigation, irrigation dates and water quantities can be adjusted to the labour economy of the holding and vice versa. This is especially apparent in intensive wet-rice systems. Neither drought nor rainfall essentially affects the production process, so that planting and weed control can be undertaken even on rainy days. The time of the labour input is dependent on dry weather only during the harvest. By regulating the water, levelling the land, maintaining a high degree of fertilization, and controlling diseases by the application of pesticides, the farmer has obtained almost complete control of all growth factors except natural catastrophes. Thus the irrigation farmer is not as dependent as the upland farmer on his daily assessment of the soil and plants, and on choosing the type of labour operations and when to terminate them. Instead, the production process is similar to that in manu-facturing, which permits relatively exact planning.

7.4. Examples

The features of irrigation farming discussed above vary in importance in the different farming systems, and the following examples are chosen to illustrate this point. Farm-management data from Mali and several parts of Asia supply information about the characteristics of wet-rice systems. The examples demonstrate one point of crucial importance: irrigation farming, in particular with all-the-year-round reliable irrigation, allows the concentra-tion of production on a small area with the productive employment of great numbers of workers. This tendency towards the concentration of production on the best land is more pronounced in the tropics than in the subtropics. This is largely due to the fact that biological innovations (new varieties, mineral fertilizer, pesticides, etc.) which are yield-increasing and labour-saving in temperate or semi-arid upland farming tend to be yield-increasing and employment-creating in irrigation farming. Examples of holdings with various arable crops in the Yemen, India, Pakistan, and Singapore illustrate how irrigators adapt land use to the varying availability of water. In the example from the Gezira, attention is drawn to the special factors that are important for organization and land use in large-scale irrigation projects.

7.4.1. *Wet-rice systems*

(a) *Characteristics of wet-rice cultivation.* In semi-humid and humid climates rice is one of the usual crops in upland farming, particularly among shifting cultivators. Much more important than this upland or dry rice is wet rice, or rice cropped on plots that are covered with water permanently, or at least for a lengthy period, during growth. The main types of wet-rice cultivation are as follows:

1. *Deep-water rice* is grown, for example in Bengal, Burma, Thailand, and Mali. Seeding or planting takes place at the beginning of the rainy season and the stem elongates—as much as 300 mm per day—adjusting to the rising water table. Along some coasts of south-east Asia floating rice is grown, which is sown from a boat, develops rhizomes 6–7 m long, and continues to grow even if uprooted.

2. In upland farming, rice is normally sown broadcast. This practice is still sometimes retained in irrigation farming, e.g. in Sri Lanka and some parts of India, when water is impounded in the paddy fields after the sowing. This is a procedure that we term the *traditional sowing method.*

3. In south-east Asia, rice is traditionally started in a nursery bed, and 6–8 weeks after sowing it is transplanted into the puddled mud of the field. The production is carried out without many purchased inputs. This method is called the *traditional south-east Asian method*, but it is common also in Madagascar and is spreading in Africa.

4. The modernized transplanting method, which is largely influenced by the *Japanese method* of rice cultivation, is common in Taiwan, and is being introduced into African and other south-east Asian countries, and this is the most intensive method. It differs from the traditional method in its high input of purchased materials and in the special care taken with the nurseries and with cultivation in general.

5. Two modern sowing methods have to be distinguished: (i) In the USA, northern Australia, southern Europe, and such tropical low-population countries as, for instance, Surinam, the sowing method is characterized

by extensive use of machinery. Rice is sown with drills or from aeroplanes and reaped with combines. (ii) Sowing methods for smallholder farming which lend themselves to double cropping of rainfed rice have been developed by IRRI (see Table 7.12).

No other agricultural product in the tropics and subtropics has as great an economic importance as wet rice. It constitutes the basis of the economy in most Asian countries, and it is expanding in Africa. The expansion of wet-rice cultivation in the tropics and subtropics is based principally on the following facts.

1. It permits constant land use, even for centuries, without decline in yields.

2. It produces high, certain yields per hectare and consequently allows the concentration of population.

3. It enjoys a relatively favoured position within traditional agriculture; manufactured inputs like mineral fertilizer, herbicides, and implements make wet-rice cropping more productive, but even the rational use of the traditional production factors of soil, work, and water can produce a relatively high yield.

The really important phenomenon is the permanence of rice cropping. Increased population densities in traditional upland systems result sooner or later in a deterioration of the habitat. In a wet-rice region the habitat remains undamaged. Some areas of Java, for example, support 2000 persons per square kilometre, without any significant decline in rice production per hectare (Geertz 1963, p. 33). Many of the famous terraced slopes in the Philippines have been planted to rice annually for more than 2000 years. Wet rice, therefore, fosters economic stability, a dense population, and an advanced civilization. It forms the technical basis for the formation of permanent *Kulturlandschaften*† in the wet valley bottoms, which contrast noticeably with the shifting cultivation of the adjoining forest landscapes. Experience in tropical Asia shows that these two types of land use represent the two forms of arable farming that are most suited to the traditional production conditions of a warm, humid climate. On the one hand there is irrigated rice culture, with high population densities, permanent housing, high labour inputs, and high yields per hectare, conquering the difficulties nature presents by conscious manipulation of the countryside. On the other hand, there is shifting cultivation, with low population densities, semi-nomadic populations, low labour input, a low yield per hectare of the total area, and adaptation to nature.

In order to practise balanced cultivation where there is a shortage of land it is often essential to make the transition from the most extensive system of land use to one of the most intensive systems, and, in addition, to concentrate production in the valleys and plains. The advantage of this transition is not

† Landscapes strongly influenced by human efforts to make them useful for man.

really that there is a higher average return on the labour input. The point is rather that, whereas intensified shifting cultivation is accompanied by rapidly falling marginal returns from extra labour input, intensified irrigated rice, at least up to a high level of yield, seems to have almost constant marginal returns from extra labour. Low average but comparatively high marginal returns characterize the farm-management situation in rice holdings, wherever the water supply is sufficient and controlled. Irrigated rice culture is thus in a position to absorb productively much more labour. At the same time it permits food production to be concentrated in narrowly circumscribed areas, and, with comparatively short transport routes and the concentration of population, the formation of towns can occur. In this way, irrigated rice has laid the foundation for Asian culture.

Rice can be adapted to suit most types of agronomic conditions. It is grown at altitudes of 2500 m (Nepal) and at sea level, in the tropics and in the warmer regions of the temperate zones, on good and on poor soils (with differences in yield of course), in the desert and where there is a heavy rainfall, provided that solar energy and water are available to a sufficient degree. Irrigated rice is not as strongly affected by waterlogging, floods, and saline soils as other crops. By the cropping of rice, areas that would otherwise be waste land can be opened up for agricultural production. In fact as long as there is a suitable supply of water, some rice can be produced on soils that are too infertile to produce any other grain crop (Jordan 1967, p. 3). Rice grows well on its own—and monocropping is therefore possible—but it can also be included in crop rotations without any difficulty. The various labour operations, like preparation of the soil, planting, and weeding, are largely independent of the weather, provided there is enough water. There are many varieties, with different growing times (100–200 days), differing demands on soil and climate, differing abilities to benefit from fertilizer, and a range of other characteristics.

In handling and transporting, rice shares the advantage of all cereals over the root crops, while in storage it suffers smaller losses than practically any other tropical food crop. Raw paddy is resistant to insect attack in storage, and parboiled milled rice is similarly resistant. Another advantage of rice as a food is its ease of preparation, and its low requirement of fuel in comparison with bread. Rice also has a high nutritive value, and its protein is of an exceptionally high quality and digestibility for vegetable protein, while it is also exceedingly palatable (Jordan 1967, p. 5).

Furthermore, wet-rice production can be fitted into most social and economic structures. Rice-cropping is practised in large and small holdings, in countries with high and low wage levels, almost without implements or with a very high investment in machinery. In short, irrigated rice-growing is more adaptable than practically any other type of agricultural production. Rice is also an ideal crop for a rapidly developing country, as the technique for its cultivation can be made to keep in line with the development of the country as a whole, and its extent is unlikely to be restricted due to rising economic levels, as may happen with other crops (Jordan 1967, pp. 4–5). It is this adaptability, not only to agronomic conditions but also to social and economic conditions, that makes wet rice so extremely attractive as a crop. The basic features in the organization of various types of rice holding can be demonstrated with a few examples taken from Africa and Asia.

(b) *Holdings with deep-water rice in the Niger Delta, Mali* (Table 7.7). Upland crop cultivation in the area is marginal. Wet-land rice cultivation occurs in plains and valleys without control of water. The result is a highly diversified system of land use (Gallais and Sidikou 1978, p. 18).

1. Relatively large areas are planted with millets and sorghums. Husbandry practices are extensive, yields are low, and crop failures due to drought are frequent. However a poor upland crop may be balanced by a better lowland crop.

2. Wet-rice production occurs on plots with different elevation, and each plot is seeded with a different mix of rice varieties so that a sufficient crop is harvested in most years irrespective of flood levels and the distribution of floods over time. Cultivators know that years with ample and short water supplies alternate in 4–7 years cycles. They conequently shift their cropping area in village blocks according to the *ex-ante* expectation of the season's flood. The result is a wet-rice shifting system (Gallais and Sidikou 1978, p. 18).

3. Cultivators increasingly tend to keep livestock. Traditionally livestock production was with the pastoral Peul in the area, but the sedentary herd owned by cultivators is gradually growing.

Incomes are low in spite of relatively large areas of fertile land with seasonally ample water. Adult males tend to migrate into higher rainfall areas of neighbouring countries. A fuller use of the potential requires the control of water.

(c) *Rice–jute holdings in Assam.* The data from these farms provide an idea about the organization of traditional lowland production in a high rainfall area with an extended rainy season. The farms are relatively large and land use is extensive:

1. Rain-fed seeded rice is the major crop during the main rainy season. The rice is broadcast, not weeded, and yields are low (average of $1 \cdot 6$ t ha^{-1}).

2. The second rice crop is transplanted and weeded and small plots of rice are planted in the third season. These two crops receive small amounts of fertilizer and yields are somewhat higher.

3. Jute is the major cash crop and occupies the land during most of the year.

The potential for agricultural production remains largely unutilized. Output

TABLE 7.7

Farm-management data of wet-rice holdings

	Mali Mopti	India Assam	Bangladesh Old Brahamaputra Plain	Indonesia Java	Nepal Western Hills	Taiwan Southern	Thailand Chanasutr Project	Philippines Cotabatu
Year	1977	1970–1	1974	1975	1974	1965	1974	1975
Rainfall	520	2641	1500–1700	2000	1700–3500	1679	1200	1772
Technique	Flood rice	Rain-fed rice[a]	Rain-fed rice[a]	Irrigated rice	Irrigated rice	Irrigated rice	Irrigated rice	Irrigated rice
		Seeding	*Seeding*	*Transplanting rice*	*Transplanting rice*	*Transplanting rice*	*Transplanting rice*	*Seeding*
Method	Case Study	Sample	Sample Model	Sample Model	Sample Model	Sample	Sample	Accounts Model
Number of holdings	1	80	—	n.a.	—	50	220	n.a.
Persons per household	15·2	8·35	6·00	—	5·00	7·86	6·00	744[d]
Labour force (ME)[c]	6·2	2·78	3·40	2·50[b]	2·00	2·00	3·20	—
Size of holding (ha)	12·0[e]	2·83	0·81	0·73	0·60[e]	1·23	3·5	500
of which lowland (per cent)	78	99	98	38	33	89	96	100
Lowland crops (ha)								
Floating rice	—	—	0·28[f]	—	—	—	—	—
Deep-water rice	9·3	—	0·64[f]	—	—	—	—	—
Rain-fed rice 1. season	—	2·02	—	—	—	—	—	—
2. season	—	1·01	—	—	—	—	—	—
3. season	—	0·17	—	—	—	—	—	—
Irrigated rice 1. season	—	—	—	0·28	0·20	1·05	3·5	500
2. season	—	—	—	0·28	—	1·05	3·5	500
3. season	—	—	0·32[f]	—	—	—	—	—
Other lowland crops	—	0·60[g]	0·16	—	0·10[a]	0·26	—	—
Upland crops and homestead	2·70	0·07	n.a.	0·45[i]	0·70[i]	0·93	—	—
Total crop area (ha)	12·00	3·87	1·40	1·01	1	3·29	7·00	1000
Cropping index	n.a.	137	173	138	167	267	200	200
Livestock (numbers)								
Draught animals	4	2	n.a.	1	2	n.a.	—	—
Other livestock[k]	3	—	n.a.	—	3	n.a.	—	—

Yields (paddy, t ha⁻¹)								
Floating rice	1·10	—	1·10	—	—	—	—	—
Deep-water rice	—	—	1·30	—	—	—	—	—
Rain-fed rice	—	1·61	—	—	—	—	—	—
Irrigated rice								
1. season	—	—	—	4·55	2·50	6·36	3·4	3·15ˡ
2. season	—	—	—	3·93	—	5·38	3·6	2·35ˡ
3. season	—	—	4·50	—	—	—	—	—
Economic analysis ($ per holding)								
Gross return	660	576	370ᵐ	379	337	1985	2695	393 000
Purchased inputs								
Seedⁿ	80	27		14	20	30	90	11 940
Fertilizer	—	14	27	54		217	205	89 851
Pesticides	—			10				40 299
Herbicides	—							26 119
Mechanization	32	62°	1	2		256	214	13 507
Irrigation charges	—	—	21			n.a.	125	47 015
Other	23	2					125	5149
Income	525	471	321	299	317	1482	2061	159 120
Wages for hired labour	—	39	91	89	—	156	404	28 881
Taxes and rents	36	8	n.a.	n.a.	n.a.	118	160	n.a.
Family farm income	489ᵖ	424	230	210	317ᵠ	1208	1497	—
Net return before interest and land rent	n.a.	n.a.	n.a.	n.a.	n.a.	n.a.	n.a.	130 239
Profit (–loss)	—	—	—	—	—	—	—	–23 642ʳ
Productivity								
Gross return ($ ha⁻¹)	55	203	457	519	562	1613	770	786
Gross return ($ ME⁻¹)	106	207	109	152	169	992	842	5311
Income ($ ha⁻¹)	44	166	396	410	528	1205	589	318
Income ($ ME⁻¹)	85	169	94	120	159	741	644	2150
Labour input (man-hours)	8500	3338	2032ˢ	2505	3400	5090	2890	148 000
Income per hour ($)	0·06	0·14	0·16	0·12	0·09	0·29	0·71	1·08
Man-hours per ME	1371	1201	598	1002	1700	2545	903	2000
MJ per man-hour	15	n.a.	10	21	6	27	73	171

ᵃ 1. season: seeding, 2. and 3. season: transplanting. ᵇ Estimate. ᶜ Agricultural labour force only. ᵈ 2000 hours per labourer and year. ᵉ Crop land only. Pasture is communal. ᶠ Floating rice from April to November. 0·32 ha deep-water rice from November. ᵍ Floating rice from March to July (aus), 0·32 ha from November/January to April/May. ʰ Wheat. ⁱ Mainly jute. ʲ Various crops on terraces on steeply sloping land. ᵏ Including sheep and goats (1:7). ˡ From Cabanilla and Herdt, p. 52. ᵐ Without returns from fishing, gardens, and manioc. ⁿ Including farm-produced seeds. ᵒ Including on-farm costs of oxen. ᵖ Plus $238 from fishing and off-farm work. ᵠ Plus $150 from off-farm work. ʳ Loss after deduction of interest and land rent. ˢ Field work only. *Sources*: Mali: Lagemann (1978, personal communication); India: Goswami and Bora (1977); Bangladesh: Khan *et al.* (1974); Thailand: Thailand (1977); Indonesia: Gauchon (1976, pp. 94–108); Nepal: Feldman (1977, personal communication); Taiwan: Department of Agriculture (1966); Philippines: Herdt and Lascina (1975) and Cabanilla and Herdt (n.y.).

could be significantly increased by higher intensities per crop and by multiple cropping.

(d) *Rice holdings in Bangladesh*. In the Brahamaputra flood plain 1500–1700 mm of rain fall in summer on relatively fertile soils. At the beginning of the rainy season the water accumulates in the fields, supplemented by flood water from the rivers. At the end of the rainy season the land is covered with water up to 1 m in depth. Thus nature supplies water free, and with it silt and nutrients are introduced. But nature does not supply the water regularly and in optimum amounts: sometimes there is too much water and sometimes too little. Consequently production conditions have been developed that in many respects combine the features of irrigation farming and upland farming, and can therefore be regarded as an intermediate stage between these two forms of land use. The data in Table 7.7 indicate how farming is organized under conditions of high risk in the area which has one of the highest agricultural population densities in the world. Various measures are taken to adapt production to an unreliable water supply.

1. A holding has numerous plots on different soil types and at different altitudes. Some can be expected to yield reliable returns in most seasons.
2. Rice-growing in the area is normally a diversified undertaking.

Highland plots that are not normally flooded are seeded with quick-growing rice (aus) in March/April and harvested in July/August. Yields are low (about 1 t ha^{-1}) because of moisture stress during dry spells, cloudiness, heavy weed growth, much pest damage, and high humidity during the harvesting time. The main crop is sometimes followed by a catch crop, mostly a legume. Impermeable soils usually produce a second rice crop (*aman*) that is transplanted in July/August and harvested in December. The crop relies on rainfall and residual moisture. Yields often suffer from drought at the end of the vegetation cycle.

Medium-high land normally receives 0·3–1·0 m of flood water. Broadcast *aus* rice or jute are grown in summer and are followed by a catch crop during the *rabi* season or by a second crop of transplanted rice (*aman*) on impermeable soils.

Medium-low land (1–2 m of flood water) is either planted with jute or, more often, with a mixture of *aus* and *aman* rice. The mixture is broadcast at the onset of the rains in March/April. The early-maturing *aus* is harvested before the heavy rains in July. The deep-water *aman* continues to grow in the rising flood water and is harvested after the rains. Some of the land is planted to a catch crop, usually a legume.

Low land receives more than 2 m of flood water during the summer and produces a slow-maturing crop of deep-water rice (*aman*) that is seeded in March/April and harvested in November/December.

Undrained depressions that hold water in the dry season and *all types of land that can be irrigated* (mostly pump irrigation) may produce a *boro* rice crop during winter. Improved *boro* rice receives a complete artificial irrigation, is transplanted, obtains heavy fertilizer inputs and yields 3–5 t ha^{-1} of paddy compared with 1–3 t ha^{-1} with the other types of rice production.

3. The higher the risk, the greater the preference for seeding rice instead of transplanting. The flood rice and rain-fed wet rice in the example is broadcast but the irrigated rice (*boro*) is transplanted.

4. Husbandry practices in nursery establishments are tuned to the reliability of water supplies. They are more intensive the higher the chances that sufficient water will be available when seedlings of an optimum size are available for transplanting.

5. Transplanting occurs only after the heavy rainfall which indicates that the monsoon has definitely started. This may mean a loss of time and yield in average years, but risks to production are lower.

6. Mineral fertilizer is often not applied at the times that would be optimum with reliable water supplies. Many farmers wait until rainfall has been ample enough to indicate that sufficient water will be available. (This can be observed particularly in areas with unreliable monsoons, as for instance in Madras State.) In the example, only the irrigated rice receives fertilizer.

7. Rice farmers tend to supplement their diversified rice economy by growing crops which are needed by the household and which fit into the cropping pattern. In the example grain legumes are planted as catch crops after rice, and vegetables are grown in small plots on irrigated land during the dry season. Jute used to be the major cash crop in the flood plains, but it has largely been replaced by rice because of depressed prices for jute combined with firm markets for rice and the great number of innovations which have made it, in particular irrigated rice, a much more productive crop.

The example indicates that farming techniques in the Brahamaputra flood plains are advanced by comparison with other rice systems, but high risks with flood and rain-fed rice prevent intensification except during the dry season. The area per labourer is very small, the return per hour of work is low and the employment capacity of the system is also low.

(e) *Rice holdings in Thailand.* The difficult situation in overcrowded Bangladesh contrasts with the favourable one in the fertile central plain of Thailand, where rice, produced on 3–8-ha holdings, is the major export crop. In much of the area the traditional Asian transplanting technique is applied:

1. The raising of seedlings in nursery beds.
2. Careful cultivation and levelling of the field.
3. Transplanting of seedlings 4–6 weeks after seeding—weeding is rare.
4. Regulation of water supplies by letting the water flow from one plot to the other—extended areas are fertilized by silt.

5. Harvesting of the crop with a knife or sickle, the straw being used for feeding or left on the fields.

6. Threshing by hand or by the trampling of animals.

7. Drying the rice in the sun.

8. Fallowing and grazing by buffaloes during the dry season.

It is, however, rare to find holdings with the transplanting technique only, because the peak in labour demand would be too pronounced. As illustrated in the model in Table 7.7, the holdings mostly have the following structure:

1. There is a sizeable homestead, planted with coconut-palms, mangoes, many bananas, and a few vegetable and root crops; it supplies a significant amount of food to the household.†

2. Deep-water rice is produced on extended flood plains. Contractors plough the land in the dry season and buffaloes prepare the seed bed. Rice is broadcast with the onset of the rains (May). There is no weeding or any other work except harvesting in December (by hired labour). The slow-maturing varieties tiller strongly and yield well, and labour productivity is high thanks to low labour requirements. Floods are rarely a threat to the crop, but a dry spell after broadcasting may seriously reduce yields. Deep-water rice does not lend itself, however, to the user of fertilizer or any other intensification, except possibly to the introduction of the newly developed high-yielding deep-water rice varieties.

3. Rain-fed wet rice is the second major activity. Crop production depends on rain and on water flowing from higher paddy fields to lower ones. The problem of the activity is a peak in labour demand at planting time. Farmers adapt by choosing different techniques:

(i) Some of the plots—even half of the area in cases when the moonson is late —are seeded, because transplanting seedlings into them all would take too much time. The crop is not weeded and is often heavily damaged by weeds. Thorough cultivation and harvesting are the major inputs. The later the rains the greater the percentage of the land that is seeded. In years with early and ample rain most of the land is planted. Transplanting and seeding alternate on a given plot to reduce the weeding problems which arise with seeded rice.

(ii) Most plots are transplanted when the monsoon is early. Producing seedlings in nurseries saves time for proper soil preparation and thus weed control. Moreover, transplanting yields more. Transplanting stretches from July to September, and some of the rice is usually planted too late. Nurseries have to be established before the farmer knows where he can transplant, so the seedlings are often too old. Rain-fed wet rice is not weeded or fertilized, and traditional varieties are grown.

4. Some of the land—but still only a small part in the central plain of

† The typical homestead in south-east Asia has 5–6 tall-growing tree species, 5–6 medium-height tree species, 5–6 bushes or banana clumps, 4–5 types of root crops, and up to 30 shade-tolerant, short-statured, or vine-type annuals (Harwood and Price 1975, p. 11).

Thailand—can be irrigated by major schemes or minor pumps. Here two rice crops are grown, all the rice is transplanted, the nursery techniques are intensive, seedlings tend to be younger when transplanted and 2–3 weedings are carried out. More and more farmers plant in rows and apply fertilizer and pesticides. In the wet season traditional varieties are preferred, but high-yielding ones are used in the dry season.

Farms of the above type yield high incomes for the farm families. Most conspicuous is the high labour productivity, which to a great extent is due to the availability of much very fertile land and the low labour input requirements of deep-water rice. The example in Table 7.7 gives information about such farms.

(f) *Rice farms in Indonesia, Nepal, and Taiwan* (Table 7.7). These three examples are from high-population-density areas. In the case from Java, land pressure has led to very intensive modes of rice production. Two crops are grown and yields are high. The upland is cropped with manioc mainly to provide starch for food. Land for livestock production is hardly left. All crop activities are modernized and yield levels are high, but only a fraction of the available labour force can be productively employed.

The situation in Nepal is even more difficult. Long-term rural over-population has led to intensively farmed, very small holdings. The effort to survive resulted in a visible ecological decline: permanent cropping of impoverished soils, mostly on steep, terraced slopes with much danger of land slides. Overgrazing has devastated the forests and led to the destruction of much of the pasture land, and erosion is heavy.

The farms which developed under these conditions are subsistence-oriented, while most of the cash originates from non-farm income. The farm consists of (1) some low-lying irrigated plots which carry rice in summer and wheat or fallow in winter. The terraces are planted with maize in summer and with mustard, buckwheat, and millets in summer. Relatively high yield levels are maintained by very high labour inputs in terrace maintenance and in crop husbandry generally. Fertilizer use is not yet feasible, given the price relations of the area. Crops are heavily fertilized with manure and compost, and much labour goes into the procurement and the application of compost and manures.

The maintenance of a very high livestock density is thus essential to the system. The livestock density with about 833 livestock units per 100 hectares is probably one of the highest in the world. Livestock is fed with crop by-products, communal grazings, and fodder trees. The remaining forest in the area is essential for the system and forest loppings are collected to feed cattle, while leaves are collected as bedding (compare Fig. 7.7).

The system is characterized by very high labour inputs with low productivity. One hectare of transplanted rice normally requires in Asia 800–

1200 hours of work. In this system it is about 5000 and much goes into terrace maintenance and compost fabrication. Also the livestock economy is highly labour-demanding, because the feed and the bedding has to be collected mainly from non-farm areas.

The example from Taiwan shows all those characteristics which are typical for a fully modernized version of intensive wet-rice farming: a long-term land shortage, skilled and efficient labourers and farmers, easy access to new technologies, and rising purchasing power. In addition there are, in the Taichung District, fertile soils, relatively warm winters, and irrigation systems with effective water control. Consequently, methods of rice production have been developed that combine high productivity per hectare with high productivity per man-hour.

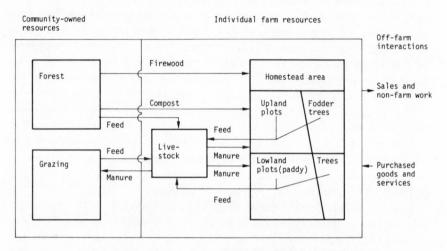

Fig. 7.7. Relations between activities in a Nepalese hill farm.
(From Harwood 1979, p. 18.)

The highly intensive rice-growing method in Taiwan, which benefited from Japanese experience, has traditionally been distinguished by:

1. Intensive organic fertilizing (night soil, animal manure, fish, oil-cake, green manure, ash, etc.).

2. Two rice harvests a year where the water supply is sufficient.

And in more recent times by:

3. Varieties that are suited to heavy applications of fertilizer.

4. A high level of mineral fertilizing in the nursery, at transplanting, and during growth.

5. Treatment of the seed with fungicides.

6. The employment of pesticides and herbicides.

7. Early transplanting three weeks after sowing.

8. Particular care in nursery techniques and in preparing the field.

9. Planting of rows from north to south to take full advantage of the sunshine.

10. Optimum distances between rows and between plants.

11. The use of improved implements.

12. Several hoeing and ridging operations (4 times in 40 days).

13. The alternation between drainage of water and a high dose of water during the growth period.

14. The threshing of the harvest by means of small, stationary threshing machines.

Intensive rice production of this type is practised in the holding depicted in Table 7.7. The total farm area of 1·23 ha is about average for the island. 1·09 ha are irrigable and the rest is upland. Multiple cropping explains the index figure of 267, which is typical for the southern rice area. Spring rice, summer rice, and a winter crop (wheat, cabbage, sweet potatoes, seed rape) are usually grown on each plot. The total area cropped thus amounts to 3·29 ha. In addition, poultry and livestock activities contribute 14 per cent of the total gross returns. One buffalo is kept for cultivation work. Mineral fertilizer and concentrates are the important purchased inputs. 320 kg N, 200 kg P_2O_5, and 200 kg K_2O are applied per hectare each year. Multiple cropping, high yields per hectare or each crop, and a high level of livestock production are the key factors behind the high return per hectare and per holding, and this in spite of relatively low prices for the products.

(g) *Large mechanized rice farms in the Philippines.* The data from this model which is based on actual performance of large enterprises in the area (Cabanilla and Herdt) provide an idea of the modes of wet-rice production which would be feasible on tropical lowlands, if population pressure were low and prices high. Gross returns per hectare and incomes per hour of work are higher than with smallholder wet rice. However, mechanized inputs imply very high capital investments, and the resulting interest costs are such that the farm produces at a loss, given the relatively low yield level which is actually achieved on such farms.

7.4.2. *Water availability and irrigation farming: examples from the North Yemen, India, Pakistan, and Singapore* (Table 7.8)

Most irrigation farming in the tropics is rice farming, but there are other systems with different arable crops. The example from the Yemen shows the characteristics of flood irrigation with an unreliable supply of water. In Uttar Pradesh rain-fed wet rice is increasingly being replaced by irrigation farming. The Pakistan and the Indian Punjab examples show well-developed canal and tube-well systems. The examples from Coimbatore show that intensive irrigation farming with traditional technology is in a position to provide a

TABLE 7.8

Farm-management data of irrigation holdings

	Yemen Tihama	India Deoria, Uttar Pradesh	Pakistan Sind	India Punjab	India Coimbatore
Country / Location					
Rainfall (mm)	80–300	1022	178	327	591
Year	1971–2	1966–9[a]	1971–2	1969–70[a]	1972–3
Type of farming	Sorghum-cotton	Sugar-cane–rice	Sugar-cane–wheat	Wheat-cotton	Rice-groundnuts
Irrigation technique	Seasonal flood irrigation	Seasonal well irrigation	Canal irrigation	Canal and tube-well irrigation	All year well irrigation
Irrigable land as percentage of crop land	n.a.	39	100	87	52
Method	Sample	Sample	Model	Sample	Sample
Number of holdings	10	53		37	30
Persons per household	4·8	12·19	6·10[b]	10·23	4·60
Labour force (ME)[c]	5·14	7·74	3·71	5·08	3·45
Size of holding (ha)	33·19	6·81	n.a.	n.a.	1·30
Homeplot (ha)	n.a.	1·23	n.a.	n.a.	0·06
Crop land (ha)	15·19	5·58	6·84	17·50	1·24
of which rented (per cent)	22	16	n.a.	16	24
Season 1[d] irrigated rice (ha)	—	0·20	—	—	0·30
irrigated cotton (ha)	3·24	—	0·61	1·25	0·01
irrigated sorghum (ha)	12·27	—	—	—	—
irrigated other crops (ha)	0·29	—	0·40[e]	2·84	0·21
rain-fed rice (ha)	—	1·87	—	—	—
rain-fed other crops (ha)	—	1·66	—	2·27	0·41
Season 2[d] irrigated wheat (ha)	—	1·42	2·11	14·23	—
irrigated rice (ha)	—	—	—	—	0·18
irrigated fodder (ha)	—	—	—	1·39	0·13
irrigated other crops (ha)	—	0·30	0·43[e]	1·73	0·13
rain-fed crops (ha)	—	0·50	—	—	0·58
Season 3 irrigated crops (ha)	—	—	—	—	0·21
rain-fed crops (ha)	—	—	—	—	—
All seasons sugar-cane (ha)	—	1·19	1·26	—	—
All irrigated crops (ha)	15·80	2·27	4·81	23·71	1·41
All rain-fed crops (ha)	—	4·87	—	3·61	0·75
Total crop area (ha)	15·80	7·14	4·81	27·32	2·16
Cropping index (per cent)	104	128	70	156	175
Livestock (number)					
Draught animals	n.a.[f]	3·00	1·90	3·83	0·87
Other cattle and buffaloes	2·4	1·90	2·30	4·40	2·37
Sheep and goats	24·7	n.a.	n.a.	n.a.	n.a.

Farm capital ($)					
Land	18 828	4435	2388	20 972	1990
Livestock	392	274	363	645	122
Wells	—	127	—	416	360
Implements and machinery	n.a.	132	66	1036	143
Farm buildings	7	154	77	338	47
Yields					
Rice (t ha^{-1})[g]	—	1·21	—	—	2·50
Rice ($ ha^{-1})	—	105	—	—	243
Cotton (t ha^{-1})	1·11	—	1·30	0·81	1·50
Cotton ($ ha^{-1})	261	—	330	270	483
Wheat (t ha^{-1})	—	1·81	1·90	2·45	—
Wheat ($ ha^{-1})	—	184	172	283	—
Sorghum (t ha^{-1})	1·23	—	—	—	—
Sorghum ($ ha^{-1})	148	—	—	—	—
Sugar-cane (t ha^{-1} plant crop)	—	42·00	57·00	—	—
Sugar-cane ($ ha^{-1})	—	631	818	—	—
Milk (kg per cow and year)	—	297	862[h]	1017[h]	—
Economic return ($ per holding)					
Gross return	2695[i]	1373	1698	3895[j]	598
Purchased inputs					
Seed	14	113	27	168	22
Fertilizers	93 }	26	43	284	27
Irrigation costs		9	46	161	15
Mechanization	55 }	52 }	46	191	8
Depreciation	n.a.		—	207	n.a.
Other inputs	—	—	43	42	12
Income	2533	1173	1493	2842	514
Wages for hired labour[k]	356	117	40	482	66
Taxes, rents, etc.	n.a.	14	51	144	17
Family farm income	2177	1042	1402	2216	431
Productivity					
Gross return ($ ha^{-1})	177	202	248	223	460
Gross return ($ ME^{-1})	524	177	457	767	173
Income ($ ha^{-1})	167	172	218	162	395
Income ($ ME^{-1})	493	152	402	559	148
Labour input (man-hours)[l]	5900[m]	5280	3464	10 702[m]	2464
of which hire (per cent)	64	51	5	46	n.a.
Income per man-hour ($)	0·43	0·22	0·43	0·27	0·21
Man-hours per ME	1147[n]	682	934	2107	714[n]
MJ per man-hour[o]	38	31	47	57	n.a.

[a] Average of 3 years. [b] Adult equivalents, not persons. [c] Including hired labour on the basis of 200 working days per ME. [d] In the examples from Uttar Pradesh, Sind, and the Punjab season 1 is the *kharif* (summer) season and season 2 the *rabi* (winter) season. [e] Fodder sorghum in the *kharif* season and berseem in the *rabi* season. [f] Ox-plough services are usually hired in the area. [g] Irrigated rice in Coimbatore and rain-fed wet rice in Uttar Pradesh. [h] Mainly buffalo cows. [i] Crops only. [j] Without by-products. [k] Including wages in kind given to labourers living in the farm household. [l] 8-hour working day in India and 7-hour working day in the Yemen. [m] Crops only. [n] Not including cotton, straw, gardens, etc. *Sources*: Yemen: Mann (1973); India (1): Lavinia (1974); Pakistan: McConnel (1972); India (2): Kahlon and Migliani (1974); India (3): Studies in the Economics of Farm Management in Coimbatore 1972–3 (in preparation).

living for a family even on a very small piece of land, and the Singapore example exemplifies one of the most intensive types of tropical land use.

(a) *Sorghum–cotton holdings in the Tihama, Yemen.* These holdings are an example of uncontrolled irrigation. Water supplies depend on floods in the 'wadis' resulting from rains in the hills upstream. The rains are irregular, and irrigation crops are produced under the risk of drought and floods. Irregular water supplies combined with ample land lead to extensive forms of irrigation farming. Sorghum is the main food crop. Cotton is grown for cash. Much of the sorghum produces a ratoon crop, provided water is available, which is used for feeding the livestock. The labour input is very low (55 man-days ha^{-1}). The income per hour of work is therefore comparatively high. We thus have a situation with low land productivity but high labour productivity.

(b) *Sugar-cane–wheat holdings in Uttar Pradesh, India.* This farming system is an example of the combination of rain-fed farming in the rainy season (*kharif*) and irrigation in the drier season (*rabi*) that is very common in the Indian subcontinent. Traditionally only rain-fed crops were grown. Irrigation in the *rabi* (winter) season came in as a supplement to summer rain-fed cropping. Paddy is the major *kharif*-season crop and two types of paddy are grown in order to make best use of the rainy season. 'Early rice' (spring planting) is traditionally broadcast and relies on rain only (1·27 ha). Irrigated rice production (0·20 ha) with high yielding varieties is, however, increasing. Transplanting and supplementary irrigation are expanding.

'Late rice' (summer planting) is a risky crop and yields good returns only provided there is a prolonged monsoon (0·60 ha). Irrigation and the availability of improved wheat varieties have changed the emphasis within the system from the kharif to the rabi season. Wheat has become a major crop and has replaced much of the low-yielding rain-fed pulses which were traditionally grown in winter. Sugar-cane is the dominating cash crop and is still mainly grown without irrigation, but irrigation is expanding. Most farmers produce the plant crop and the one ratoon crop, and thus integrate sugar-cane in a rotation with wheat and paddy.

The greater part of the crop area devoted to annuals still produces one crop per year. The bottleneck is water supply. There is not yet enough water when needed and the animal and human labour force is insufficient to distribute the water effectively over the farm area. Another central problem of the system is the livestock economy. Cattle densities are 284 head per 100 ha, and the output of this herd in terms of draught power and milk is extremely low. Milk yields are negligible, and bullocks perform not more than 91 days of work per year.

(c) *Sugar-cane–wheat holdings in the Sind, Pakistan.* In the Sind rainfall is so low that rain-fed cropping is a very marginal affair, and is normally absent. All the land on the farm is irrigable, but water supplies are insufficient to irrigate all of it. During the *kharif* season only one-third of the land is cropped, and most of the water goes to sugar-cane. During the *rabi* (winter) season evaporation is less. Mainly wheat is grown which can mature with relatively little water, and two-thirds of the land are cropped. Livestock is closely related to cropping. Bullocks supply draught power. Milk is a major activity, and manure is required for the sugar-cane. The cattle are fed on sugar-cane tops, berseem grown in the winter season, and fodder sorghum in the summer season.

Fig. 7.8 shows the relations between the activities within the holding. Five cropping and two livestock activities (dairying and draught animals) are distinguished. Each of these activities receive inputs: seeds, fertilizer, water, hired labour, family labour, and bullock labour. The feed and manure input is farm-produced and therefore not expressed in monetary terms. The inputs flow into the various activities and produce outputs which are partly consumed by the household and partly sold. The various activities are closely related by the joint use of the family labour, the production and consumption of draught power, feeds, and manure, as is indicated by the flows going in and out of the 'bullock', 'feed', and 'manure' pools (McConnel 1972).

(d) *Wheat–cotton holdings in the Punjab, India.* The average farm size in the area from which the example is taken is 5 ha, but most of the land belongs to farms in the 15–20 ha size group, and the example is taken from this group. Irrigation farming is highly developed. Water from canals is supplemented by water from tube-wells, and two irrigated crops are grown per year. Almost all land is planted to wheat in the *rabi* (winter) season, and in the *kharif* season mainly cotton and maize are grown. The rain-fed land produces fodder in the *kharif* season. Farming techniques have been modernized. High-yielding varieties of wheat, maize, and cotton are planted, and the input in mineral fertilizer is substantial. The traditional organization of livestock, which is still to be found in the other two irrigation examples from India, no longer applies. Bullocks are in the process of being replaced by tractors, and commercial milk production is moving in. Milk yields average 1000 kg per cow, which is several times the Indian average.

Most conspicuous is the employment of the available labour force. Work with crops only, not including work with livestock, amounts to 2107 hours per labourer per year. There is thus clearly full employment of the farm labour force. Full employment applies not only to human labour but also to bullocks which work on average 148 days per year, compared with 60–100 days in most other Indian farming systems. The important aspect of the Punjab farming system is not so much a high return per hour of work, but the great number of working hours per labourer and year. This reflects the general tendency in irrigation farming with multiple cropping.

(e) *Holdings with diversified multiple cropping in Coimbatore, India.* The

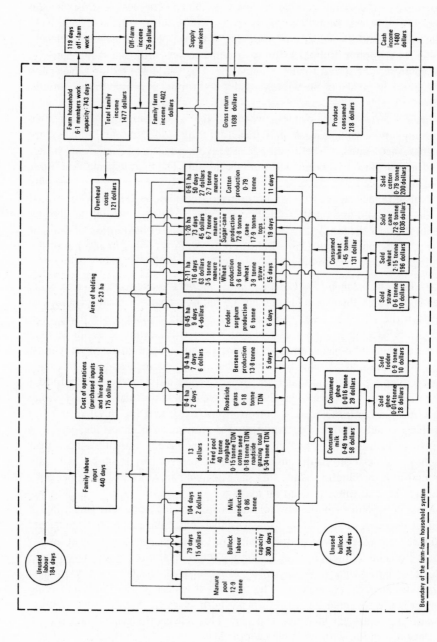

Fig. 7.8. Structural model of a sugar-cane–wheat farm, Sind, Pakistan (from McConnel 1972 p. 47. The data have been slightly modified. The flow chart has been significantly changed).

example from Coimbatore demonstrates the farming pattern which tends to develop in conditions of great land scarcity, a tropical climate, a combination of rain-fed and irrigation farming, reliable irrigation from wells, and favourable price relations for the producer: low wages in relation to producer prices. This situation induces a high intensity in irrigation farming. The cropping index of the total cultivated area is 182 and of the irrigable area it is 237. Three crop seasons are distinguished. Wet rice is the major crop, but several types of oil-seeds, pulses, and fodder crops are also grown.

The high intensity makes it possible to sustain a family on a very small area at an income which is higher than in most rain-fed systems. Incomes are, however, much lower than in larger irrigation farms reflecting the influence of diminishing returns to labour. However, farming techniques in the Coimbatore example are still traditional The mineral fertilizer input is negligible and the fertilizer economy still relies mainly on a large herd of cattle (261 head per 100 ha) which produces little else but draught power and manure.

(f) *A Chinese-type vegetable holding in Singapore.* There is hardly a more intensive farmer or gardener in the world than the Chinese vegetable producer, who has adapted the subtropical Chinese system of market gardening to the humid tropics of south-east Asia. Table 7.9 shows the farm-management data of a holding of 1·21 ha in Singapore, run by three brothers. The vegetable gardens occupy the flat land of the holding, covering not more than 0·6 ha. The garden land is divided into 127 beds. Dispersed among the beds are five man-made ponds. These not only provide water for the vegetables, but also function as sites for growing water hyacinth, an important source of pig fodder. In addition, they are stocked with fish. The ground between the gardens and the hill slope is occupied by dwellings and farm buildings. The hillside of the holding is left unused, except for some fruit trees.

The bulk of the vegetables produced are of the quick-growing, leafy type, mainly Brassicaceae. Most of them are grown in stages—first in the nursery beds, and later in the main production beds. The vegetables usually occupy the production beds for 4–6 weeks. In the afternoon or the morning following the harvest, the beds are planted again. The fallow beds amount normally to not more than 5–10 per cent of the total. The average number of crops per year and per bed hovers around 8. Fig. 7.9, taken from farms in Hong Kong, shows the sequence of cropping that is typical for this type of farming. The rainfall, amounting to 2500 mm annually, is too irregular and inadequate to rely on. The watering of the vegetables is roughly equivalent to an additional rainfall of 1400 mm per year.

The land provides hardly more than the physical medium for plant roots. The gardens are fertilized with both farm-produced and purchased fertilizers, including diluted pig manure, chicken droppings, mud from the ponds, compost, prawn dust (a product of the shrimp industry), and mineral fertilizer.

TABLE 7.9

Farm-management data of a Chinese-type vegetable holding in Singapore

Year	1965
Rainfall (mm)	2500
Method	Case study
Persons per household	30
Labour force (ME)	10
Size of holding (ha)	1·21
Cultivated land	0·60
Total crop area	4·80
Cropping index (per cent)	800
Livestock	
Pigs	300
Poultry	440
Economic return ($ per holding)	
Gross return	
Crops	3559
Livestock	16 726
Total	20 285
Purchased inputs	
Concentrates	12 356
Fertilizer	1054
Other	712
Income	6163
Productivity	
Gross return ($ ha^{-1})	16 764
Gross return ($ ME^{-1})	2028
Income ($ ha^{-1})	5093
Income ($ ME^{-1})	616

Source: Compiled from Fong, Lian, and Wikkramatileke (1966). This publication does not contain a farm-management balance sheet. The figures in this table in some instances had to be derived from various data, and they give only a rough idea of the economic state of the holding.

Night soil is traditionally used in the farming system, but is no longer allowed for vegetable production.

The rearing of pigs and poultry, livestock activities which lend themselves to a high turnover, is an important part of the farming system. The pigs are fed on concentrates and on a mash prepared from vegetable waste. Apart from supplying much of the income, livestock also provide an important source of manure. Most of the nutrients are not imported as purchased fertilizers, but as feeding stuff.

Intensive techniques thus enable the Chinese market gardener to cultivate vegetables successfully in an environment where natural soil fertility is low, rainfall irregular, natural soil aeration often inadequate, solar radiation often too intense, and harmful organisms numerous.

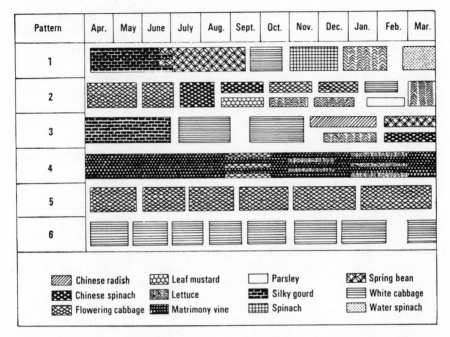

Pattern	Apr.	May	June	July	Aug.	Sept.	Oct.	Nov.	Dec.	Jan.	Feb.	Mar.

Chinese radish   Leaf mustard   Parsley   Spring bean
Chinese spinach   Lettuce   Silky gourd White cabbage
Flowering cabbage   Matrimony vine   Spinach   Water spinach

FIG. 7.9. Multiple-cropping system on some Hong Kong farms (from Wong 1968).

Even more intensive are vegetable enterprises in Southern China. Over a dozen crops may be grown in a single year, and this includes every conceivable type of intercropping such as the combination of early-maturing crops with long-duration crops (Chinese cabbage among aubergines), tall-growing with short-growing and shade-tolerant crops (celery under tomatoes) and relay planting (wheat and maize) (Harwood, Plucknett, and Romanowski 1977, pp. 14–17). These enterprises produce a gross return of about $6500 per hectare and year with relatively low amounts of purchased inputs of about $330. Labour inputs, however, are very high and average around 24 000 hours per hectare and year. Roughly half of the labour input is due to such time-consuming tasks as pig-tending for meat and manure and compost-preparation (Wiens 1977, pp. 38–40).

7.4.3. *Farming systems in large irrigation schemes*

(a) *Characteristics of production under close supervision.* Recently, small-holder settlements have been increasingly established in areas that are being opened up for arable farming in large-scale irrigation projects. It is a characteristic of large-scale irrigation developments that there is a great measure of physical interdependence between different users of the water, and, therefore, to achieve maximum benefit, the resources must be planned and operated as part of an integrated system (Carruthers 1968, p. 7). In several countries, large-scale projects have been created by the government and are operated by government-backed agencies. The farm-management position of small-holders in these irrigation schemes is different from their position in holdings that obtain water from minor irrigation works. The minor schemes have been set up in the course of time not so much by the use of capital as by work on the part of the local population. For this reason, they rarely entail high fixed costs.

Large irrigation schemes, on the other hand, necessarily involve high fixed costs asking for high productivity and high water charges. The farmers, who are not as a rule highly experienced in modern irrigation farming, are not in a position to realize the production potential within a reasonably short period after the water has been made available. The answer to the problem is production under close supervision, that is, the obligatory application of modern techniques by all water users. The State, which has the right to distribute land to settlers and to grant water rights, has a 'power key' by which it can introduce better techniques. This is selective pressure in the sense that it is used only on those smallholders who are interested in becoming irrigation settlers. They must agree to abide by a series of principles that guarantee technically and economically efficient production. At the same time, participants may retain ownership of the means of production that they possessed previously, or which have come into their possession, with the proviso that they may not dispose of this property just as they please. They are dependent in this respect on the rules laid down by the scheme management.

In large irrigation schemes the management functions are usually not in the hands of one man, but partly in the hands of the scheme agency and partly in those of the cultivators.

1. The project manager runs the framework of the scheme. He sees that production is carried on in an organized fashion; he arranges the acquisition of water, he organizes tractor ploughing, the supply of purchased inputs, and the marketing. His function is that of a strict but benevolent and technically progressive landlord. Co-operative management is not feasible in irrigation schemes. The task of distributing water and supervising production is too delicate to be left to a group of people who are the representatives of the irrigators. The successful running of large irrigation schemes obviously requires a high degree of centralized authority (see Wittfogel 1931).

2. The smallholders do the work delegated to their plot. Within the rules, which are laid down by the project agency and can be organized and enforced on a broad or narrow basis according to the knowledge and the drive of the irrigators, the smallholders operate at their own discretion.

Their position is often closer to that of sharecroppers than of independent farmers. Frequently they are tenants.

3. The irrigators can, if necessary, join forces and co-operate in certain undertakings, as for example raising credit or selling their produce. The activities of the co-operative are again supervised by the project management.

Giglioli (1965, p. 202) describes production under close supervision in one of the most successful irrigation schemes in Africa, the Mwea-Tebere irrigation scheme in Kenya. He writes:

> Tenants are allocated four one-acre plots as a rice holding. Only one crop of transplanted rice is grown per year . . . No rotation is practised . . . The rules are very comprehensive . . . and . . . particularly detailed on matters relating to cropping practices, water control, absentee ownership, and disposal of the crop. Management has three disciplinary tools to deal with recalcitrants. These are, in order of increasing severity: written warnings, prosecution and termination of licence . . . Selection (of tenants) has been limited to the disinherited, and the Clan Committee have a very human tendency to unload onto the Settlement the less desirable members of their community . . . (Nevertheless) in the three years that the Irrigation Rules have been in force only 19 tenants have had their licences terminated (Mwea section only). The basic organizational unit . . . is a block of 2400 acres of irrigated rice, supporting 600 families . . . The Assistant Agricultural Officer is in charge of all activities on the block . . . The day to day supervision . . . is in the hands of the Field Assistants. Each man is responsible for 150 tenants . . . All water control is in the hands of the Settlement staff. A tenant is not allowed to interfere in any way with the water regime in this field . . . the Head Water Guard and his Guards are responsible for execution . . . The last link in the chain between the Manager and the tenants is constituted by the Head Cultivator . . . a tenant selected for his farming and leadership qualities . . . The introduction of mechanical cultivation has made possible a degree of planning and discipline and extension unthought of in the past. The orderly progression of cultural operations made possible by mechanization of puddling has produced an atmosphere in which strict discipline can be enforced without opposition . . . It is now possible to meet every group of tenants . . . before the beginning of each stage of the cultivation and to indoctrinate them on the measures. Repetitive indoctrination made possible by mechanization is showing its effectiveness by both the increase in yield and the even distribution of the increase over all the tenants on the Settlement.

Irrigation farming under close supervision combines a number of advantages that characterize small-scale and large-scale farming. The stimulus for efficient production, which is generated by the direct relationship between the individual's efforts and the returns of the land, is maintained. Every irrigator receives an income according to the yield of his land; since the quality and amount of land and other inputs provided for each settler are the same, the yield is mainly a direct reflection of the amount and skill of the labour input. Large-scale operations, such as ploughing or spraying with pesticides, can be organized by the management on a large-field basis. The plots of the irrigators can be arranged in such a way that large machinery can be used efficiently. Innovations like new varieties or crops can easily be introduced by the scheme management.

(b) *A case study: the Gezira scheme, in the Sudan.* The biggest example of large-scale irrigation farming under close supervision is the Gezira scheme in the Sudan. In the project area, which including the Managil and Guneid extensions comprises 862 500 ha and 96 000 tenants, the land was leased or bought from its owners, reorganized, and developed for cropping with cotton, sorghum, groundnuts, wheat, fodder crops, and vegetables. The division of operations between the government, the scheme management, and the tenants is well defined.

The government has four main responsibilities:
1. It buys or leases the land from its owners and pays small rents.
2. It builds and maintains dams and the primary and secondary distribution and drainage channels.
3. It finances the acquisition of water and land development.
4. Finally, it establishes the scheme management as an autonomous body —the Gezira Board.

The scheme management is responsible for the following functions:
1. It distributes the land among the leaseholders and enters into lease contracts with them. In the original Gezira area, the smallest irrigation unit is a rectangle of 1420 m by 292 m. Each of these blocks is divided up into nine plots, each of 4·22 ha. Each leaseholder receives one plot of land in four blocks, giving 16·8 ha altogether. Further subdivisions and amalgamations are permissible. The holdings in the Managil extension are laid out on a smaller scale (6·3 ha), because the original holding size was too great for the tenants' labour force to manage.
2. Land use is not determined by the settler but is prescribed by the management. The usual crop rotation in the Gezira has an 8-year cycle as follows:

      lubia   Phillipesara
cotton—groundnuts—sorghum— fallow —fallow—cotton—wheat—fallow.
          vegetables

From 1975–6 on the crop rotation will be:
               cotton—wheat—groundnuts—fallow.

In the Managil extension, a six-field crop rotation was introduced: cotton—fallow—cotton—lubia—sorghum—fallow. Fig. 7.10 shows the plot division in the Gezira rotation and illustrates the connection of smallholder farming with large-field management. Each field is 2·1 ha in size and forms, with seven others, a unit of 16·8 ha (see Table 7.10). Each tenant uses his plots in obligatory rotation. Thus, for example, tenant A1 uses a cotton field of 2·1 ha, half a sorghum field, and half a lubia–groundnuts–vegetable field. In the

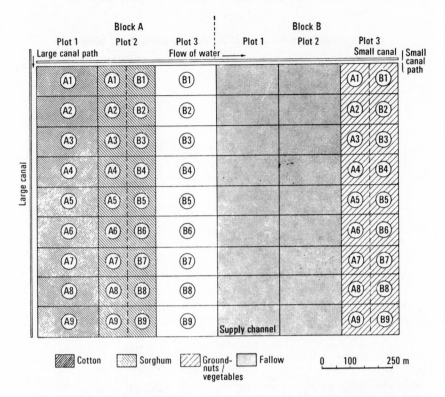

FIG. 7.10. Organization of water delivery, land use, and field layout in the Managil extension of the Gezira scheme, Sudan. Two blocks, each with nine holdings are shown (three plots of 2·1 ha per holding). A1 and B1 are the cultivators of the first strip, each farming 2·1 ha cotton, 1·05 ha sorghum, and 1·05 ha lubia. (A new layout will be introduced in 1975–6, but the principle of an obligatory rotation will be adhered to.) (Wörz 1966, p. 33).

following year his cropping is adjusted according to the rotation, but only within the strip that he has been allotted together with the tenant B1.

3. It organizes the water supply, maintains the canal system, and levels the land. The days when the water is supplied and the amount are determined by the Board. Furrow irrigation is supplied 12–15 times in the growth period of cotton, i.e. 1200 mm, which is supplemented by 200–400 mm of rain in a year.

4. The annual preparation of the soil with tractors is also carried out by the scheme management and presents the tenants with land ready to be sown—the arrangement of cropping in large blocks with one crop facilitates mechanization.

5. Seed, mineral fertilizer, and other material inputs are supplied.

6. Pest control is centrally undertaken by the management.

TABLE 7.10

Farm-management data of tenants in the Gezira, Sudan

Location	Gezira
Year	1973–4
Type of farming	Cotton–wheat
Method	Farm accounts
Number of holdings	95
Persons per household	7·0
Labour force (ME)	4·4
Size of holding (ha)	18·90[a]
Cotton	4·91
Wheat	2·42
Groundnuts	2·73
Sorghum	2·36
Total crop area	12·42
Cropping index (per cent)	66
Irrigated (per cent)	100
Livestock (number)	
Cattle	3
Sheep and goats	7
Yields	
Cotton (t ha^{-1})	1·84
Cotton ($ ha^{-1})	1169[b]
Wheat (t ha^{-1})	1·36
Wheat ($ ha^{-1})	216
Groundnuts (t ha^{-1})	1·34
Groundnuts ($ ha^{-1})	184
Sorghum (t ha^{-1})	1·98
Sorghum ($ ha^{-1})	101
Economic return ($ per holding)	
Gross return	
Crops	7012
Livestock	355
Total	7367
Purchased inputs	
Seeds, fertilizer, etc.	547
Mechanized services	228
Other	82
Income	6510
Costs carried by Gezira Board[c]	616
Wages for hired labour	1292
Share going to Gezira Board	3450
Family farm income	2384
Productivity	
Gross return ($ ha^{-1})	390
Gross return ($ ME^{-1})	1674
Income ($ ha^{-1})	344
Income ($ ME^{-1})	1479
Labour input (man-hours)	6048
Income per man-hour	1·08
Family farm income per man-hour	0·39

[a] The size of holding is not entirely uniform. The accounts are taken from farms which are somewhat larger than in Fig. 7.10. [b] In 1973–4 farmers received the very high price of $0·63 kg^{-1} compared with $0·31 in 1971–2. [c] The Board's share in joint costs for fertilizer. It has to be added to arrive at the farm family income. *Source:* Sudan Gezira Board (1975).

7. It also ensures that operations delegated to the tenants are carried out according to plan.

8. The transport, preparation, and sale of cotton is centrally organized.

9. The scheme management organizes the distribution of small amounts of credit for the payment of seasonal labour for cotton picking.

10. It is responsible for the collection, ginning, sorting, storing, and sale of cotton.

11. It also finances research.

12. It supports the social institutions in the project area.

The tenants are assigned special tasks:

1. They are responsible for all manual work—seeding, weeding, harvesting, uprooting old cotton stalks, ditch clearance, etc. The work must be carried out at specific times. Each block is under the supervision of a block inspector, who checks the timing of operations in cotton growing, but the tenants are left to do the work themselves, or to employ seasonal workers.

2. The cotton has to be surrendered by them to the scheme management.

3. Tenants can keep their cattle in the project area. The fallow is used as communal grazing. Cropping of fodder legumes bridges the fodder shortage in the dry season.

4. Farmers are members of a tenants' organization, which represents the interests of the tenants at the scheme management level.

The harvested crops of sorghum, other crop products, and animal products belong to the tenant in their entirety (see Table 7.10). The dues of the project management are deducted from cotton receipts. First of all, those cost items are deducted that have to be carried by the project and the cultivator jointly. These include: mineral fertilizer, seed, pest control, tractor ploughing for uprooting cotton stalks, the wages of field guards, the costs of transport, processing, and marketing of the cotton, and applied research. The amount remaining after these deductions is distributed as follows:

36 per cent to the Government;

49 per cent to tenants, 2 per cent of which go to the Tenants' Reserve Fund in order to maintain it at a maximum level of 0·25 per feddan sown with cotton;

2 per cent to the Local Government Councils within the irrigated area;

3 per cent to the Social Development Department;

10 per cent to the Sudan Gezira Board.†

7.5. Problems of the system

Agricultural innovations during the last decade have increased so rapidly that the gap between actual and optimum output is usually high, and particularly so in irrigation farming. The main problems of irrigation systems are not inherent in the system as such, as is the case in shifting systems and generally to a lesser extent in upland farming. The major weaknesses lie rather in the inadequacy of water control, husbandry practices, and irrigation institutions.

† The distribution is varied from time to time.

7.5.1. *Insufficient control of water supplies*

An inherent problem of irrigation farming is the irrigators' dependence on others whose land-use practices influence his water supplies. As a rule the irrigation farmer has to reckon with insufficient water at certain times and waterlogging at others. This is particularly true for terraces with wet rice and for minor irrigations. Upland and irrigation farming are not only related within the holding but also within the wider framework of the watershed or river basin. Growing frequencies of floods and droughts may be due to over-grazing or the extension of cropping into forested areas. A farmer downstream may not get the accustoned amount of water because others upstream use more of it. Drainage for one may mean more waterlogging for another. Decision-making on each holding usually occurs without regard for external economies and dis-economies, and co-operative or Government institutions are, as a rule, not yet very effective in caring for the common good. Land use in the watershed or river basin and water control and distribution should be under the control of one effective agency. In minor irrigation this was traditionally organized within the framework of a landlord–tenant system. More recent institutions which have taken over the landlord's role are rarely adequate in their performance.

Most of the wet rice in Asia and Africa is grown during the rainy season without a reliable source of water. Wet-rice production relying mainly on impounded rainfall rarely exceeds $1 \cdot 5$ t ha^{-1}, while twice the yield can be expected with controlled irrigation, even without improved practices. The major reasons why wet-rice producers harvest low average yields are primarily outside the control of the individual farmer. The modifications in the environment necessary to achieve effective water control, irrigation, and drainage during both the wet and the dry season will have to come primarily from investments in water-distribution structures. Public investments and communal work are required to narrow the gap between typical and potential yields in wet-rice production (Ruttan, Soothipan, and Venegas 1966, pp. 32–3).

The failure to develop an effective water storage, transportation, and drainage system in large parts of the tropics is partly due to the physical geography of the areas concerned. Japan and Taiwan, for instance, are characterized by short river valleys and narrow coastal plains, which lend themselves to locally organized, small-scale, labour-intensive irrigation, and drainage work. Most other rice-growing countries in south-east Asia and Africa are characterized by broad river valleys and plains. Under these conditions, the physical geography requires the organization of large systems of water distribution on a national basis. The construction of such systems requires high levels of capital investment. Accomplishment of this scale of investment is held up by the generally low profitability of large-scale irrigation schemes, which is to be attributed mainly to low product prices, and difficult organization and administration.

The rational use of water in minor irrigations, which in terms of area and production potential are even more important in the tropics than large schemes, suffers for various reasons, the most conspicuous being the following: inappropriate water and land rights, administrative inadequacies, difficulties in the organization of underemployed rural labour without payment in communal work, difficulties in the organization of efficient co-operatives for water use, lack of small amounts of credit, and lack of incentives for farmers who are prepared to undertake development.

7.5.2. *Wasteful irrigation and husbandry practices*

Wasteful irrigation practices, which prevail in most of the minor irrigation areas, are another drawback of irrigation farming. Much water is lost in field canals because of seepage.† Attempts to irrigate land with an uneven surface generally result in low efficiency of water use and low crop yields. Sufficient drainage is a prerequisite for high yields and the prevention of loss of land due to salting. Misuse of irrigation water, in particular too high application of water, can cause soil erosion, waterlogging, and a buildup in soil salinity, and can thus lower efficiency in terms of output per cubic metre of water and per unit of investment in irrigation facilities. In particular, the sharing of water on a rotational basis without water meters, as is customary in minor irrigations that supply several irrigators, often leads to the uneconomic use of water. The users will tend to take their full share at each turn, irrespective of the water needs of their crops.

Wasteful irrigation practices encourage the tendency to continue traditional farming, although new varieties, mineral fertilizers, and pesticides have created a new situation with attractive possibilities. In this connection, it is necessary to point out that irrigation farming, because of its production potential, depends more on the willingness of farmers to change and on knowledge, skill, and drive than do other land-use systems. Multiple cropping makes special demands. Preparation of the land, planting, and harvesting can be accomplished several times within a year only if the cultivators are prepared to work hard and to discipline themselves to observe the necessary schedule of work.

7.5.3. *Dependence on inadequate institutional arrangements*

Shortcomings due to uncontrolled irrigation and inadequate husbandry practices are, however, only partly a matter of lack of knowledge, skill, and drive. In irrigation systems, closing of the gap between actual and optimum

† Case studies in India showed that for every 100 m³ at the head of the canal only about 40–50 m³ actually reached the field. About one-third of the water applied to the field percolates to the water table (Randhawa, Cheema, and Dev 1972, pp. 493 and 496).

farming depends more than in any other land-use system on institutional arrangements that promote production. This is particularly true of land and water rights.

Irrigation is a practice that relatively few farmers are able to organize on their own. Water storage, supply, and drainage are complex processes that can be carried out only through joint action that will affect a large number of individual farmers. Consequently, the way in which these things are organized can be a crucial factor in the progress of irrigation farming. Often the functions of land ownership and water supply on the one hand, and their use in farming on the other, are divided between different people: the private or public landlord and the smallholder. A practice detrimental to irrigation farming in such a situation is sharecropping, which reduces the incentives for intensification. On the other hand, a high fixed monetary charge per unit of land, in the form of either a land or a water charge, although highly effective in introducing intensive farming, is politically difficult to implement. In some regions, where the normal rainfall is adequate, the farmers tend to regard irrigation only as an insurance against years with insufficient rainfall, because water is free or the water charge is not sufficiently high to induce more intensive farming.

A number of projects that have been operating for years—the Gezira scheme in the Sudan is the most famous example—show that irrigation farming under close supervision allows modern farming techniques to be used by unskilled smallholders where required. Production under close supervision, however, is not a generally applicable principle. The approach shows several weaknesses, the most important of which seem to be the following.

1. High costs of supervision and mechanization have to be borne either by the tenants or the government. Projects must be large enough to work economically. Since the scheme managements only rarely have any competition to contend with, there is the danger that top-heavy bureaucracies might develop.

2. The layout has to be simple and uniform. Concentration on one or two major cash crops is advisable. The patterns of land use are necessarily rigid. Smallholder farming under close supervision is not characterized by variety in cropping like other types of irrigated smallholder farming, though this variety helps to increase yields and reduce risk.

3. The tenants are inclined to employ labourers and to become miniature landlords. The more compulsive and complete the supervision is, the smaller the scope for individual initiative and the less willing farmers are to invest in their holdings.

4. Production under close supervision is usually unpopular, and the success of production depends, therefore, upon political backing. The participants very rapidly feel themselves to be capable of running the

production properly without supervision, an opinion that is not as a rule confirmed in actual practice.

Minor irrigation areas are too small to be organized along the lines of production under close supervision. The traditional landlord, who had the authority to distribute or sell water economically, has largely been abolished by land reforms. Co-operatives may fill the gap, but experience shows that it is very difficult indeed to organize irrigation properly without the backing of some outside authority, and this runs contrary to the principle of co-operative action.

7.6. Development paths of irrigation farming

Further development of holdings with arable irrigation farming will lead to fundamentally different land-use systems only in exceptional cases. The change to irrigated tree-crops or to irrigated pasture are two relevant possibilities. The return to extensive grazing is another possibility where land is lost due to salting. As a rule, the real issue for farm management is how to increase production within the framework of arable irrigation farming. The following basic possibilities present themselves:

1. Increase of the irrigated area, by increase of water supplies: or obtaining a more efficient use of the available water.

2. Expansion of the hydrologic growing season by (i) supplementary irrigation (often pump irrigation), (ii) early crop establishment techniques (nurseries), and (iii) growing drought-tolerant crops in the unirrigated season, using residual soil moisture (Zandstra and Price 1977).

3. Better utilization of the growing season by (i) higher yielding varieties, (ii) higher chemical inputs, (iii) reduced 'turnaround' times, (iv) harvest at physiological maturity, and (v) relay cropping, intercropping, and multiple cropping (Zandstra and Price 1977).†

7.6.1. *Extension of irrigation systems*

The extensions of farming systems with irrigation is an obvious way to increase output and productive employment. There is much scope for more irrigation farming, in particular wet-rice farming, in semi-humid and humid Latin America and Africa. Where land is ample and labour scarce, the tendency is towards very large and fully mechanized operations as they are operating in Para State, Brazil (two rice crops with 14 t ha^{-1}, year^{-1}), or in Ivory Coast. Also much valley-bottom land can be developed for wet rice in Africa.

In semi-arid Asia the tendency is towards increasing the irrigation capacity by economizing on the use of water.

---

† 'Turnaround time': number of days between harvest and next planting.

1. Irrigation techniques with higher water efficiency (the lining of canals, underground pipes, sprinkler and trickle irrigation, etc.) are likely to gain in relation to other irrigation techniques.

2. The return per unit of water applied can be increased by techniques summarized by Carruthers:

   (i) Adaptation of cropping pattern to the availability and the reliability of water supplies.

   (ii) Allowing crops to draw upon stored soil moisture.

   (iii) Allowing crops to suffer water stress for some time, if this can contribute to increasing over-all production with the given resources of water, labour, and capital.

   (iv) Fulfilment of leaching requirements at times of the year when there is no peak demand for water.

These are economic rather than technical criteria for determining water application. Furthermore, there is a good chance that important water-saving innovations might be developed in the near future. Water development on the watershed is likely to gain in importance. Water-harvesting (channelling run-off on to fields) is an age-old technique which may be significantly improved by modern water-collecting techniques. Water stored on the watershed can be applied as 'life-saving' irrigation to crops which mainly rely on rainfall. The large-scale extension of this type of irrigation farming would mean the gradual substitution of classical irrigation farming, which concentrates much water on a limited surface, by the application of small amounts of water on extended areas. The potential gain in water-use efficiency is obviously great.

7.6.2. *More reliable water supplies*

Even more important, in terms of output, than the extension of irrigation seems to be the provision of more reliable and controlled sources of water to the farmers. With unreliable water supplies little intensification in irrigation farming can be expected. In the Philippines, for instance, Herdt and Wickham (1974, p. 12) estimated that lack of control over water is the biggest single yield constraint. It is responsible for about 25 per cent of the difference between the potential and the actual yield. In south-east Asia only about 20 per cent of the wet-rice area is under artificial irrigation, but it produces about 40 per cent of the total rice crop.

7.6.3. *High-yielding crops and varieties*

In most tropical irrigation areas yields per hectare and crop can be doubled with modern production techniques: high-yielding varieties, mineral fertilizer, pesticides, improved irrigation, timely planting, and effective weeding. Particularly important is the possible impact on rice production. In traditional wet-rice farming it is common for average yields to hover around 1–2 t ha^{-1}

in the wet season and 2–3 t ha⁻¹ in the dry season. The new varieties seem to have a yield potential under field conditions of 4–6 t ha⁻¹ in the rainy season and 5–7 t ha⁻¹ in the dry season, and this with only 80–110 days of field occupation.

The general principle is demonstrated by Fig. 7.11, which shows the interaction between the application of nitrogen and dwarf varieties. High-yielding varieties not only produce more, but they mature earlier, losses due to drought are lower, yields are more reliable, and these varieties may be grown at any time of the year as they are sensitive to the photoperiod. The first high-yielding varieties had several shortcomings: poor cooking qualities, less

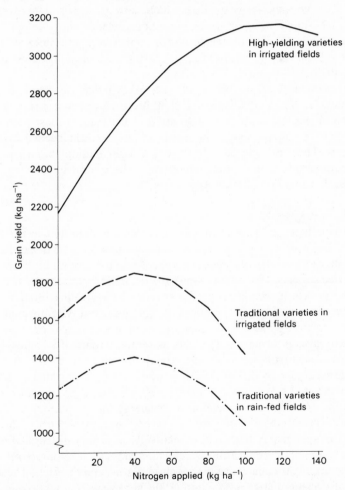

FIG. 7.11. Responses of different varieties of rice to nitrogen fertilization under field conditions in the Philippines (from Hayami, Bennagan, and Barker 1977.)

tolerance to insects and disease than traditional varieties, and higher weeding requirements, because there was less shading at ground level. The new varieties became, however, the starting point for the following sequence in rice development:

1. Varieties that yield more, mature earlier, and are not sensitive to the photoperiod (IR 8).
2. Varieties with quality (1) and better cooking qualities (IR 5).
3. Varieties with quality (2) and greater resistance to insects and disease (IR 26).
4. Varieties with quality (3), but which mature even earlier and are more resistant to a greater number of insects and diseases (IR 28, 29, and 30).
5. Varieties with the above qualities which show greater drought resistance, which are suited to deep-water conditions, which show tolerance for adverse soils and extreme temperatures, which compete more effectively with weeds, and/or which produce a grain with more and better protein (IRRI 1975).

Similar sequences are to be found with other irrigation crops. It seems reasonable to assume that advances in plant breeding will continue to increase yields also in less favourable areas, to assist in making the most economical use of purchased inputs, and to improve the return to the needed irrigation investments. There is evidence indicating that high-yielding varieties increase employment even though their introduction may be accompanied by a growing interest in motorization (Barker and Anden 1975, p. 28).

7.6.4. *Multiple cropping*

The prospects for an increase in multiple cropping, in particular for double cropping, are of no lesser importance than for increasing yields per hectare. Fig. 7.12 provides an idea of three basic approaches to more multiple cropping in Asian rice systems. Fig. 7.12(a) shows the possibilities for a fuller use of the rainy season with a second crop. Traditionally low-yielding mungbeans or cowpeas are grown after transplanted rice. The improved system would imply earlier transplanting and a full second crop of maize (corn) or vegetables. Fig. 7.12(b) gives information about the potential impact of a change in rice technology. The first rice crop is directly seeded and can be followed by a second transplanted crop. Fig. 7.12(c) applies to a fully-irrigated area where three transplanted rice crops are followed by a winter crop of wheat.

The most effective way of turning solar energy into organic matter is the all-the-year-round cover of the soil by several storeys of leaves which are organized in such a way that light is relatively uniformly distributed among all layers of the plant association which is maintained, if required, by irrigation. Growing sugar-cane comes rather close to this, and the highest recorded amount of organic-matter production (150 t ha^{-1} $year^{-1}$) comes from sugar-cane in Hawaii (Chang 1968, p. 348). A similar output could be achieved with a

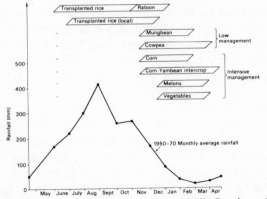

(a) Cropping patterns in rainfed lowland rice areas of Iloilo Province, the Philippines.

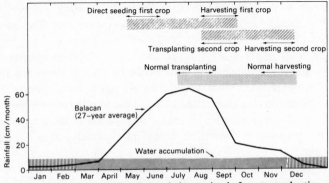

(b) The possible combination of directly seeded wet rice before transplanting wet rice at Balacan, the Philippines.

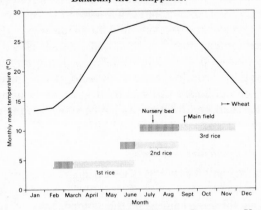

(c) Three rice crops per annum in the Tai Lai People's Commune, Kwantung Province, China.

FIG. 7.12. Three examples of land-use intensification through multiple cropping in Asian wet-rice systems (from (a) Palada, Tinsley, and Harwood 1976, (b) IRRI 1976, p. 29, (c) IRRI 1978.)

multi-storey canopy of various interplanted crops which are tailored to each other regarding their requirements of light, nutrients, and water. Such a system is not yet likely to be an economic proposition from the point of view of the farmer, but between sole cropping and high intensities in multiple cropping there are numerous intermediate steps which are likely to become economic where there is ample availability of rural labour and favourable price relations for the farmers.

Deductive reasoning, based on what is known about plant production, leads to the hypothesis that increasing intensity in multiple cropping would result in increased gross returns per hectare but decreasing net returns per hour of work. A great number of experiments with various types of multiple cropping have been carried out in India (Indian Society of Agronomy 1973) and most of the results indicate that the point of diminishing returns has not yet been reached. In Table 7.11 the more intensive rotation is clearly the more profitable, although labour is costed at the going wage rate. The information collected by Dalrymple (1971, pp. 50–2) indicates the same: net returns per hectare and returns per hour of work increase with an increasing cropping index even up to the index level 300. However, this information should be considered with caution. Most of the data are not from operating farms but from experiments, and it is reasonable to assume that the discrepancy between the return in experiments, and in commercial farms is the greater the more complex the process, and multiple cropping is clearly more complex than sole cropping.

Much research is being undertaken to improve the productivity of multiple cropping: varieties can be 'tailored' to fit intercropping, relay-planting, and ratooning techniques. Crop combinations have to be found which reduce disease problems. The leaf-area index of crops has to be developed in a way to use solar energy more effectively and to shade weeds (Harwood 1973, pp. 6–11). Intercropping may be organized in such a way that nitrogen fixed by leguminous crops is readily available for nitrogen-demanding crops. By understanding the crop interactions involved, it may be expected that a technology will be developed which is more effective in using the available resources of solar energy and water than has been the case in known multiple-cropping systems (Herrera and Harwood 1973).

7.6.5. Changes in the labour economy

Most tropical irrigation farming relies on high inputs of labour, supplemented by animal and tractor power. This is likely to remain so, but there are indications of significant changes in the composition of the labour and support energy inputs.

(a) The available evidence indicates that more labour is employed if high-yielding varieties are grown under smallholder conditions, but the pattern of labour use tends to change:

TABLE 7.11

Costs and returns of multiple cropping in Hissar District, India[a] (per hectare per year)

	Crop duration (days)	Man-hours[b]	Yields (tonne)	Gross return[c] ($)	Variable costs[d] ($)	Net return ($)	Gross return ($ per man-hour)
Rotation 1							
Lady finger	80	966	0·52	308	122	186	0·32
Potato	95	1290	15·22	708	381	327	0·55
Tomato	120	880	14·96	796	343	453	0·90
Total	295	3136	——	1812	846	966	0·58
Rotation 2							
Cotton	200	602	1·32	371	145	226	0·62
Wheat	150	517	3·68	402	167	235	0·78
Total	350	1119	——	773	312	461	0·69
Rotation 3							
Maize	95	873	3·56	290	172	118	0·33
Potato	70	1214	10·18	474	349	125	0·39
Wheat	120	476	3·01	329	158	171	0·69
Green gram (Mung)	65	363	0·49	85	71	14	0·23
Total	350	2926	——	1178	750	428	0·40
Rotation 4							
Pearl millet	105	482	2·85	260	126	134	0·54
Wheat	150	502	3·59	392	165	227	0·78
Green gram (Mung)	65	377	0·47	81	72	9	0·21
Total	320	1361	——	733	363	370	0·54

[a] Results of multiple-cropping rotations in a sample of 30 farmers. Averages of 2 years (1969–71). [b] 8 hours per man-day. [c] Including by-products. [d] Value of labour, purchased inputs, and farm-produced inputs (bullock-labour, seeds, manure, etc.). *Source:* Singh, Nandal, and Singh (1973, p. 130).

(i) New nursery techniques require less labour. This applies in particular to the mat (dapog) technique for rice seedlings.†

(ii) Tractors reduce labour requirements for land preparation.

(iii) Fertilizer and pesticide application requires additional labour hours and more time is needed for weeding. With herbicides, however, the labour-saving effect is very pronounced.

(iv) Much more labour is required for harvesting and threshing.

† Traditionally seedlings for the transplanting of rice are produced in nurseries with roughly 1 ha of nursery for 10 ha of rice. With the 'mat' technique seedlings are produced on 'mats' which may be kept on the homestead. Land and labour requirements are much lower than with traditional nurseries.

There is thus reason to assume that labour inputs per crop will tend to decline. Cordova and Barker (1977) assume that the appropriate wet-rice production technique for modernized smallholder farming of transplanted rice would involve (i) the mat technique to the production of seedlings, (ii) tractor ploughing, (iii) buffalo harrowing, (iv) use of fertilizer, (v) pesticides and herbicides, (vi) hand harvesting, and (vii) machine threshing. This would require about 430 hours of work only. Most of the additional employment which is required to absorb a growing rural labour force, is likely to come from multiple cropping.

(b) Labour requirements tend to become more adaptable to changing price relations. Traditionally low labour inputs implied low yield levels. Modernized irrigation farming allows high yield levels with very different levels of labour input. Labour inputs can more easily be substituted by support energy, and they therefore vary widely between countries with different resource endowments.

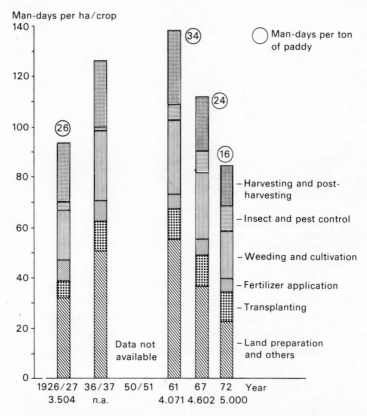

FIG. 7.13. The development of labour use per rice crop in Central Taiwan (from Lih-Yuh Shy Tsai and Herdt 1976.)

(c) Labour requirements per hectare increase with land-use intensification and tend to decline once the rural labour surplus has been absorbed by economic growth, as is indicated by the data from Taiwan (see Fig. 7.13). Land-use intensification brought increasing outputs per hectare, but declining labour productivity. From 1961 on output per hectare and labour productivity increased.

(d) Table 7.12 illustrates the employment potential of modern rice production without mechanization. However, there are signs that wet-rice farmers tend towards more mechanized inputs for a variety of reasons:

TABLE 7.12

Input–output relations with double- and triple-cropping of rain-fed wet rice in the Philippines (per hectare and year)

Location	Iloilo	Iloilo
Year	1976–7	1976–7
Technique	Rain-fed, modern Two seeded crops	Rain-fed, modern Three transplanted crops
Method	Case study[a]	Case study[a]
Technical data		
Labour input (man-hours)		
Land preparation	311	503
Nurseries	3	65
Planting or seeding	92	560
Fertilizing	13	22
Plant protection	22	45
Weeding and herbicide application	52	35
Harvesting and threshing	2000	3339
Other work	11	10
Total	2504	4579
Yield (t paddy)		
1. crop	4·88	5·06
2. crop	1·93	3·81
3. crop	—	2·93
all crops	6·81	11·80
Days of field occupation[b]	240	300
Yield (kg day^{-1})	28	39
Economic analysis ($)		
Gross return	1111	1854
Material inputs		
Seed	67	49
Fertilizer	106	68
Pesticides	47	36
Herbicides	18	23
Mechanization[c]	1	1
Income	872	1677
Income per man-hour	0·35	0·37

[a] Trial fields inspired by IRRI, but carried out by the farmers. [b] Estimate. [c] Hand labour only. *Source:* Magbanua, Roxas, Raymundo, and Zandstra (1977).

(i) Mechanization is essential for the extension of the irrigation area and for off-season irrigation. This is obvious in the case of pumps which replace human or animal labour or with innovations in the field of water distribution (plastic pipes) and water application (trickle irrigation). Mechanization is essential for the development of valley-bottom land.

(ii) The widespread use of pesticides and herbicides requires spraying equipment.

(iii) Tractorization is essential for the success of multi-cropping systems. Tractor ploughing has little direct impact on yields, but it is essential to reduce 'turnaround' periods. Under the traditional conditions of wet-rice farming at Iloilo (Philippines) it takes 21 days between the harvest of the first rice crop and the planting of the second, and 40 days for the non-rice crop to be planted. In modern multiple-cropping systems a field that is 'yellow' (ripe) in the morning should be 'green' (replanted) in the evening, having been harvested, cultivated, and replanted during the day.

7.6.6 *Introduction of intensive types of animal husbandry*

As the tractor replaces the draught animal, the tendency in irrigation holdings is increasingly to introduce intensive types of animal husbandry, dairying in particular. In any case, harvest residues and grass growth on the edges of tracks and canals usually provide fodder that cannot be used in any other way. There are the additional possibilities of multiple cropping with fodder crops, and the introduction of fodder plants into the rotation. The potential for producing beef and milk with irrigated tropical grasses is very great, and can be realized if price relations move in favour of animal products. It might perhaps be easier to meet the rapidly growing demand for animal protein in tropical countries by intensification of animal activities in irrigation farming (fish, ducks, pigs, and milk) rather than by improved grazing in hill country or dry areas (ranching).

8. Systems with perennial crops

8.1. Definition and genesis

8.1.1. *Definition*

A FORM of land use in the humid and semi-humid tropics, and as typical as shifting cultivation and wet rice, is the cultivation of perennial crops, primarily tree and shrub crops with growth cycles of several decades. Crops like sugar-cane, pineapple, and sisal, which occupy the land for several years but not for as long as shrubs or trees, are akin to arable crops from a farm-management point of view, and are best regarded as perennial field crops. Bananas are also included in this category, although the farm-management characteristics of bananas are distinct. Grasses on artificial pastures are also perennial plants, but the farm-management characteristics of livestock systems based on arti-ficial pastures are very different from the systems dealt with in this chapter.

Palms, trees, bushes, bananas, and various perennial field crops are grown in gardens and fields under almost all tropical farming systems. This chapter, however, is concerned with those farming systems in which perennial crops are the main activity, in terms of their contribution to the gross return. Perennial crops are grown in estates and smallholdings. The terms 'estate' and 'plantation' usually designate large units (Jones 1968). To avoid confusion, any land that is planted with perennial crops will be called a plantation and any holding that is largely or entirely planted with such crops will be called an estate, provided it is large and operated by a planter or manager command-ing a large number of paid workers.

8.1.2. *Genesis*

Plantations of perennial crops, in particular of tree and shrub crops, may come into existence in various ways. Large estates prefer to clear and plant, and, where land is ample, estates tend to shift their plantations either within the estate's boundary or by acquiring virgin land, i.e. to clear new land and to abandon run-down sites. Until recently, shifting of this kind was practised with coffee in southern Brazil, and it spread into Paraguay. Intra-farm shifting was customary for Central American banana estates and East African sisal estates that had sufficient land in reserve. But as land becomes scarce, plantations tend to become stationary within the estate boundary.

The genesis of smallholder plantations does not usually follow the 'clear and plant' sequence. Two main types of development in smallholder plantations are distinguishable. The development sequence:

1. exploitation of fruit trees or palms that grow naturally in forests or exploitation of infillings (irregularly planted seedlings) in the forest;

2. exploitation of systematically established plantations under the cover of some remaining forest trees;

3. exploitation of systematically established plantations with no forest trees left;

is characteristic of the way coffee cropping has developed in Ethiopia, West Africa, and Madagascar, and oil-palms and cacao in many areas of West Africa have evolved in a similar manner. Plantations are developed directly from forest vegetation mainly where there are extensive forest areas. However, more frequently the evolution of smallholder plantations has followed a different sequence, stemming from fallow farming. It usually includes four phases.

1. The planting of perennial crops around the house slows down the rotation cycle of shifting cultivators.

2. The area devoted to perennial crops is extended. Cleared plots with arable crops are interplanted with perennial plants.

3. In time, the perennial crops predominate in the mixed-cropping system, and young stands of tree crops are usually interplanted with arable crops.

4. As the perennial crop creates more and more shade, intercropping becomes less important.

When this occurs, the cycle arable cultivation–fallow is replaced by the sequence arable cropping–perennial crops, and at the same time shifting systems become stationary. Perennial-crop cultivation does not, however, always lead to stationary farming. The abandonment of old plantations and the establishment of new ones, i.e. the slow shifting of tree-crop production, can also be found. The slow movement of cacao holdings in forest areas of Ghana is a case in point. Generally, however, the introduction into small-holdings of commercial production with perennial crops is associated with stationary farming. Groeneveld (1968, p. 222) and Meillassoux (1964) illustrate how cultivation of shrub or tree crops is integrated into an arable system.

Fig. 8.1 shows the establishment of smallholder coconut-palm plantations. Manioc, maize, peas, or beans, and also coconut-palm seedlings are planted or sown on one plot simultaneously. The annual crops, maize, beans, and peas are harvested after 3–5 months. Their harvest also provides the final weeding for the manioc. The manioc harvest commences 13–15 months after planting and continues on the same plot up to the twenty-fourth month. The planting of manioc, peas, and beans is repeated on two to three consecutive occasions. Meanwhile, the young coconut-palms are growing. By weeding the arable crops and thereby preventing the bush from regenerating, the farmer also protects the palms. After 4–6 years, arable cultivation comes to an end. The plot is left to the coconut-palms, which at this time yield their first nuts. A fallow ensues after 3–4 manioc cycles and can extend over several decades. In the meantime (8–10 years after the first planting), the coconut-palms have grown so large that the regenerating bush can no longer endanger their development. The establishment of smallholder coconut plantations is thus a by-product of arable subsistence farming. The stands of coconut-palms are set up without expenditure for clearance and care, since the planting proceeds almost without cost and effort as a consequence of the manioc cultivation. On the other hand, palm development is slower, owing to interculture. It should be noted, however, that the area of coconut-palms is dependent upon the extent of manioc cultivation, since young palms are planted only where manioc is cultivated on freshly cleared bushland with soils suitable for coconut-palms. Planting of additional coconut-palms would give rise to high expenditure for clearing and tending the palms solely for their own sake.

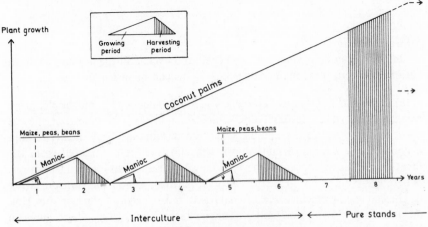

FIG. 8.1. Interculture of coconut-palms and various arable crops, Tanga, Tanzania (from Groeneveld 1968, p. 223).

Fig. 8.2 provides an example of this type of cropping from Ivory Coast. Whereas in the manioc–coconut system the permanent crop is planted right at the beginning of the cropping cycle, the smallholders of Ivory Coast plant their coffee later. After clearing the land, yams and manioc are grown, interplanted with short-term crops like groundnuts and beans. About a year after clearance, coffee is interplanted, and occupies the plot for 15–30 years, together with spontaneous bush.

8.2. Types of system with perennial crops and their geographical distribution

The various ways in which plantations are established give rise to different farming systems with perennial crops, which can be conveniently classified according to their cropping and exploitation systems.

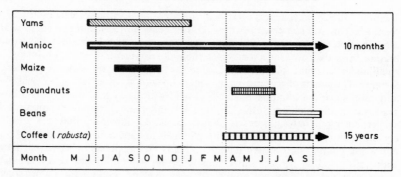

FIG. 8.2. Interplanting of coffee into arable plots of shifting cultivators with stationary housing, Ivory Coast (from Meillassoux 1964).

8.2.1. *Cropping systems*

On the basis of the length of the vegetation cycle and the amount of cultivation involved, three main cropping systems can be distinguished:

(a) *Perennial field crops.* This class includes sugar-cane, pineapple, sisal, bananas, and other similar types, which do not have the characteristics of shrubs or trees. With the notable exception of bananas, they mostly require a considerable degree of cultivation, and can thus be considered from the farm-management angle as something between arable farming and tree crops. Where these crops are grown for export, cultivation and the transportation of the harvested produce are often largely mechanized. Crops of this kind are therefore grown principally on large estates. The regions where they can be grown are determined not so much by climate and soil conditions as by the availability of markets, transport, labour, and technical knowledge.

(b) *Shrub crops.* Coffee and tea are typical shrub crops. The vegetation cycle lasts longer than that of the perennial field crops mentioned above. Shrub crops usually differ from tree crops in the high input of manual labour required for weeding, pruning, and harvesting. They usually yield an end-product that requires early processing and can thus be transported subsequently at relatively low costs. Tea-cropping and coffee-cropping areas are therefore determined principally by soil and climatic conditions.

(c) *Tree crops.* Finally, tree crops like cacao, rubber, coconut-palms, and oil-palms constitute another group. The difference between them and other perennial crops is their particularly long vegetation cycles, and the lower labour input required for their maintenance and weeding. Work in a tree plantation is usually less strenuous than with shrub crops, and takes place largely in the shade. Harvesting is the main task.

8.2.2. Exploitation systems

Whereas arable farming in the tropics has been from time immemorial a domain of the smallholder, until the twentieth century cultivation of perennial crops for export was carried out mainly in large estates. However, this distinction no longer applies. Plantations vary in size from very large estates of over several thousand hectares down to only a few shrubs and trees. Almost all types of perennial crop are grown in large estates as well as in smallholdings, although some are better suited to large-scale production than others, and are mainly found in estates. The distinction between cropping systems is, therefore, in many respects a distinction between large- and small-scale production.

(a) *Estates*. Although virtually all perennial crops are grown by small-holders, whose relative importance tends to increase steadily, there is still important large-scale production, of which the following are the most important examples:

1. Sugar-cane estates in almost all tropical regions.
2. Banana estates in Latin America and West Africa.
3. Sisal estates in East Africa and Madagascar.
4. Tea estates in India, Sri Lanka, and East Africa.
5. Rubber estates in Liberia, Malaysia, Sri Lanka, and Indonesia.
6. Oil-palm estates in West Africa, Malaysia, and Indonesia.
7. Coffee estates in El Salvador, Brazil, Kenya, Colombia, etc.
8. Coconut estates in Malaysia, Sri Lanka, Indonesia, India, and the Philippines.

In estate production, a distinction can be made between individual planters and firms, the former usually operating fairly small estates. In the normal way, the owner of the small plantation is also the estate manager, and he owns only one estate. This form of undertaking in the tropics is typified on the one hand by the shortage of working capital, exposure to risks, and often obsolete technology, and on the other by adaptability and low overheads. Firms, however, usually each own several estates. The tendency to cultivate very large areas springs from the necessity to use industrial processing and transport to their utmost capacity, in order to recoup the high costs of overheads through economies of scale. The motive underlying this kind of large-scale production is frequently to obtain a regular supply of goods of constant quality, whether for delivery to a factory (tea and sisal) or for delivery to an established market (export bananas). Almost always, production of this kind is associated with well-developed reliable marketing outlets. The producing firms often regard their farming activities as a necessary evil, undertaken only because other farmers are not in a position to provide the raw materials in the desired way. On estates, emphasis is often laid more on industrial processing and the organization of transport than on producing the raw material.

Large-scale cultivation of perennial crops is found usually in the form of pure stands, and there is an obvious trend towards monoculture. This is for a number of reasons.

1. The producer is interested only in obtaining large quantities of a single raw material.

2. Where monoculture is carried out, unskilled labour can be more easily supervised and brought to a relatively high degree of productivity.

3. A relatively large part of the cost consists of paying the wages that managers can command, and diversified production is not worthwhile in many cases because additional skilled personnel would be required.

4. The production conditions favour monoculture. With a number of perennial crops, labour is required so regularly that it is not necessary to diversify production in order to find productive employment all the year round for the labour force. Monoculture does not inevitably lead to soil erosion. Indeed, with several perennial crops monoculture can even be termed soil-conserving compared with arable farming.

5. Whereas smallholders usually have to operate where they are settled, and adapt to the natural habitat, and are thus compelled to diversify production, the firm can select the most favourable economic and natural location, which is chiefly on land suitable for monoculture.

In addition to estates with monoculture of perennial crops, diversified large-scale estates occur that are engaged in other activities to supplement the main crop. The following forms can be distinguished.

1. The predominant tree crop may be supplemented by one or several other perennial crops, which can be integrated either by mixed cropping— e.g. cacao under coconut-palms—or by part of the land carrying another perennial crop.

2. The perennial crop may be supplemented by arable farming. Managers of estates choose only rarely to grow arable crops between the rows of tree or shrub crops. Instead, they prefer to use separate plots, as for example in the coffee holdings in Brazil, which grow maize, rice, cotton, groundnuts, and onions (see § 8.5.1 (a)).

3. Perennial crops may be supplemented by stock-keeping, particularly with cattle, e.g. dairying in the coffee estates of Brazil and dairying among the coconut-palms in Sri Lanka.

The motives behind diversification in estates are different in kind from those in smallholdings. Subsistence needs do not play a great part, since in general the workers are allocated some land to cover their food requirements. Hardly any draught animals are kept, which means that there is no need to grow fodder. Similarly, the idea behind diversification is rarely to employ the available labour force more regularly and to its capacity, since many of the estates have seasonal labour at their disposal in the surrounding countryside. Other factors are usually the main reasons for diversification such as (i) the use of land which is not claimed by the major permanent crop, (ii) the use of by-products for livestock operations, and (iii) the reduction of risk.

(b) *Smallholdings*. A large and growing proportion of perennial-crop cultivation is in the hands of smallholders. Of particular importance for the world's economy are the semi-commercialized holdings, e.g.:

1. The cacao holdings in Ghana, Nigeria, and Trinidad.
2. The rubber holdings in Malaysia, Nigeria, and Sri Lanka.
3. The coffee holdings in Colombia, Ivory Coast, Kenya, and Madagascar.
4. The oil-palm holdings in West Africa.
5. The coconut-palm holdings in the Philippines, Sri Lanka, Venezuela, and Samoa.
6. Cashew holdings in India, Tanzania, and Mozambique.
7. Vanilla holdings in Madagascar.
8. Sisal holdings in north-east Brazil.
9. Banana holdings in Central America, South America, and Asia.

The primary cash crop is usually supplemented by yams, manioc, rice, maize, beans or other crops for household use. A strikingly large proportion of the cash production is in the hands of a numerically small but economically important group of smallholders, who employ seasonal labour, and who might be termed small entrepreneurs.

The development of these holdings from pure subsistence farming brings about a situation where the household's food requirements compete with the desire to earn money from cash crops. The competition for land takes two main forms.

1. In some cases, perennial crops are grown on separate plots as supplementary activities. The husbandry techniques are usually similar to those on large estates, and operations are often carried out by seasonal, paid labour. However, as cash cropping increases, there is less land available for meeting food requirements. At the same time, arable farming is undergoing a change, which as a rule is a process of degradation, because the fallow is shortened and fertility declines. Farmers are obliged to reduce fallows and pastures and to turn from maize or millets to root crops, in particular manioc (Hurault 1970, p. 225). Manioc is eaten with less relish but it does supply a greater amount of energy per hectare and per hour of work. Not infrequently the stage is reached where 'dualistic' peasant holdings evolve, so that within a small-scale holding the interrelation between the two branches of production is as weak as that between the large estate and the traditional subsistence holdings in a dualistic society. Arable farming is carried on by the family in the traditional way, although with reduced fallows and declining yields. The plantations of the same holding are regarded as a special undertaking, and new husbandry and economic standards apply (von Blanckenburg 1964).

2. In other cases, cash crops and subsistence crops are grown in mixed stands. The degree of competition is reflected by the composition of the mixed stand. Thus, when coffee and plantains are grown in mixed stands, the proportion of the subsistence crop (plantains) tends to be greater the more people per hectare there are to be fed, and vice versa. Compared with the dualistic smallholdings, mixed cropping has the advantage that the progressive degradation of the arable branch is avoided. On the other hand, mixed cropping has proved to be an obstacle to such technical advances as use of mineral fertilizers and plant protection.

Besides the semi-commercialized holdings that have developed from subsistence farming, smallholdings are found that have come into being in consequence of large-scale undertakings or in relation to non-farm work. This applies primarily to the great number of part-time farmers who frequently cultivate perennial crops. People work for a few years in towns, on large-scale holdings and in mines as seasonal workers, or on a regular wage-earning basis. Their own holding provides a base and a measure of support, and a home in old age, because, through the investment of savings in shrub or tree crops, the holdings can offer a relatively high degree of security.

Moreover, smallholder plantation farming has sometimes been evolved on land where it is no longer profitable to operate on a large scale. The smallholdings command the labour of their families, for whom there is usually no alternative but to work their own land. It pays them to buy land from run-down estates and to establish small plantations. In some areas this procedure has led to the spread of smallholder farming on land that has been purchased from run-down estates, in which the new producers take advantage of the marketing channels created by large-scale production. Their organization of land use is very similar to that in estates.

Finally, a trend of increasing importance is the establishment of plantations by middle-class entrepreneurs who aim at a reliable return on capital and a secure farm as a base for their old age.

8.3. General characteristics of farming systems with perennial crops

Although perennial crops are a heterogeneous group, grown under various circumstances by various types of farmer, some common characteristics exist, which are mainly relevant to shrub and tree crops and which to a lesser degree apply also to several perennial field crops (see Table 8.1). As a first step, the farm-management characteristics of perennial cropping can be compared with those of arable farming, and then some general principles, applied by farmers to fit perennial crops into their farming systems, can be outlined. The various aspects of the maintenance of soil fertility, of animal husbandry, and of the labour economy are discussed within the case studies.

TABLE 8.1

Some features of the principal perennial crops

Crop	Years to first crop	Years to maturity	Years of production	Harvested portion	Urgency for processing
Perennial field crops					
Sugar-cane	1–1·5	1–1·5	4–6	Stem	High
Bananas (export)	1–2	3	5–50	Fruit	High
Pineapples	1·5	2	3–5	Fruit	Fair
Sisal	3	5	8	Leaf	Fair
Shrub crops					
Coffee	3	5–6	12–50	Fruit	High
Tea	3	6	50	Leaf	High
Tree crops					
Oil-palm	3–4	7–9	35	Fruit	High
Rubber	4–7	8–11	35	Sap	Fair
Cacao	8–11	15–20	80–100	Seed	Low
Coconuts	4–6	8–15	60	Fruit	Low
Cloves	8–9	20	100	Flower buds	Fair

Source: MacArthur (1969).

8.3.1. *A comparison with arable farming*

The culture of perennial crops involves a number of farm-management characteristics that in many ways can be regarded as advantages compared with those of arable farming.

1. Perennial crops give rise to stationary living and to investment in permanent improvements like houses, tracks, irrigation systems, etc. Where perennial crops are grown, the individual ownership of land tends to become established.†

2. Most perennial crops are characterized by a comparatively high productivity per hectare. This encourages the use of fertilizer and thus the conservation of soil fertility. On account of their high productivity per hectare, some perennial crops actually economize on land use. Thus in Bukoba, Tanzania, 0·3 ha of plantains mixed with beans are sufficient to cover

† In tropical Africa, for instance, by planting trees the planter establishes an almost permanent right to the land. The traditional land-tenure situation consequently provides powerful incentives for an individual to plant many trees and to claim larger pieces of land. The grower's main interest is in the extent of his area under trees and not with well-cared-for plantations which yield much per hectare, and which are recommended by the extension service. The individual's gain is not so much with the yield of the trees, but with his claim to the land which is getting scarce. The rapid expansion of the tree-crop area in many parts of Africa and the poor husbandry which prevails, may be explained in this way.

the basic food requirements of one family, whereas in arable farming the annual cultivation of about 1·2 ha is necessary to meet the same need. Most perennial crops produce by-products that are valuable to the household, can be fed to livestock, or can be sold in local markets: fuel for the kitchen, timber for construction purposes, branches for covering roofs, tops and by-products for fodder, etc.

3. There is a good chance of conserving fertility when perennial crops are grown. The frequency and intensity of cultivation is disproportionately less than in arable farming. Perennial crops shade the soil and permit or require a permanent cover of grass or leguminous vegetation. Some tree and shrub crops influence the soil in the same way as the forest, and moreover permanent planting encourages the construction of terraces, the control of water courses, and other permanent land improvements. The destruction of fertility in the cultivation of some perennial crops, like coffee in southern Brazil or tea in Sri Lanka is very serious, but the fact remains that the damage is usually less than in arable farming under similar conditions.

4. Various perennial crops allow the farm-management advantages of monoculture without a reduction in soil fertility. Properly husbanded, pure stands of rubber, oil-palms, or tea may well be considered as soil-conserving types of land use.

5. For some perennial crops, land can be used that is either out of the question for arable farming, or would be usable only if high costs could be borne. Examples are steep slopes (bananas, tea, rubber), rocky terrain (tea, rubber), land with unreliable rainfall (sisal), and slightly saline land (date-palms).

6. Important perennial crops like coffee, tea, and rubber require a great deal of manual labour, but repay it with a relatively high income. They are therefore suited to the production conditions of developing countries, which are characterized by the availability of low-cost labour. Smallholder systems with tree crops usually show a higher employment content than arable systems. There are more working days per worker and more hours per working day (Cleave 1974, p. 32). Particularly conspicuous is the high additional labour input required by high-yielding clones.

7. The labour required for many perennial crops can be spread fairly evenly over the year. This facilitates the expansion of monoculture. Moreover, most operations required by perennial crops tend to be disproportionately easier and more agreeable than in arable farming, particularly hoe farming.

8. The fluctuations in the yield of crops like sisal, rubber, and coconut-palms are small—at least smaller than in arable farming under similar conditions. Thus, the sisal hedges in north-east Brazil and in the smallholdings on Lake Victoria are cut chiefly when arable crops suffer harvest failure; in other words they can counterbalance such risks. Similarly, fodder bushes and trees are a source of fodder when required. The greater certainty of yields

from perennial crops is important in small-scale farming, because the yields from arable rain-fed farming are often unreliable.†

9. The products are chiefly goods that can be transported and stored, and have a high value per unit of weight, or they acquire these qualities through industrial processing. Perennial export crops make it possible to develop areas for commercial production where only subsistence farming could otherwise be practised.

10. Perennial crops present considerable scope for intensification according to variety, raising and planting method, density, fertilization, pruning, harvesting, and processing. With some crops, irrigation is economic and increases the yield appreciably. Several perennials are relatively efficient converters of solar energy into products useful to man, and they are efficient in the use of chemical support energy, such as fertilizers, pesticides, and herbicides.

11. In the case of most perennial crops, cultivation can begin with only a few plants. The enterprising individual can progress without having to overcome any particular technological threshold, nor is he dependent on the co-operation of neighbours to the same extent as in irrigation farming, for instance.

12. Wealth in the form of plantations, and therefore security in old age, can be accumulated by planting shrubs and trees. This is particularly important where there is no private ownership of arable land, and where investment in cattle as security is not possible.

13. Perennial crops give rise to social differentiation—perhaps more than any other cash crop. A family's income is based on the number of trees and shrubs it has and their yield. This forms an outlet for the common human desire to excel, and at the same time provides an additional incentive towards economic activity.

On the credit side we find that perennial crops have many agro-economic advantages, which naturally do not obtain in every case, but which characterize the group as a whole. But on the debit side there are certain important farm-management requirements that have to be met.

1. Creation of plantations usually requires a high initial investment per hectare. Yields do not reach their capacity until some years after planting (see Table 8.2). To cultivate perennial crops it is necessary to finance the establishment of the plantation, maintain plants during the years when there are no

† The ability of perennial crops to produce a reliable yield and to store the starch is indicated by biennial manioc and is even more pronounced with ensete bananas (*ensete edulis*) and the sago palm. The ensete banana, grown in Ethiopia, stores the starch in the stem and may be harvested after 3 or 4 years when it yields about 17 kg of fermented and dried produce per stem. The Gurage, a tribe in Ethiopia, who mainly live on banana-stem starch, never experienced famine (Shack 1969, Westphal 1975). The sago palm is grown on rice field borders in Malaysia and Indonesia. According to Flach (1973) wild palms produce 7–11 t of water-free starch per hectare, and much more can be expected with improved husbandry. This perennial crop fits well into the ecosystem, allows interculture, and utilizes swampy land, and harvesting is not confined to a definite season.

TABLE 8.2

Some examples of input–output relations of perennial crops[a] (per ha)

Age of plantation (years)	0[b]	1	2	3	4	5	6	7	8	9	10	15	20	25	30	50
Vanilla (Madagascar, 1975, modernized technique in commercial farming)																
Labour (man-days)	198	150	150	215	274	274	274	262	230	—	—	—	—	—	—	—
Wages ($0·7 per day)	139	105	105	150	192	192	192	183	161	—	—	—	—	—	—	—
Material inputs ($)	60	1	1	1	1	1	1	1	1	—	—	—	—	—	—	—
Total inputs ($)	199	106	106	151	193	193	193	184	162	—	—	—	—	—	—	—
Yield (t, green vanilla)	0	0	0	0·6	0·9	0·9	0·9	0·7	0·5	—	—	—	—	—	—	—
Gross return ($)	0	0	0	576	864	864	864	672	480	—	—	—	—	—	—	—
Net return before interest and overheads ($)	−199	−106	−106	+425	+671	+671	+671	+488	+318	—	—	—	—	—	—	—
Arabica coffee (Colombia, 1978, small estate)																
Labour (man-days)	174	108	104	154	187	179	166	141	129	129	129	—	—	—	—	—
Wages ($2·50 per day)	435	270	260	385	467	447	415	352	322	322	322	—	—	—	—	—
Material inputs ($)	717	314	230	230	230	230	230	230	230	230	230	—	—	—	—	—
Total inputs ($)	1152	584	490	615	697	677	645	582	552	552	552	—	—	—	—	—
Yield (t, dried beans)	0	0	0·5	1·5	2·0	1·9	1·7	1·3	1·1	1·1	1·1	—	—	—	—	—
Gross return ($)	0	0	835	2505	3340	3173	2839	2171	1837	1837	1837	—	—	—	—	—
Net return before interest and overheads ($)	−1152	−584	+345	+1890	+2643	+2496	+2194	+1589	+1285	+1285	+1285	—	—	—	—	—
Cacao (Ivory Coast, 1973, intensive technique in smallholder farming)																
Labour (man-days)	212	72	73	65	68	85	115	115	125	135	155	155	155	155	155	—
Wages ($1·24 per day)	263	90	91	81	84	105	143	143	155	167	192	192	192	192	192	—
Material inputs ($)	19	34	48	68	88	20	14	14	20	14	16	16	16	16	16	—
Total inputs ($)	282	124	139	149	172	125	157	157	175	181	208	208	208	208	208	—
Yield (t, dried beans)	0	0	0	0	0·15	0·3	0·5	0·6	0·7	0·8	1·0	1·0	1·0	1·0	1·0	—
Gross return ($)	0	0	0	0	43	91	144	173	202	230	288	288	288	288	288	—
Net return before interest and overheads ($)	−282	−124	−139	−149	−129	−34	−13	+16	+27	+49	+80	+80	+80	+80	+80	—
Oil-palms (Sumatra, 1972, smallholder farming)																
Labour (man-days)	185	100	39	39	48	56	59	58	68	68	68	63	62	62	n.a.	n.a.
Wages ($ 0·90 per day)	166	90	35	35	43	50	53	52	61	61	61	57	56	56	n.a.	n.a.
Material inputs ($)	159	35	29	12	12	12	12	12	12	12	12	12	12	12	n.a.	n.a.
Total inputs ($)	325	125	64	47	55	62	65	64	73	73	73	69	68	68	n.a.	n.a.
Yield (t, fruit bunches)	0	0	0	4	7	11	12	14	18	18	18	14	14	14	n.a.	n.a.
Gross return ($)	0	0	0	64	112	176	192	224	287	287	287	226	224	224	n.a.	n.a.
Net return before interest and overheads ($)	−325	−125	−64	+17	+57	+114	+127	+160	+214	+214	+214	+157	+156	+156	n.a.	n.a.

Rubber (Sumatra, 1975, intensive technique in smallholder settlement)

	1	2	3	4	5	6	7	8	9	10	11	12	13	14	15	16
Labour (man-days)	204	49	47	49	29	25	118	118	114	114	114	114	104	104	—	—
Wages ($1·29 per day)	263	63	61	63	37	32	152	152	147	147	147	147	134	134	—	—
Material inputs ($)	199	13	90	125	131	179	116	116	111	111	111	111	28	28	—	—
Total inputs ($)	462	76	151	188	168	211	268	268	258	258	258	258	162	162	—	—
Yield (t, dry rubber)	0	0	0	0	0	0·4	1·0	1·2	1·4	1·5	1·6	1·7	1·7	1·0	—	—
Gross return ($)	0	0	0	0	0	157	392	469	549	588	626	783	639	392	—	—
Net return before interest and overheads ($)	−462	−76	−151	−188	−168	−54	+124	+201	+291	+330	+368	+525	+477	+230	—	—

Coconuts (Ivory Coast, 1969, intensive technique in smallholder farming)

	1	2	3	4	5	6	7	8	9	10	11	12	13	14	15
Labour (man-days)	38	24	17	12	7	5	5	8	17	23	27	27	27	27	n.a.
Wages ($1·24 per day)	47	30	21	15	9	6	6	10	21	29	33	33	33	33	n.a.
Material inputs ($)	78	10	10	14	14	14	14	24	24	24	24	24	24	24	n.a.
Total input ($)	125	40	31	29	23	20	20	34	45	53	57	57	57	57	n.a.
Yield (t, copra)	0	0	0	0	0	0	0	0·2	1·2	1·8	2·1	2·2	2·2	2·2	n.a.
Gross return ($)	0	0	0	0	0	0	0	19	114	176	200	209	209	209	n.a.
Net return before interest and overheads ($)	−125	−40	−31	−29	−23	−20	−20	−15	+69	+123	+143	+152	+152	+152	n.a.

Coconuts interplanted with maize and cowpeas (Kenya, 1977)

	1	2	3	4	5	6	7	8	9	10	11	12	13	14	15
Labour (man-days)															
Coconuts	5	2	3	4	4	5	6	9	10	10	11	11	12	12	13
Maize	40	38	36	34	34	31	29	25	22	22	22	22	22	22	22
Cowpeas	23	22	21	17	17	18	17	14	13	13	13	13	13	13	13
Total	68	62	60	55	54	52	50	48	45	45	46	46	47	47	48
Material inputs ($)															
Coconuts	3	—	—	—	—	3	—	—	—	—	—	—	—	—	—
Maize	44	42	40	37	35	32	30	27	25	25	25	25	25	25	25
Cowpeas	20	19	17	16	15	14	13	13	13	13	13	13	13	13	13
Total	67	61	57	53	50	46	43	40	38	38	38	38	38	38	38
Gross return ($)															
Coconuts	—	—	—	—	—	—	37	50	62	75	87	100	112	122	122
Maize	115	108	101	94	88	85	74	67	67	67	67	67	67	67	67
Cowpeas	34	32	30	28	26	24	22	20	20	20	20	20	20	20	20
Total	149	140	131	122	114	133	133	137	149	162	174	187	199	209	209
Income ($)	90	88	83	78	72	68	90	93	99	111	124	136	149	161	171
Income per man-day ($)	1·32	1·42	1·38	1·42	1·33	1·31	1·80	1·94	2·20	2·47	2·70	2·96	3·17	3·42	3·56
Net return before interest and overheads ($1.25 per day)	+5	+10	+8	+9	+5	+3	+27	+33	+43	+55	+66	+78	+90	+102	+111

* These are planning data for smallholder development. Costs exclude overheads, land rent, interest, taxes, etc. A man-day in Ivory Coast and Kenya has 6 hours, in Sumatra 7 hours, and in Colombia 8 hours. [b] Planting material or nursery costs are registered as material inputs in year 1. Material inputs do not include interest, extension fees, etc. *Sources:* Vanilla: Refeno (1975); Arabica coffee: Varela (1977); cacao: Satmaci (1973); oil-palms: (Personal enquiries); rubber: Ptsipef (1975); coconuts: Sodepalm (1973); coconuts interplanted: Thorwart (1977, personal communication).

yields, and bear the burden of accruing interests or—which is more important in smallholdings—forgo for some years the return in food or cash that the plot would have yielded under arable farming.

On the other hand, a plantation is a form of investment, which a family can build up by supplementing small amounts of cash with much unpaid family labour. Machine inputs are usually not necessary for plantation establishment, and labour requirements can substantially be reduced by arboricides and herbicides. Plantation establishment primarily requires labour and chemicals.

2. For many of these crops, it is very important that processing should take place within a relatively short period after harvesting. In the absence of early treatment, which is designed to effect some change in the composition of the produce, chemical or biological deterioration would set in and either reduce the quantity of valuable constituents extractable from the harvested material—oil-palm and sugar-cane especially—or else spoil the quality of the final product, as with bananas, coffee, and tea (see Table 8.1).

3. Where perennial crops are the raw material for a local industry, in addition to the initial costs of the plantations it is necessary to invest in a local processing plant, to invest in transport for bringing the product to the factory, and to arrange delivery very carefully, so that the processing plant can work to capacity.

4. Consequently, holdings with perennial export crops, like irrigation holdings, operate with high fixed costs, but, unlike arable irrigation holdings, they are committed to one type of production for a long time.

5. Plantation husbandry demands different kinds of work and skills from those of arable farming. Cultivation is of less importance, and with several crops not even necessary. Weeding is labour-demanding only in the early phases of plantation development. Instead, much work is required for raising plants in nurseries, planting, pruning in the case of shrubs, spraying, mulching, harvesting, and so on. The labour input is particularly high when crops are being established and with mature crops. The intervening period usually requires little attention. In comparisons of arable production with shrub-crop and tree-crop production, it is also important to realize that the influence of husbandry measures may extend over years and decades. Poor plant development, unskilled pruning, and delay in pest control may damage the crops for years to come, while in arable cropping, last year's failure may be avoided next year (MacArthur 1969).

6. Another disadvantage of perennial crops is that the work is difficult to mechanize. Where cultivation, weed control, and transport in bulk strain the labour economy, as in sugar-cane and sisal estates, mechanization is more likely to follow. With tree and shrub crops, where the main emphasis is on tending the crops and harvesting, operations cannot be mechanized at all, or only with rather poor rates of substitution.

In assessing the various advantages and requirements of perennial crops, it is clear that, provided market outlets are available, they are relatively well suited to tropical agriculture, and several of the shrub and tree crops are particularly suited to smallholder production.

8.3.2. *Some principles of crop management*

(a) *The spatial and vertical organization of cropping.* Large estates are usually uniformly organized, but in smallholdings, planting patterns near the house or village are mostly different from those some distance away, so that more or less pronounced spatial patterns are found in the arrangement of the various perennial crops. In this respect they are not basically different from arable crops, where distinct spatial arrangements of cropping in smallholder farming are common.

A specific feature of plantains, some shrub crops, and most tree crops is that they lend themselves to 'multi-storey' cropping. The planting of several crops which differ in height, root development, and light requirements allows a more efficient use of solar energy, soil nutrients, and water (Nelliat, Bavappa, and Nair 1974, p. 263). In estates we occasionally find 'two-storey cropping', where two perennial crops, e.g. rubber and cacao, are interplanted. In smallholdings we find a great variety of 'multi-storey cropping', including not only perennial crops, but also various annual crops. A typical case in point is the land use on some Polynesian islands, as depicted by Barrau (1961). Fig. 8.3 (a), (b), and (c) shows distinct spatial arrangements in land use combined with several two-storey and three-storey cropping patterns.

(b) *Monoculture, interculture, and combination with grazing.* The spatial and vertical organization of cropping leads naturally to various crop combinations within a holding and within a field. Large estates usually prefer to cultivate pure stands, to make optimum use of the husbandry techniques specifically required by one plant. Where production is concerned with only one crop, pure stands result in monoculture, of which two types have to be distinguished. From the standpoint of farm management, there is monoculture where only one crop is grown in the holding. This is usually the case in oil-palm or banana estates. From the standpoint of agronomy, there is monoculture where one field is planted with the same crop for each succeeding vegetation cycle, which is usually the case with sugar-cane or sisal. Both types of monoculture are identical in some but certainly not in all cases.

While estates tend to prefer pure stands and monoculture, smallholdings usually prefer interculture, because it helps them to obtain higher short-term returns, higher returns per hectare (lower returns per unit of labour being accepted), and the benefits of a more diversified production.

However, some types of interculture are also prevalent in estates. The following types of interculture with perennial crops are important.

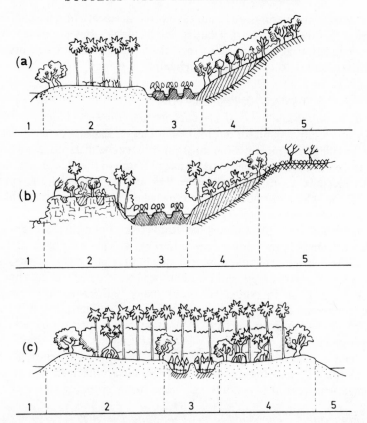

FIG. 8.3. Various types of land use on Polynesian islands. (a) High Island, Rarotonga:
(1) ocean; (2) coral calcimorphic soils, growing coconuts, sweet potatoes, tomatoes;
(3) hydromorphic soils, taro gardens; (4) colluvium clay loam, citrus fruits and gardens;
(5) lateritic soils, scrub and secondary forest. (b) High island, Atiu: (1) ocean; (2) uplifted
coral plateau with clay loam in pockets, a few gardens and coconuts; (3) hydromorphic
soils, taro gardens; (4) colluvium clay loam, gardens; (5) lateritic soils, ferns and *Casuarina*.
(c) Atoll, Tarawa: (1) ocean; (2) coral calcimorphic soils, coconut and *Pandanus*; (3)
Cyrtosperma pits; (4) coral calcimorphic soils, coconuts and *Pandanus*; (5) lagoon. From
Barrau (1961, p. 26).

1. Perennial crops and cover crops. In large estates, the planting of
leguminous creepers (e.g. *Centrosema pubescens, Pueraria javanica*) as cover
crops is a useful practice for protecting the soil in the early stages of plant
development (oil-palm, cacao, rubber) or to help in weed control (sisal).
2. Tree crops and shade plants. Some tree crops need shade in their early
stages. The shade plants frequently produce a yield of their own, for
instance bananas shading cacao.
3. Perennial crops of various sizes. A widespread practice, in estates as well
as in smallholdings, is the combination of various perennial crops of

differing heights (see § 8.4.3 (b)). This multi-storey physiognomy is closer to the natural climax vegetation than are pure stands. In some cases, interculture of this kind is limited to the early stages of plantation development.

4. Perennial crops and annual cropping. In large estates, perennial and annual crops are not usually combined, whereas in smallholdings their combination is common. With some crops, like rubber, oil-palms, and coffee, interculture is limited to the early stages of plant development, and arable crops are grown only until the main crop branches out and forms too dense a shade for annual crops. With other crops, like coconut-palms, arable farming continues to be practised under mature stands.

5. Perennial crops and grazing. In addition we find various combinations of perennial crops with grazing. In some cases (sisal, oil-palms), grazing is limited to the early stages of plant development, when ample grass grows between the rows. In other cases (as frequently with coconut-palms) cattle may graze under mature stands.

The importance of interculture varies with the type of crop, the farm size and price relations. Interculture is generally of importance in smallholdings where the farmer, by practising arable farming in the early stages of shrub- or tree-crop development, may significantly reduce the costs of plantation establishment. This may be more important economically than the reduction in future returns which is likely to result from poorer development of the dominant crop.

(c) *Vegetation and production cycles.* Another variable peculiar to perennial crops is the length of the production cycle and the distribution of input requirements and returns over several years. MacArthur (1969) summarizes this as follows:

During the early stages there is a slow but accelerating build-up towards maturity, when maximum yields are obtainable. A high level of production is thereafter obtained for several years, but ... age ... begins to affect the levels of production, and yields fall off. With tree crops a fairly distinct type of production pattern can be discerned, but with crops whose natural form is that of a bush or shrub, productivity can be renewed through pruning back almost all of the growth above the ground and permitting new main stems to grow up from the capitated main stem.

This generalized scheme is subject to many exceptions. Sisal follows a peculiar pattern wherein leaves grow and can be harvested for a given period until the plant puts up a stem with a flowering head, after which no new leaves are produced and the plant dies. Pineapple and sugar-cane grow in a series of 'ratoons' or successive waves of new shoots which grow up from the root stock, and the yield level falls steadily after the first harvest.

... this ... usually implies that there is a period during the early stages of growth (which can last for several years) when annual costs will exceed the value of harvested produce, but eventually a stage will be reached when new costs each year will be less than the product value. This phase may persist for a very long time, but there will come a stage . . . when production will fall to such a low level that its value is insufficient to cover annual production costs. At this stage, the crop's useful life will be over. With sisal, there is a firm limit fixed by the botanical behaviour of the plant, whilst with other crops like pineapples . . . and sugar-cane the point of unprofitable diminishing output is encountered after only a few years. Theoretically, it may be possible for naturally or artificially self-regenerating crops like banana, coffee and tea to remain in profitable production almost indefinitely, but most plants of these species, if they do not . . . fall victim to a disease . . . are eventually superseded by new varieties of the crop whose yields exceed those of established plants by so much that they merit uprooting and replacement, despite the interruption in production that this means.

Consequently, in the case of shrub and tree crops we can usefully distinguish the following stages in plantation development.

1. The period between clearing and planting and the beginning of the crop's productive life (in the calculations in Table 8.2 for cacao, rubber, and oil-palms, this period is 4–7 years). In this period farmers usually seek to reduce costs by interculture. The example with interculture under coconut palms at the Kenya coast indicates the essential points: (i) costs of plantation establishment are reduced often to such a degree that the balance is positive from year 1 onwards. (ii) There is a decline in net returns when the plantation crop is increasingly shading the others but not yet fully producing. (iii) Net returns reach a higher level when interculture is practised under a fully producing stand of palms. (iv) Incomes per man-day are relatively low with interculture and reach a much higher level with full production of the palms.

2. The time between planting and the year after which the annual costs are covered by the annual yield (in the examples 3–8 years).

3. The period between planting and the year after which the capacity yield is produced (10 years with coconut).

4. The period of capacity production, which begins when the first full yield is produced. In the following years, a regular yield, sometimes rising or falling slightly, is produced. This period comes to an end when the yield tends to decline more rapidly.

5. The length of land occupation. According to the level of running costs, the availability of land for planting, the cost of replanting, and the returns of new plantations in relation to old ones, it is sometimes advisable to continue using the land beyond the period of full production, in spite of rather low yields. The period of land use thus extends from the first to the last harvest.

6. The growth cycle, which is the time between creation of the plantation and clearance of it or its abandonment to run wild.

Plantation farmers may influence the economic length of the production cycle of their various perennial crops. Extension of the time in nurseries, special husbandry techniques, proper weeding, and manuring may reduce the initial period before a crop is obtained, while interculture may delay development. Pruning and harvesting techniques may hasten or delay returns. Husbandry practices in general influence the economic life of perennial crops. With sugar-cane the grower has to decide the number of ratoons. With shrub and tree crops in general the grower has to choose between production cycles of various lengths, and has to decide when to cease cropping and to plant afresh. Short production cycles are frequently, but not always, an indication of intensive farming.

(d) *Rotations*. When a stand of perennial crops is created, the way in which the plot is used is fixed for a long time. An annual change in the use of the plot, such as we find in arable farming with crop rotations, is only possible to the extent that interculture can be practised. In perennial cropping long-term rotations do occur, and the following are the basic types.

1. Perennial crops may be followed by a lengthy forest or bush fallow. Thus old cacao plantations in West Africa revert to bush vegetation, which is then cleared after one or two decades. The rotation runs: fallow–food-crop farming–cacao–fallow (von Blanckenburg 1964, p. 38). Planned reforestation after coffee cultivation is practised occasionally in Latin America (Sick 1969, p. 164).

2. Perennial crops may be followed by grassland. Thus in southern Brazil old coffee plantations are chopped down and the ground becomes covered with grass and serves as rough pasture. In tropical high-rainfall areas, an even more negative sequence can be observed: exhausted coffee plots gradually turn into fields of useless *Imperata* grass.

3. Perennial crops may be followed by arable farming. In the coastal lowlands of Ecuador, for example, banana land which becomes available through the change from relatively extensive production of Grande Michel to intensive Cavendish banana production changes to continuous cultivation of soya and upland rice (see § 6.4.3).

4. One perennial crop may be followed by another. The change may occur by uprooting and replanting. On some banana plantations in Central America, for instance, cacao or oil-palms are planted after bananas. It may also occur by underplanting the outgoing crop with a new one. Thus, in some estates in Sri Lanka, old rubber is underplanted with cacao. The rubber slowly gives way to a pure stand of cacao. With declining cacao yields the area is cleared and planted with rubber again. But more often than not estates practise monoculture and rubber is planted after rubber or oil-palms after oil-palms. Plantations showing the third consecutive rubber or oil-palm stand are no rarity and yields tend to increase each time, because of technical innovations.

5. A perennial crop, or a mixture of perennial crops, may be renewed by

continuous planting, as in the coffee plantations of El Salvador, or the coconut plantations in Sri Lanka.

6. A perennial crop may continuously renew itself through suckers or self-sown seedlings (as is the case with some smallholder sugar-cane in Sumatra), with permanent banana groves anywhere, or, often, with smallholder rubber in south-east Asia.

The sequence is determined partly by the characteristics of the particular crop and partly by the availability of land and price relations. Where there is plenty of land, one or several growth cycles of perennial crops is followed by many years of forest or bush fallow. As land becomes more scarce, there is a tendency to replace the fallow by use of the land for pasture or arable crops, or perennial crops may follow each other. Where it is technically possible, continuous replanting (infilling) is a sign that land is particularly scarce.

(e) *Spatial distribution of crops*. Tree-crop production usually leads to a pronounced spatial pattern of land use. In estates with monoculture and low population pressure we normally find the crop on the upland and the valleys left as wasteland. With increasing population density upland tree crops are combined with valley-bottom wet rice. In smallholder farming the forest-type land produces cash crops, while the more open savanna-type land supplies food (Sautter and Mondjannagni 1978).

8.4. Estates and smallholdings with perennial field crops

The basic farm-management characteristics of perennial-crop production are best depicted by reference to existing types of farming in both the estate and the smallholder type of economy.

8.4.1. *Sugar-cane holdings*

Sugar-cane production is carried out partly in diversified holdings where cane is one of the crops in the rotation (see § 7.4.2), but it is mainly grown in specialized sugar-cane plantations. The latter group comprises three distinct types: first, millers who combine sugar-cane production and a factory for processing; secondly, planters with plantations of large and medium size who produce mainly or exclusively cane for sale to factories; thirdly, smallholder producers who grow sugar-cane in addition to food crops. Large estates are often surrounded by a belt of small-scale producers, who sell sugar-cane to the factory and provide extra labour for cane-cutting.

Table 8.3 shows input–output data of sugar-growing under smallholder and estate conditions.

1. Sugar-cane production is an intensive form of agriculture. Gross returns in the example vary between $400 and $1000 per hectare and year.

TABLE 8.3

Some examples for input–output relations in sugar-cane production (per ha)

	India	Kenya	Peru	Brazil
Country	India	Kenya	Peru	Brazil
Location	Uttar Pradesh	Mumias	Lambayeque	Rio de Janeiro
Year	1966–9	1977	1977	1977
Type	Smallholder	Outgrowers	Co-operative	Company
Technique	Irrigated	Rain-fed	Irrigated	Rain-fed
Method	Sample	Accounts	Accounts	Sample
(a) *Plant crop* (months)	18	22	20	18
Labour input (man-hours)				
Land preparation	288	36[a]	42	13
Planting	456	185	105	57
Fertilizing	84	18	10	2
Irrigation	372	—	95	—
Weeding	576	480	60[c]	4
Harvesting	1416 ⎱	1116	211	414
Transporting	648 ⎰			
Total	3840	1835[b]	523	490
Yield (t)	85	128	200	90
Economic analysis ($)				
Price ($ t)	10·87	16·22	8·59	11·00
Gross return	924	2076	1718	990
Material inputs				
Seed	33	141	36	68
Fertilizers	12	102	109	98
Pesticides and herbicides	—	—	15	56
Irrigation	n.a.	—	137	—
Mechanization	64	375	439	224
Other costs[c]	18	127	—	—
Income before interest and overheads	797	1331	982	544
Income per man-hour	0·21	0·72	1·88	1·11
(b) *1st Ratoon crop* (months)	15	20	16	11
Labour input (man-hours)				
Crop production	620	432	213	n.a.
Harvesting and transports	1270	850	152	n.a.
Total	1890	1282[b]	365	n.a.
Yield (t)	46	97	144	68
Gross return ($)	500	1573	1236	748
Chemical inputs	5	102	110	n.a.
Irrigation	n.a.	—	117	n.a.
Mechanization	62	235	122	n.a.
Other	15	106	15	n.a.
Income before interest and overheads	418	1130	872	n.a.
Income per man-hour	0·22	0·88	2·39	n.a.
(c) *2nd Ratoon crop* (months)	n.a.	20	16	11
Labour input (man-hours)				
Crop production	n.a.	432	213	n.a.
Harvesting and transport	n.a.	712	152	n.a.
Total	n.a.	1144[b]	365	n.a.
Yield (t)	n.a.	80	135	55
Gross return ($)	n.a.	1298	1160	605
Chemical inputs	n.a.	102	110	n.a.
Irrigation	n.a.	—	117	n.a.
Mechanization	n.a.	193	122	n.a.
Other	n.a.	88	15	n.a.
Income before interest and overheads	n.a.	915	796	n.a.
Income per man-hour	n.a.	0·80	2·18	n.a.
(d) *All crops*	2 crops in	3 crops in	7 crops in	4 crops in
	3 years	5 years	10 years	4·5 years
Gross return per year ($)	475	989	875	655
Income before interest and overheads per year ($)	405	675	583	n.a.
Labour hours per year	1910	852	271	n.a.
Income per hour ($)	0·21	0·79	2·15	n.a.

[a] Mechanized company operation. [b] Farm labour and company labour. [c] Including herbicide application.
Sources: India: Lavinia (1974, pp. 152–93); Kenya: Mumias Sugar Company (1977, personal communication); Peru: Hatzius (1977, personal communication); Brazil: Carvalho and Grace (1976), Estado de Sao Paulo (1977).

2. Sugar-cane estates are systems which show only one major internal relation, namely the one between the fields and processing factories. There is no need for diversified cropping to maintain soil fertility. A dense vegetation covers the soil for most of the year and large amounts of organic matter are conveyed to the soil in the form of trash and by an extended root system. Traditionally oxen were employed to produce draught power and manure and these oxen require fodder production on the estate. However, manure, fodder, and oxen have been replaced by tractors and mineral fertilizers. Sugar-cane is therefore mostly grown as a monoculture, which does not endanger the long-term stability of the systems. Some areas (West Indies and Mauritius) have been practising sugar-cane monoculture for centuries without any damage to soil fertility.

3. Labour costs are the most important cost item. This applies not only to smallholder production but also to estate production in most tropical countries. 50–70 labourers per 100 ha were traditionally employed in sugar-cane estates, and even more in smallholder farming. The example from Brazil reflects the tendency for sugar-cane production in high-wage economies (Hawaii, Queensland, Guadeloupe) to be mechanized. Mechanized estates require less than half the labour force (15–20 labourers per 100 ha).

4. The economy of sugar-cane is highly sensitive to transport costs because cane is heavy, bulky, and perishable. Large sums must be invested in field tracks, roads, tractors, lorries, cranes, etc. In Guyana, transport costs largely take the form of age-old investment in canals and the costs of canal maintenance.

5. Sugar-cane cropping has shown itself able to benefit from yield-increasing innovations to which sugar-beet is far less responsive: improved varieties, improved disease control, sprinkler irrigation, increased application of mineral fertilizer, improved cultivation practices, etc. Sugar-cane yields in the West Indies are reportedly 10 times those of 150 years ago, and there seems to be no end to further yield-increasing innovations. The subtropical cane-growers are mostly ahead of the tropical producers in applying these techniques. Thus, if the trend continues, the yields per hectare are likely to carry on rising, and proven mechanical methods will be adopted where the wage level makes it necessary. Sugar-cane cropping tends towards monoculture, although yields from rotations are usually higher. The practice of incorporating a few years of field grass between the cane-cropping cycles, which was common when the estates depended on large numbers of oxen and therefore on a fodder and manuring system, has become rare now that tractors and mineral fertilizers have been introduced. Where the cropping cycles are interrupted by arable crops, we find either green-manure crops on large estates or intensive crops in smallholdings, like tobacco in Puerto Rico and rice in Taiwan.

The trend towards monoculture is closely linked to the development of larger processing plants and the transport problems that they entail. Thus, in Mauritius, for example, the sugar factories in 1832 had an average processing capacity of 200 t, in 1903 of 30 000 t, and in 1961 of 24 000 t. As the capacity of factories has been increased, and when the plots are near the factory and transport is easy, other crops are even less able to compete with sugar-cane.

6. Sugar-cane is definitely a crop that is most suitable for estates, because it is bulky and perishable, and costs limit the distance over which it can be carried. The modern factory should therefore stand in the centre of a relatively compact area, where cane is grown in monoculture. Monoculture, however, rarely fits into the working pattern of smallholdings. Cane cutting creates peaks in labour demand that can hardly be tackled by the labour capacity of a smallholding. The organization of co-operative cutting is usually a very difficult and intricate undertaking. Also, cane production must be accompanied by thorough cultivation, which is best carried out with heavy tractors, and cane is very susceptible to poor husbandry and numerous diseases. It is therefore more important than with other crops that technical advances should be available to the estate manager. Moreover, the smallholder cannot establish a sugar plantation as cheaply as he can a tree or shrub plantation. The superiority of large-scale producers of cane was so great in, for example, former Dutch Java that the smallholders recognized the advantage of leasing land to estates.

Smallholder production of cane has proved to be competitive only:

1. In satellite holdings near an estate—often on adjoining marginal land, as, for example, in Mauritius, where this type of holding produces cane and is also a source of seasonal labour.

2. In an area with a good transport system, where the smallholders are settled as tenants and are subject to closely supervised contract farming. as it is the case in the successful outgrower scheme at Mumias, Kenya; or

3. In places where the smallholders have technical training and farm very intensively, as in Taiwan.

Thus if estate production is not feasible or desired, sugar-cane can be grown in medium-sized holdings (cultivated by tractor) more economically than in smallholdings. Such medium-sized holdings are typical of production in, for instance, Queensland, Australia.

8.4.2. *Pineapple holdings*

With pineapples the economies of scale are less pronounced than with sugar-cane or sisal. A growing proportion of the supply of processing factories comes from commercial planters and from smallholders with diversified holdings. Table 8.4 illustrates some characteristics of the production process.

1. Commercial pineapples are an intensive crop yielding high gross and net returns per hectare and per hour of work. Particularly striking is the high

TABLE 8.4

Some examples of input–output relations in pineapple production (per hectare per vegetation cycle)

	Sri Lanka	Ivory Coast	Ivory Coast	Cameroun
Country	Sri Lanka	Ivory Coast	Ivory Coast	Cameroun
Location	Colombo	Bonoua	Bonoua	Nyombe
Year	1973	1974	1974	1974
Technique	Semi-intensive smallholders, for processing	Smallholder contract farmers with shifting[a] cultivation, for processing	Smallholder under close supervision in a permanent pineapple block,[b] for processing	Estate intensive techniques for fresh-fruit export
Method	Sample	Accounts	Accounts	Project plan[c]
Vegetation cycle[d]	3 years	2 years	2 years	2 years
Technical data				
Labour input (man-hours)[e]				
Land preparation	920[f]	—	1260[g]	100[h]
Planting	600	—	600	1500
Fertilizer application	584	—	144	75
Application of herbicides and pesticides	160	—	—[i]	450
Weeding	1240	—	—[i]	600
Application of catalyst	136	—	108	400
De-crowning ⎫	560	—	240	400
Harvesting ⎭		—	420	2250[j]
Other work	336	—		1100
Total	4536	1500	2772	6875
Yield (t)	49	50	60	45 (+ 15 per cent unexported)
Economic analysis ($)				
Gross return[k]	2217	2262	2715	9654
Material inputs				
Planting material	185	257	179	n.a.
Fertilizer	334	724	808	245
Herbicides and pesticides	74	—	152	⎫ 165
Catalyst	47	86	52	⎭
Mechanization ⎫	106	388	210	1095
Other ⎭		—	—	3992[l]
Administration overheads	—	216	989	—[m]
Total	746	1671	2390	5497
Income before interest	1471	591	325	4157
Income per man-hour	0·32	0·39	0·12	0·60
Wage costs	—	—	—	2096

[a] Independent smallholders who produce on a contract basis for a processing factory which supplies inputs and buys the produce. [b] Layout as in an estate. Smallholders receive allotments which have to be farmed according to rules laid down by the scheme management. [c] Based on estate accounts in Ivory Coast. [d] The production cycle usually includes one more year. [e] 6 hours per day in Ivory Coast, 5 hours in Cameroun, and 8 hours in Sri Lanka. [f] Hand-clearing. [g] Clearing of last cycle's vegetation. [h] Mechanized clearing. [i] Use of plastics. [j] Including packing of fresh fruit. [k] In Ivory Coast and Sri Lanka: $45·25 per tonne at packing station. Cameroun: $214 per tonne of fresh fruit net of transport costs. $394 per tonne at French wholesale market. [l] Mainly packing material. [m] Administration overheads are not specified in the source and included in the other positions. *Sources:* Sri Lanka: Kuhonta, Wijekoon, and Ariyaratnam (1973); Ivory Coast: Dietz (1975, personal communication). Cameroun: Gaillard and Lossois (1974).

chemical input which, in the case of Ivory Coast, amounts to about half of the total costs of material inputs.

2. The product is bulky and perishable and the return is sensitive to transport costs, but not to the same degree as sugar-cane. Transport costs are not prohibitive even if the supply originates from shifting cultivators growing pineapple and food crops.

3. Crop production is the least costly part of the total production process. Most of the costs are in packing and transporting in the case of fresh fruit export or in processing. The costs of a 100 000-tonne factory are distributed as follows: purchase of fruits 22 per cent; factory operations 68 per cent; depreciation 5 per cent; administration 5 per cent (planning figures, SALCI, Ivory Coast 1970).

The Sri Lanka growers (Table 8.4) produce for the local markets and their production process is comparatively extensive. The Bonoua growers (Ivory Coast) are wealthy farmers with extensive coffee and cacao plantations under a remaining forest cover. Plantations are interplanted with plantains and manioc which makes it difficult to define the cash-crop areas. Pineapples are more productive per man-day than traditional coffee or cacao. Table 8.5 shows the change in the farm-management data if pineapples are introduced. Growers participate in contract schemes which imply very modern husbandry practices. The Cameroun example shows the rather different cost and return structure of a plantation producing fresh fruit for export. The produce receives a much higher price per tonne (net of transport costs to the French market), but much more labour is required, mainly for packing, and packing material is the most important cost item.

8.4.3. *Banana holdings*

Plantains are an important food crop of the tropics and they are increasingly grown for the supply of local markets. Fruit bananas rank among the major export products of humid tropical lowlands. Both crops combine many of the advantages of annual arable and of permanent crop production:

1. The gross return per hectare, measured in energy or money, is usually high. The input for land clearance is low, establishment of the plantation simply implies fire clearance, planting, and weeding, and the first yield is obtained 10–12 months after planting. Bananas and plantains can be grown under a wide variety of environmental conditions. They produce reliable yields even under relatively dry conditions (down to 1000-mm rainfall per annum and sometimes less) provided there is a bimodal pattern of rainfall, and they can be grown on soils of poor fertility and in cooler highland climates. Export bananas, however, are grown in humid tropical lowlands and usually require supplementary irrigation.

2. The adaptability of the banana stems from the great number of varieties, which not only are suitable for different purposes (for cooking, beer, and use as a fresh fruit), but also have varying location requirements, and grow to different heights (the tall-stemmed bananas and the dwarf banana). Bananas also yield valuable by-products. In southern India, for instance, green leaves are sold to urban households, where they serve as plates. Elsewhere, dry banana leaves serve as roofing material.

3. By providing plenty of shade, trash, and permanent cover on slopes,

TABLE 8.5

Farm-management data of pineapple–coffee–cacao and coffee–cacao–banana holdings at Bonoua, Ivory Coast

	Bonoua	Bonoua	
Location	Bonoua	Bonoua	
Rainfall (mm)	1587	1587	
Year	1972	1972[a]	
Type of farming	*Robusta* coffee–cacao–banana	Pineapple–*robusta* coffee–cacao	
Method	Project records[b]	Survey	
Number of holdings	n.a.	n.a.	
Persons per household	9·1	12·2	
Labour force (ME)	3·4	4·2[c]	
Size of holding (ha)	n.a.	n.a.	
Pinaepple	—		1·95
Coffee	2·82		2·27
Cacao	2·77		3·83
Mixed tree crops	1·95		0·27
Food crops (mainly plantains)	0·26		0·60
Fallow land[d]	n.a.	n.a.	
Crop land	7·80	8·92	
Yields			
Pineapple (t ha⁻¹)	—	55	
Pineapple ($ ha⁻¹)	—	1236	
Coffee (t ha⁻¹)	0·57	0·57[e]	
Coffee ($ ha⁻¹)	216	216[e]	
Cacao (t ha⁻¹)	0·15	0·15[e]	
Cacao ($ ha⁻¹)	45	45[e]	
Economic return ($ per holding)			
Gross return	1083	2074	
Purchased inputs	55	601	
Income	1028	1473	
Wages for hired labour	133	176	
Family farm income[f]	895	1297	
Sales as percentage of gross return	74	91	
Productivity			
Gross return ($ ha⁻¹ crop land)	139	233	
Gross return ($ ME⁻¹)	318	494	
Income ($ ha⁻¹ crop land)	132	165	
Income ($ ME⁻¹)	302	351	
Labour input (man-hours)	3288	n.a.	
Income per man-hour	0·31	n.a.	
Man-hours ME⁻¹	967	n.a.	

[a] 1966 survey data of man-equivalents and crop areas. 1972 data of yields, prices, returns and costs. [b] Project information based on SEDES surveys and case studies. [c] Including two non-family members working with the family. [d] Fallow is not included. 60–85 per cent of the land claimed by the family is cropped. [e] 1972 yields and prices taken from Z. Nagel (1972). [f] Including non-family labour staying with the family. *Sources:* Nagel (1972, personal communication); Ripailles (1966).

the banana helps to maintain soil fertility and reduces erosion. It makes the application of large amounts of fertilizer worthwhile and responds well to fertilization with mulch, manures, and mineral fertilizers. Bananas may occupy a plot permanently, provided plenty of mulch and fertilizer are applied.

4. Banana bunches mature throughout the year and provide a regular source of food, income, and employment.

5. Labour requirements are relatively high even in modernized enterprises, varying from 0·5–3·0 ME ha^{-1} in smallholdings to 0·4–0·8 ME ha^{-1} in estates, but there are no pronounced peaks in labour demand, and the ratio between labour input and output in terms of energy is usually better than with most short-term food crops.

6. Finally, the banana is compatible with other crops, facilitating inter-planting and the incorporation in a crop rotation.

(a) *Banana estates.* Export bananas, however, are a sensitive crop in several ways. The fruits have to be in a refrigerated boat within 8 to 12 hours after cutting. This implies a well-organized (contract) market, the concentration of production close to harbours, all-weather roads, and efficient transport. The costs of disease protection are high. The change from the Grande Michel variety to Cavendish banana, which became necessary in the major Latin American exporting countries because of the Panama disease, implied a jump in land-use intensification. Much of the Grande Michel was grown without fertilizer, irrigation, and pesticides. Cavendish banana production is a factory-type production process which varies little between the major producing areas, and includes high and regular applications of fertilizer, nematocides, herbicides, and pesticides, irrigation in areas with several dry months, a cable-way to transport the bunches to packing units where fruits are sorted, washed, and packed. The machine input is limited to some spraying equipment and trucks. Rising wages are not absorbed by more mechanization but by increasing yields (mainly bunch weight) with higher chemical inputs (up to 60 t of fruits with about 450 kg of nitrogen ha^{-1}). Traditionally bananas used to be replanted and to shift within a larger estate area. Most of today's commercial banana production represents a permanent type of land use, based on fertilizers and nematocides. Whereas sugar-cane estates tend towards one-crop enterprises only, banana production usually leads to more diversified farming. Estates often keep cattle to utilize surrounding grassland and reject bananas, or swine are fed with rejects.

Table 8.6 presents the farm-management data of export banana enterprises in Ivory Coast, Ecuador, and Martinique. Costs of production are lowest with producers in Ecuador, who have soils of high fertility (volcanic origin), low cost gravity-irrigation, and an efficient labour force with lower wages than in Central America. Enterprises in Ivory Coast require much

TABLE 8.6

Input–output relations of plantain and banana production in Uganda, Colombia, Ivory Coast, Ecuador and Martinique (per hectare and year)

Country	Uganda[a]	Colombia[b]	Ivory Coast[c]	Ivory Coast[d]	Ecuador[d]	Martinique[e]
Year	1969	1976	1970	1973	1976	1976
Crop	Plantain	Plantain	Banana	Banana	Banana	Banana
Length of cycle	Permanent	7 years	3 years	3 years	Permanent	5 years
Labour input (man-hours)						
Land preparation	n.a.	29	216[f]	213[f]	n.a.	5
Planting	n.a.	33	92	—	n.a.	62
Plantation maintenance	556	282	588	123	196	200
Irrigation and drainage	—	—	—	73	146	—
Mulching	99	—	798	—	—	—
Fertilizing	—	56	112	81	18	73
Pesticide and herbicide application	—	43	108	229	151[g]	74
Fruit protection and support	—	—	378	—	75	87
Harvesting and processing	99[h]	304	240	211	258	220
Total	754	747	2532	930	844	721
Yield (t)	9	20	22	30	30	40
	fruits	bunches	fruits	bunches	fruits	bunches
Economic analysis ($)						
Gross return[i]	225	672	2283	1737	1778	5040
Mechanical operations	1	10	261	17	n.a.	67
Fertilizer	—	119	333	368	146	665
Pesticides and herbicides	—	101	177	155	183	700
Other inputs[j]	—	73	54	—	278	128
Administrative overheads	—	14	203	442[k]	61	838
Income before interest	224	355	1225	755	1110	2642
Income per hour of work	0·30	0·48	0·50	0·81	1·32	3·66
Wages	n.a.	161	712	176	300	1708
Return to land and capital	n.a.	149	513	579	810	934

[a] Smallholder production for subsistence food. [b] Small entrepreneurial production for local markets. [c] Accounts of a greater number of estates with motorization. [d] Estate records. [e] Costs published by Planters' Association. They are probably somewhat overestimated. [f] Forest clearance. [g] Pesticide application by plane. [h] Underestimate because of continuous harvest. [i] Loco farm or estate, net of transport costs to harbour. [j] Mainly plastic for bunch protection, support poles, and packing materials. [k] Thereof 64 per cent overheads proper and 36 per cent amortization of land development costs. *Sources:* Uganda: Kyeyune-Sentongo (1973b); Ivory Coast (a): Bonnefond (1977); Ivory Coast (b): Champion (1974, personal communication); Colombia: Federacion Nacional de Cafeteros de Colombia (1976); Ecuador: Programa Nacional del Banano y Frutas Tropicales (1977); Martinique: Rocheteau (1978, personal communication).

more labour and much higher chemical inputs which (including plastics for fruit protection) are about three times the wage bill. The example from Martinique demonstrates the high costs of production under the condition of a high-wage economy. Martinique, as a part of France, receives very high prices for its bananas and wage levels are those of an industrialized economy. The production of bananas is unsuitable for much mechanization. Producers in Martinique try to absorb wage costs by very high yield-increasing and labour-saving chemical inputs.

The two examples with plantains from Uganda (subsistence-oriented) and Colombia (market-oriented) provide an idea of the economics of this type of starch production. The labour input per hectare is lower than with annual food crops given hand technologies. Material inputs remain at a low level in relation to the energy output, and plantains seem to be efficient users of support energy.

(b) *Banana–coffee smallholdings: the Wahaya of Bukoba, Tanzania.* Banana production for food and local markets is one of the main activities of tropical smallholders, wherever rainfall and temperature are sufficient for banana growing. A case in point is the banana–coffee culture of the Wahaya, which is particularly interesting because of the intelligent adaptation of banana production to the difficult natural conditions of the location.

The Bukoba District receives ample and well-distributed rainfall averaging 2000 mm annually. The soils, however, are badly leached and very poor. The average holding comprises 1·4 ha, of which 0·80 are homesteads carrying mainly cooking bananas (plantains), which are usually used as subsistence food, but which are also sold to markets as far away as Dar-es-Salaam. Bananas, which yield more than half of the gross return, are interplanted with *Coffea robusta* coffee, maize and beans. In addition to the homestead, the holding includes 0·6 ha of open, infertile grassland.

Fig. 8.4 shows the layout of a Bahaya holding. The so-called homestead (1), a banana plot with coffee and several other crops in mixed stands, is situated around the hut. At the edge of the homestead there is usually a garden growing predominantly root-crops (2). At some distance from the homestead lies the grassland (3) belonging to the farm. Here small plots are cultivated on a short-term basis. The grassland serves as pasture and supplies mulch to the homestead. In addition to the grassland that belongs to the farm, everyone can make use of communal grassland. Within the homestead both a spatial and a vertical order in plant growth is recognizable (see Fig. 8.4 (b) and (c)).

1. Directly around the hut and the cattle kraal, the focal point of the holding where the soil has received the most manure, a dense growth of bananas can be found. The greater the distance away from the hut, the less manure is applied. The bananas become thinner. More and more coffee trees, maize plants, and beans grow between the bananas until the belt of

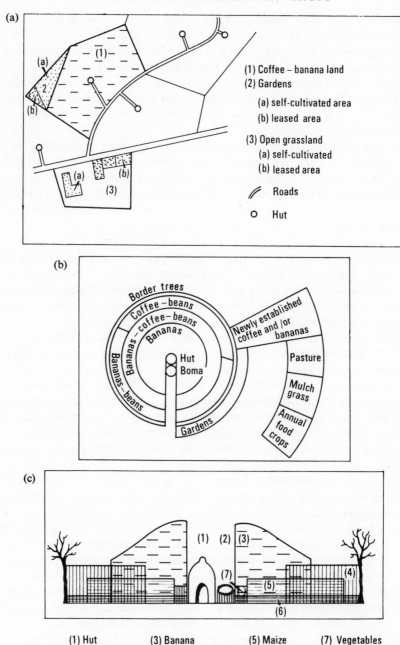

FIG. 8.4. Land use in a Bahaya holding, Bukoba, Tanzania: (a) layout; (b) horizontal order; (c) vertical order. From Friedrich (1968, p. 175).

bananas around the centre of the holding gives way to a belt of banana–coffee cultivation. Gradually the mixed stand of banana–coffee–beans is succeeded by one consisting of banana–maize–beans or coffee–beans. In some cases, the belt of mixed stands is supplemented by a plot with pure stands of new coffee. Land use is thus organized in concentric circles around the central hut, the bands ranging in width from a few feet to several dozen yards.

2. In addition to the horizontal order of plant growth, there is a vertical order. On the heavily fertilized soil close to the house, the bananas reach unusual heights, casting dark shadows upon the ground, where only some arrowroots grow. Bananas close to the centre of the farm frequently grow twice as high as those growing on the less fertile edges of the holding. With the distance from the hut increasing and the supply of manure and mulch material growing less, and at the same time the heights and shading of the bananas decreasing, more coffee trees can be found growing between the bananas. If the density of the banana–coffee vegetation permits it, the ground is covered with beans, so that, in accordance with their need for light, three crops are growing one above the other: beans, coffee, and bananas. Arrowroot, yams, tobacco, maize, and vegetables are also grown where possible, in such a way that on different plots different crops are grown which are compatible with regard to their light requirements. At the same time, the ground is given complete protection by a thick layer of mulch. Arable crops are grown within a zero-tillage technique.

Naturally, the various belts around the hut are not kept strictly separate: instead, there is a gradual transition in the combination of plants, which may vary from one location to another, from one farm to another, and even on any one farm, according to the nature of the soil and the site of the slope. The determining factors in the cultivation of the homestead are the fertility of the soil, domestic needs, the land available, and—occasionally—the influence of extension officers. The way mixed cropping is apportioned depends primarily upon domestic requirements. In the case of large families with small plots of land, bananas are given priority, and there is little room left for coffee. In contrast with this, there is proportionately much coffee where a family has at its disposal a relatively large homestead.

Bananas at Bukoba are a permanent crop. Some of the homesteads have carried bananas for more than 60 years, and this permanency of farming is based on intensive manuring. The export of nutrients is limited to some dried coffee. Instead, there is a substantial import of nutrients by the gathering of fuel, the purchase of food, the use of ample mulch collected from the open grassland, and the fact that the cattle are kept in stables and their manure is systematically applied. The buildup in soil fertility is the more marked the older the homestead. The man farming an old plot, which in the course of time has received much refuse and mulch, can rely on his bananas having a denser growth and a richer yield than a man cultivating a new plot.

One-third of the farmers own cattle, usually 3–4 head. The cattle graze on the open grassland and produce mainly manure for the banana land. The production of milk and meat is very low. Friedrich (1968, p. 159) gives the following information: calving interval 25·5 months, weaning loss 30 per cent, daily milk production during the 6-month lactation period about 1·5 kg, and 20 per cent of the cows are probably sterile. The average milk production per cow and per year, over and above calf-rearing, is thus estimated at 50 kg. The most important product of the cattle economy is obviously manure (see Fig. 8.5). On farms where cattle are kept, the gross return per hectare of banana–coffee land is on average $200 above that of farms without cattle. If we

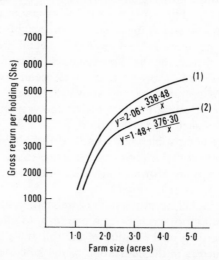

FIG. 8.5. The relation between farm size and gross return in Bahaya holdings with and without cattle. (1) Farms with cattle: $R^2 = 0·98$, $n = 16$. (2) Farms without cattle: $R^2 = 0·51$, $n = 18$. From Friedrich (1968, p. 204).

assume—certainly not entirely correctly—that the additional average gross return on farms with livestock is wholly due to the manure, we arrive at an annual manure value per livestock unit of $78. This figure may explain the Wahaya's interest in obtaining cattle, and thus manure.

Table 8.7 shows the economic return of Bahaya farming, which is higher than in most other farming systems of Tanzania, in particular in respect of returns per hour of work. The comparison with banana–coffee holdings in neighbouring Uganda reveals some interesting differences (Brandt 1971). Land around Jinja is more ample and more fertile. There is less need for intensive husbandry techniques. Because of this, bananas and coffee are grown in sole stands, and banana groves are not yet permanent. After 5–10 years the soil is exhausted and banana production moves to another plot. Jinja farmers

TABLE 8.7

Farm-management data of banana–coffee holdings in Tanzania and Uganda[a]

Country	Tanzania	Uganda	Uganda
Location	Bukoba	Jinja	Ankole
Rainfall (mm)	2041	1227	985
Year	1964–5	1968–9	1969
Type of farming	Banana–*robusta* coffee	Banana–*robusta* coffee	Banana–*arabica* coffee
Method	Survey	Case studies	Survey
Number of holdings	120	3	108
Persons per household	5·3	6·3	4·25
Labour force (ME)	1·64	3·0	n.a.
Size of holding (ha)	0·88	2·74	1·73
Garden, homestead ⎱	0·81	0·26 ⎱	1·31
Banana and coffee[b] ⎰		1·14 ⎰	
Arable crops, season 1	0·07	0·59	0·25
season 2	—	0·25	0·17
Total crop area	0·88	2·24	1·73
Livestock per holding (numbers)			
Cattle	2·00	—	3·00
Small stock	n.a.	n.a.	5·00
Yields			
Bananas (t ha^{-1} year)	n.a.	9·2	9·1
Bananas ($ ha^{-1})	n.a.	206	225
Coffee (t ha^{-1})	n.a.	1·4	0·81
Coffee ($ ha^{-1})	121	183	234
Economic returns ($ per holding)			
Gross return	230	488	471
Purchased inputs	3	5	7
Income	227	483	464
Wages for hired labour	5	38	43
Family farm income	222	445	421
Sales as per cent of gross return	41	62	37
Productivity			
Gross return ($ ha^{-1})	261	178	272
Gross return ($ ME^{-1})	140	162	n.a.
Income ($ ha^{-1})	258	176	268
Income ($ ME^{-1})	138	161	n.a.
Labour input, man-hours	1604	4135	2912
Income per hour of work ($)	0·14	0·12	0·16
Man-hours per family ME	978	1142	n.a.

[a] Most of the bananas are cooking bananas = plantains. [b] Mostly mixed. *Sources:* Friedrich (1968, p. 206), Brandt (1971, pp. II/1–III/1), Kyeyune-Sentongo (1973*b*).

use mulch, but not to the same extent as Bukoba farmers. Interculture is limited to the first 2 years after the planting of coffee. Cattle are not yet needed at Jinja to supply the manure. Jinja farmers receive higher family incomes, but labour productivity is not much higher, and husbandry practices of the Bukoba farmers show the way to adapt to growing land scarcity and depleted soils. The example from Ankole relates to a rather similar situation with Arabica coffee under lower rainfall conditions.

8.5. Estates and smallholdings with shrub crops

Coffee and tea are the main representations of a distinct group of crops that botanically are tree crops but which because of their farm-management characteristics are termed shrub crops. The main difference in cultural practice between these and tree crops is the input for the maintenance of the shrub-like appearance (pruning or stumping) and the high labour requirements of harvesting.

8.5.1. *Coffee holdings*

The production of coffee in most cases started on estates and then spread to smallholdings. A number of the farm-management features of coffee promoted this pattern of development.

1. Coffee growing is adaptable to wide differences in farming intensity and to various forms of mixed cropping and intercropping.
2. Coffee production at low levels of yield is not very susceptible to plant disease.
3. By far the most important input in coffee production is manual labour.
4. The produce may be dried on the farm or processed in small pulperies. Transport of cherries to large pulperies is a problem, but by no means comparable with that of transporting sisal leaf or sugar-cane.
5. The introduction of coffee into subsistence holdings does not usually lead to pronounced peaks in labour demand.

Because coffee cropping suits small-scale farming, expansion in various parts of the world since the 1930s took place primarily in smallholdings, as in Colombia and East Africa. On the other hand, estate production has recently gained in competitiveness, owing chiefly to the more rapid application of technical advances like improved varieties, mineral fertilizing, plant protection, pruning techniques, and processing.

(a) *Coffee–maize holdings at Jimma, Ethiopia.* In higher rainfall areas of the Ethiopian highlands, coffee grows naturally under the original forest. Traditionally there is a dense undergrowth of bush species, mainly coffee, in a thick forest with trees of varying size. In extensive areas the forest has been thinned to give more light to the coffee. This increases coffee yields, but produces a weeding problem. More slashing is required and some weeding to control couch grass. Gaps with no spontaneous coffee growth have to be planted with coffee seedlings. Most of the labour input is, however, still in harvesting. Farmers usually wait for the cherries to fall down. The collected cherries are dried and sold. Coffee yields are low, but there are no costs for plantation establishment, and farm family incomes are higher than in most other Ethiopian smallholder systems (compare Table 8.8 with Table 6.4). Subsistence food requirements of these holdings are mainly covered by maize grown in a fallow system.

Coffee smallholdings at Jimma are far from the efficient use of their natural potential. Their way of farming may be considered as the first phase in the commercialization of subsistence holdings.

(b) *Coffee estates in Sao Paulo, Brazil.* While the low intensity of coffee production in Ethiopia can primarily be explained as a 'lag' in socio-economic development, we may regard the Brazilian pattern of coffee production which is also relatively extensive, as a function of increasing wages, the seasonal pattern of labour requirements in coffee, and the low-quality produce delivered from southern Brazil.

Coffee in Sao Paulo, Brazil, is almost entirely produced on diversified large estates, and is usually combined with dairying as a relatively low-risk activity. Table 8.10 shows a specialized coffee estate from Sao Paulo and the cost structure of such enterprise. Generally, however, coffee-producing farms in southern Brazil are diversified undertakings. The diversified holdings usually show the following enterprises (Table 8.8).

1. A coffee plantation of 30 to 100 ha.
2. A large pasture stocked with either dairy or beef cattle.
3. Arable land which mostly varies between 20 and 40 per cent of the total farm area. On dairy farms it produces maize silage and sugar-cane as dry season fodder. Other farms grow soya beans, maize, upland rice, and onions.
4. Odd pieces of forest or waste land amounting to 10 to 20 per cent of the farm area.

Diversification is also explained by the high purchasing power of nearby urban markets and the need to employ the labour force regularly throughout the year.

Coffee production in southern Brazil used to be a soil-mining activity. Depleted land went into extensive grazing and new land was cleared and planted with coffee. Production techniques were significantly changed during the last decade. Rising wages imply mechanization and leaf rust requires intensification, because only highly yielding and intensively-husbanded plantations justify the necessary plant protection input (Monaco 1977). Traditionally coffee was seeded with several plants per hold, not pruned, weeded by hand, completely picked at one time, and processed through sun-drying. The prevailing production technique (1978) is characterized by seedling production in nurseries, fully motorized land preparation, and the planting of 2000 trees ha^{-1} with narrow spacing within the row and wide spacing of rows to allow mechanized spraying and weeding operations. Rows are

TABLE 8.8

Farm-management data of Arabica-coffee farms

	Ethiopia	Kenya	Costa Rica	Colombia	Brazil
Location	Jimma	Embu	La Suiza	Restrepo	Sao Paulo
Rainfall (mm)	1534	1600	2800	1200	1500
Year	1972–3	1977	1978	1977	1974–5
Type of enterprise	Smallholdings	Smallholdings	Smallholdings	Smallholdings	Estate
Type of farming	Coffee–maize	Coffee–maize–potatoes–milk	Coffee–milk	Coffee–plantain–milk	Coffee–milk
Method	Sample	Case Study	Case Study	Model	Accounts
Number of holdings	10	1	1	1	9
Persons per household	5·9	6·0	13·0	n.a.	n.a.
Labour force (ME)	2·3	2·4	4·0	3·5	77
Size of holding (ha)	6·46	2·75	14·0	9·8	567
Coffee	1·95[a]	1·00	2·0	4·7[a]	40
Arable crops	1·15	1·00	1·5	0·6	136[b]
Pasture	3·36[c]	0·75	7·0	4·0	278
Bush	—	—	3·5	0·5	113
Cattle (numbers)	2·0	2·0	9·0	4·0	228
Yields					
Coffee (t ha⁻¹)	0·24	0·74	0·8	1·0	0·97
Coffee ($ ha⁻¹)	207	2431[d]	2248	1667[d]	1003
Maize (t ha⁻¹)	1·04	2·90	—	1·0	2·55
Maize ($ ha⁻¹)	50	522	—	—	284
Milk (kg per cow and year)	n.a.	720	1800	1100	1560
Economic return ($ per holding)					
Gross return	447[e]	3412	7342	6910	331 582
Purchased inputs	16	1046	1263	1208	171 007
Income	431	2366	6079	5702	160 575
Wages for hired labour	34	88	2459	2259	96 266
Taxes		n.a.	n.a.	n.a.	n.a.
Family farm income	397	2278	3620	3443	—
Net return before taxes, interest, and land rent	n.a.	n.a.	n.a.	n.a.	64 309
Productivity					
Gross return ($ ha⁻¹)	69[f]	1241	524	705	730[g]
Gross return ($ ME⁻¹)	194	1422	1836	1974	4306
Income ($ ha⁻¹)	67	860	434	582	354
Income ($ ME⁻¹)	187	986	1519	1629	2085
Labour input (man-hours)	3974	4309	8000	7021	154 000
Income per man-hour ($)	0·11	0·55	0·76	0·81	1·04

[a] Ethiopia: A natural vegetation of coffee under a cover of forest trees. Colombia: Interplanted with plantains. 2 ha mature new coffee, 1 ha newly planted coffee, 1·7 ha old coffee yielding 0·5 t ha⁻¹. [b] Mostly bush fallow. [c] Including land cropped by workers. [d] At the height of the coffee boom. 1976 prices were about 50 per cent lower. [e] Without return from livestock and gardens. [f] Per hectare of total land. [g] Per hectare of utilized land which is 80 per cent of the total. *Sources:* Ethiopia: Friedrich, Slangen, and Bellete (1973); Kenya: Schmidt (1978, personal communication); Costa Rica: Personal enquiries; Brazil: Bemelmans (1978, personal communication); Colombia: Varela (1977, with some modifications by the author based on enquiries at Restrepo in 1978).

carefully planted on the contour and effectively control erosion. No mulch is applied to avoid damage through frost and fire. Weed control occurs mainly with herbicides and is supplemented by some mechanized and hand weeding. No pruning is practised except head cutting to ease harvesting. Fertilizer and pesticide inputs are heavy, but labour inputs for maintenance are low. In the average ten labourers per 100 hectares are required for maintenance operations. Two-thirds of the total labour input are for harvesting and processing. All the crop is stripped off. Trees are shaken and beaten to dislodge the cherries from the higher branches. Labour requirements for harvesting are highly seasonal (three months) and are mainly covered by seasonal labour. Yields are lower than in most other arabica–coffee growing areas and quality is lower, but labour productivities are very high.

Coffee production in Sao Paulo faces numerous problems. There is the risk of devastating frosts. Nematodes make it questionable whether continuous coffee can be grown. The opportunity costs of the land are high, because the area lends itself to numerous other profitable activities such as citrus, soya, maize, cotton, and dairying. A coffee harvesting machine has been invented which would reduce labour requirements for harvesting to negligible proportions, but which implies heavy capital investments, trees suited to machine harvesting, and large rectangular fields. Without such a machine coffee is probably no longer competitive under the conditions of rising wages and a highly seasonal employment pattern.

(c) *Coffee smallholdings in Colombia.* Coffee farmers in Colombia produce under favourable conditions of production: rainfall is relatively high, reliable and bimodally distributed, land is ample and the infrastructure is well developed. The examples in Tables 8.8 and 8.9 are taken from an area with less rainfall than is usual in main coffee-growing areas, but husbandry practices are similar. Coffee is grown on the best land of the farm under a shade of plantains (400 clumps ha^{-1}) and occasionally some shade trees. There are two harvesting seasons and the cherries are pulped and dried on the farm. Young coffee is interplanted with root crops and beans. Much of the production occurs on steeply sloping land, but erosion in modern plantations is effectively controlled by a very dense vegetation of coffee and plantains. Plantains are an important element of the system. They provide shade, erosion control, feed for cattle and pigs, and produce for home consumption and cash. No pruning occurs, except the head cutting of tall trees. Labour input is about twice as high as in Sao Paulo, but yields are higher and the produce is of a higher quality, receiving higher prices. Coffee is grown on land with low opportunity costs and in 1978 production was being expanded very much. Much of the coffee has been recently planted with seed-suppressing and self-mulching high densities (3000–4000 trees ha^{-1}). This coffee is heavily fertilized but pesticide inputs remain moderate.

TABLE 8.9

Some examples of input–output relations in the production of Arabica coffee (per hectare and per year of mature stands)

Country	Kenya	Kenya	Colombia	Costa Rica	Brazil	Brazil
Location	n.a.	Nyeri	Restrepo	Zone IV	Sao Paulo	Espirito Santo
Year	1976	1977	1978	1976–7	1977	1977
Exploitation system	Estate	Smallholder	Smallholder	Small entrepreneur	Estate	Estate
Technique	Modern	Modern	Modern	Modern	Average	Average
Method	Accounts	Model	Model	Sample	Sample	Accounts
Technical data						
Labour input (man-hours)						
Weeding	808	150	224	95	148	240
Pruning	552	85 ⎫	64	195[a]	—	—
Infillings	16	— ⎬		94[a]	—	—
Fertilizer application ⎫		80	80	35	48	24
Mulching, soil conservation ⎬	328	176	n.a.	67	8	104
Plant protection	96	24	16	40	34	75
Harvesting	936	1175	648	1111	373	176
Total field work	2736	1690	1032	1637	611	619
Processing	432	—	96	—[b]	40	n.a.
Total before overheads	3168	—	1128	—	651	619
Power inputs (hours)						
Tractors	n.a.	—	—	n.a.	19	24
Animals	—	—	—	n.a.	28	—
Yield (kg dried beans)	1170	810[c]	1125	1797	600	700
(kg dried cherries)	—	—	—	—	1200	—
Economic analysis ($)						
Gross return	1325	2535	1875	2946	1071	1250
Price ($ kg^{-1} beans)	1·13	3·13	1·67	1·63	1·79	1·79
Material inputs						
Fertilizer and herbicides	164	36	174	249	138	125
Plant protection	195	81	35	14	29	38
Mechanized operations	46	33	21	66	61	105
Processing materials	54	—	1	—	—	—
Other materials	18	50	—	89	—	—
Income before interest and						
overheads	848	2335	1644	2528	843	982
Overheads	203	—	n.a.	108	87	n.a.
Wages	485	357	352	715	321	233
Net return before taxes, interest						
and land rent	160	1978	1292	1705	435	749
Income per hour of work ($)	0·27	1·38	1·45	1·54	1·29	1·59

[a] Coffee and shade trees. [b] Sale of unpulped cherries. [c] 4 kg of cherries per kg of dry beans. *Sources:* Kenya: Estate: Kenya Coffee Growers Association (1976), Smallholdings: Schmidt (1978, personal communication); Colombia: Varela (1977); Costa Rica: Lizano and Jefe (1977); Brazil: Estado de Sao Paulo (Prognostico 1977).

The coffee–plantain field is supplemented by some small plots with manioc, beans, and maize. Most of the land is extensive pasture utilized by dairy cattle for subsistence milk production. Cattle obtain supplementary dry season feed from small plots with planted grass and sugar-cane. Most farms keep a pig.

(d) *Coffee smallholdings in Costa Rica.* The natural conditions of production in most coffee-growing areas of Costa Rica are exceedingly favourable with soils of volcanic origin in a temperate highland climate with about ten humid months. The farm given as an example in Table 8.8 shows the typical combination of coffee with sugar-cane and dairying, and all activities are performed at a rather low level of intensity, while the input–output relations in Table 8.9 are more representative for coffee-producing techniques in the country (Marin 1977).

Most of the coffee in Costa Rica is produced in smallholdings or by small entrepreneurs. Most farmers maintain 2600–3600 trees per hectare under a light cover of shade trees which are pruned according to the seasonally changing light requirements of coffee. Fertilizer and pesticide inputs are modest. Most coffee is grown on a three-stem method, with the eldest stem being replaced each year by a regrowing stem. Picking is selective, and ripe cherries are sold to private pulperies. The Oficina de Cafe acts as the intermediate between producers and processors. Picking is highly seasonal (3–4 months) and farmers tend to combine coffee with dairying and sugar-cane which employs labour when coffee does not. Average yields are relatively high because of the very favourable soil and climate conditions of the area. Table 8.9 gives information about the costs of production (Marin 1977).

Shade trees are a supply-regulating device in the system. With high coffee prices shade trees are heavily pruned, coffee receives more fertilizer and yields more heavily. With low coffee prices shade trees are allowed to shade more heavily. Coffee receives less fertilizer and remains under the shade in a productive state with relatively low labour and chemical inputs, yielding significantly less than in periods of less shading and heavy fertilizing.

Coffee farmers in Costa Rica tend to absorb rising wage levels through the interacting effect of higher plant densities and higher fertilizing inputs. The Brazilian mechanization-oriented approach is not feasible given the sloping nature of the terrain. Dense planting is possible because coffee rust and coffee berry disease have not yet been observed. Most farmers interplant into existing coffee, up to 5000 trees per hectare. A few enterprising ones clear and replant with 7000 trees per hectare. These high-density plantations are husbanded with rotational stumping of complete coffee rows, no shade trees are maintained and fertilizer inputs are heavy (250 kg N, 50 kg P_2O_5, 135 kg K_2O plus lime and minor elements). Little weeding is required and soil protection is perfect. Yields reach the 3000-kg level with labour inputs of about 1700 hours, mainly for stumping and harvesting.

(e) *Coffee estates and smallholdings in Kenya.* Coffee in Kenya is a highly intensive crop. Inputs of labour, fertilizers, and pesticides are higher than in any other example cited in Tables 8.8 and 8.9, and this is related to the high price of the produce, the land scarcity, the ample availability of low-cost labour and the high standards of technology in the estate sector. Chemical inputs for plant protection and soil fertility maintenance are very high in the example from the estate sector. Plantations are maintained by pruning, interplanting and rotational stumping. Much labour goes into mulching and high

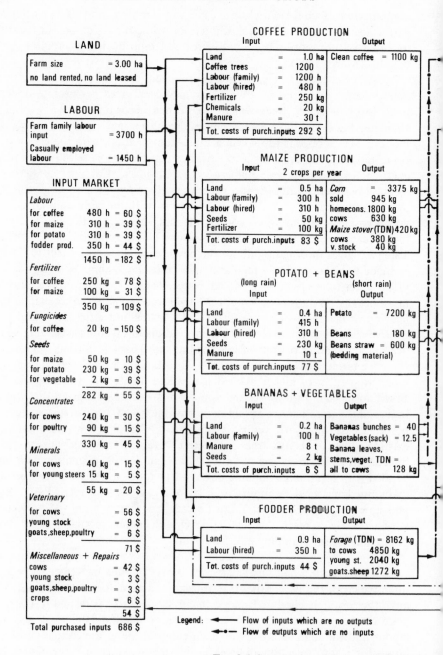

LAND

Farm size   = 3.00 ha
no land rented, no land leased

LABOUR

Farm family labour
input   = 3700 h

Casually employed
labour   = 1450 h

INPUT MARKET

Labour

for coffee	480 h	= 60 $
for maize	310 h	= 39 $
for potato	310 h	= 39 $
fodder prod.	350 h	= 44 $
	1450 h	=182 $

Fertilizer

for coffee	250 kg	= 78 $
for maize	100 kg	= 31 $
	350 kg	=109 $

Fungicides

for coffee	20 kg	=150 $

Seeds

for maize	50 kg	= 10 $
for potato	230 kg	= 39 $
for vegetable	2 kg	= 6 $
	282 kg	= 55 $

Concentrates

for cows	240 kg	= 30 $
for poultry	90 kg	= 15 $
	330 kg	= 45 $

Minerals

for cows	40 kg	= 15 $
for young steers	15 kg	= 5 $
	55 kg	= 20 $

Veterinary

for cows		= 56 $
young stock		= 9 $
goats,sheep,poultry		= 6 $
		71 $

Miscellaneous + Repairs

cows		= 42 $
young stock		= 3 $
goats,sheep,poultry		= 3 $
crops		= 6 $
		54 $

Total purchased inputs 686 $

COFFEE PRODUCTION

Input			Output
Land	=	1.0 ha	Clean coffee = 1100 kg
Coffee trees	=	1200	
Labour (family)	=	1200 h	
Labour (hired)	=	480 h	
Fertilizer	=	250 kg	
Chemicals	=	20 kg	
Manure	=	30 t	
Tot. costs of purch.inputs 292 $			

MAIZE PRODUCTION

Input	2 crops per year		Output	
Land	=	0.5 ha	*Corn* =	3375 kg
Labour (family)	=	300 h	sold	945 kg
Labour (hired)	=	310 h	homecons.	1800 kg
Seeds	=	50 kg	cows	630 kg
Fertilizer	=	100 kg	*Maize stover* (TDN)	420 kg
Tot. costs of purch.inputs 83 $			cows	380 kg
			v. stock	40 kg

POTATO + BEANS

(long rain)			(short rain)	
Input			Output	
Land	=	0.4 ha	Potato =	7200 kg
Labour (family)	=	415 h		
Labour (hired)	=	310 h	Beans =	180 kg
Seeds	=	230 kg	Beans straw =	600 kg
Manure	=	10 t	(bedding material)	
Tot. costs of purch.inputs 77 $				

BANANAS + VEGETABLES

Input			Output	
Land	=	0.2 ha	Bananas bunches =	40
Labour (family)	=	100 h	Vegetables (sack) =	12.5
Manure	=	8 t	Banana leaves,	
Seeds	=	2 kg	stems,veget. TDN =	
Tot. costs of purch.inputs		6 $	all to cows	128 kg

FODDER PRODUCTION

Input			Output	
Land	=	0.9 ha	*Forage* (TDN) = 8162 kg	
Labour (hired)	=	350 h	to cows	4850 kg
Tot. costs of purch.inputs 44 $			young st.	2040 kg
			goats.sheep	1272 kg

Legend: ◄——— Flow of inputs which are no outputs
◄—•— Flow of outputs which are no inputs

FIG. 8.6. Structural model of a coffee-milk farm

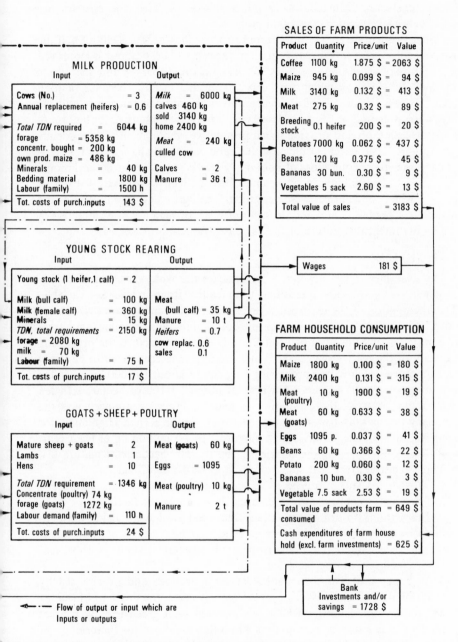

SALES OF FARM PRODUCTS

Product	Quantity	Price/unit	Value
Coffee	1100 kg	1.875 $	= 2063 $
Maize	945 kg	0.099 $ =	94 $
Milk	3140 kg	0.132 $ =	413 $
Meat	275 kg	0.32 $ =	89 $
Breeding stock	0.1 heifer	200 $ =	20 $
Potatoes	7000 kg	0.062 $ =	437 $
Beans	120 kg	0.375 $ =	45 $
Bananas	30 bun.	0.30 $ =	9 $
Vegetables	5 sack	2.60 $ =	13 $
Total value of sales			= 3183 $

MILK PRODUCTION

Input		Output	
Cows (No.)	= 3	*Milk* =	6000 kg
Annual replacement (heifers)	= 0.6	calves 460 kg	
		sold 3140 kg	
Total TDN required =	6044 kg	home 2400 kg	
forage = 5358 kg		*Meat* =	240 kg
concentr. bought = 200 kg		culled cow	
own prod. maize = 486 kg			
Minerals =	40 kg	Calves =	2
Bedding material =	1800 kg	Manure =	36 t
Labour (family) =	1500 h		
Tot. costs of purch.inputs	143 $		

YOUNG STOCK REARING

Input		Output	
Young stock (1 heifer,1 calf) = 2			
Milk (bull calf) =	100 kg	Meat	
Milk (female calf) =	360 kg	(bull calf) = 35 kg	
Minerals =	15 kg	Manure = 10 t	
TDN, total requirements =	2150 kg	*Heifers* = 0.7	
forage = 2080 kg		cow replac. 0.6	
milk = 70 kg		sales 0.1	
Labour (family) =	75 h		
Tot. costs of purch.inputs	17 $		

GOATS + SHEEP + POULTRY

Input		Output	
Mature sheep + goats =	2	Meat (goats) 60 kg	
Lambs =	1		
Hens =	10	Eggs = 1095	
Total TDN requirement =	1346 kg	Meat (poultry) 10 kg	
Concentrate (poultry) 74 kg			
forage (goats) 1272 kg		Manure 2 t	
Labour demand (family) =	110 h		
Tot. costs of purch.inputs	24 $		

Wages	181 $

FARM HOUSEHOLD CONSUMPTION

Product	Quantity	Price/unit	Value
Maize	1800 kg	0.100 $ =	180 $
Milk	2400 kg	0.131 $ =	315 $
Meat (poultry)	10 kg	1900 $ =	19 $
Meat (goats)	60 kg	0.633 $ =	38 $
Eggs	1095 p.	0.037 $ =	41 $
Beans	60 kg	0.366 $ =	22 $
Potato	200 kg	0.060 $ =	12 $
Bananas	10 bun.	0.30 $ =	3 $
Vegetable	7.5 sack	2.53 $ =	19 $
Total value of products farm consumed			= 649 $
Cash expenditures of farm house hold (excl. farm investments)			= 625 $

Bank Investments and/or savings = 1728 $

◄--·-- Flow of output or input which are
Inputs or outputs

ambu, Kenya. (From Stotz 1979, personal communication.)

capital costs are incurred for sprinkler irrigation. The ripe cherries are selectively picked and processed in pulperies.

Coffee production in the small-farm sector—which produces about half the Kenya crop—is far less intensive than in the estates. Less labour is employed per hectare of crop than in estates. The example in Table 8.8 and 8.9 reflects the average situation which was very favourable in 1977 because of high coffee prices. Yields of coffee, maize and milk in small farms are still far below the potential of the area. Land use, however, is intensive. Potatoes are a major cash crop and provide much productive employment. The tendency generally is to substitute grade cows for zebu cows and change from partial-grazing to zero-grazing types of dairying. Fig. 8.6 shows the structural model of a coffee–milk smallholding in Kiambu District, Kenya (compare Table 5.3), in 1977 when coffee prices were very high. The figure not only illustrates the high return per ha of land and per hour of work of such holdings, but also of the highly integrated nature of the various production processes.†

8.5.2. *Tea holdings*

The examples in §§ 8.4.3 and 8.5.1 indicate the wide variation of coffee-producing techniques and intensities. With tea the situation is far less variable. The assimilation of the tea plant drops rapidly once temperatures rise above 30 °C. Tea is therefore grown either in tropical highlands or under shade trees (Assam). The organization of production is closely related to the processing capacity of the tea factory. In Sri Lanka and Assam, there used to be many small factories each with a processing capacity of about 100 000 kg made tea a year which would require a tea area of about 50–100 of hectares. In tea processing, however, the economies of scale are very pronounced and the tendency is generally towards larger factories. In the Kenya Highlands, for instance, most of the factories manufacture 1 to 1·5 million kg of tea a year, and this requires, depending on yields, 500 to 1500 hectares of mature tea.

Three types of green leaf suppliers have to be distinguished:

1. Most tea is still produced by estates combining processing with green leaf production on a compact tea area surrounding the factory.

2. Innovations in transport reduce the need of having fields close to factories and external suppliers of green leaf are increasingly of importance. Besides the estate we find planter-operated commercial holdings whose leaf is sold to neighbouring factories.

3. In Japan and Taiwan smallholders still produce and process tea by traditional hand methods. In East Africa, however, roughly half of the tea originates from factories which are supplied with green leaf by a great number of associated smallholders who often also own the factories.

Tables 8.10 and 8.11 illustrate the labour inputs and costs of tea production.

† For advice on the design of this structural model I am indebted to P. Chudleigh (1979, personal information).

TABLE 8.10

Farm-management data of coffee and tea estates

Crop	Arabica coffee			Tea	
Country	Brazil	Kenya	Sri Lanka	India	Kenya
Year	1977	1975	1976	1975	1975
Method	Model built on accounts		Accounts[a]	Accounts	Accounts
Size of estate (ha)	260	162	n.a.	1196	506
Mature coffee or tea (per cent)	73	56	n.a.[b]	n.a.[b]	n.a.[b]
Immature coffee or tea (per cent)	15	—	n.a.	n.a.	n.a.
Other land (per cent)	12[d]	44[c]	n.a.	n.a.	n.a.
Labour force (ME per 100 ha planted)[e]	29	132	94	n.a.	191
Yield (t clean coffee or made tea ha^{-1})	0·8	1·17	1·0	1·70	1·98
Economic return ($ per ha Planted)					
Gross return	1428	1325[f]	n.a.	n.a.	n.a.
Operating costs	794	962	1238	1795	1440
Plantation maintenance (per cent)	42	49	23 ⎫		
Harvesting (per cent)	44	19	26 ⎬	73	69
Processing, running costs (per cent)	⎫	9	22 ⎭		
Depreciation of factory and buildings (per cent)	⎬ 14	5	5	3	6
Transport, marketing cases and other overheads (per cent)	⎭	18	24	24	25
Net return (before depreciation of plantation and interest)	634	363	n.a.	n.a.	n.a.
Wages and salaries as per cent of operating costs	50	53	52	52	54
Productivity					
Gross return ($ ME^{-1})	4924	1004	n.a.	n.a.	n.a.
Income ($ ha^{-1} planted)	1028	873	n.a.	n.a.	n.a.
Income ($ ME^{-1})	3555	661	n.a.	n.a.	n.a.

[a] Accounts of more than 100 estates. The data from India and Kenya on tea concern individual estates. [b] Almost entirely mature tea. [c] Mainly grass for mulching. [d] Roads, etc. [e] Total labour input in hours divided by 2400 hours per year. With the Brazil coffee estate the labour is mainly seasonal. With the other estates it is almost entirely permanent labour. [f] In 1977 prices and gross returns on the Kenya estate were twice as high as in 1975. *Sources:* Brazil: (personal enquiries); Kenya: (coffee): Kenya Coffee Growers' Association (1976); Sri Lanka: Ministry of Agriculture (1976); India: (personal enquiries); Kenya: (tea): (personal enquiries).

In Sri Lanka and India the industry is very old and much of the tea was planted more than 50 years ago. In Kenya the tea industry is younger and benefited from several decades of economic stability and modern management.

TABLE 8.11

Labour input into established tea (man-hours ha^{-1} year^{-1})

Country	Sri Lanka	Kenya (1)	Kenya (2)	Kenya (3)	Kenya (4)
Year	1976	1977	1977	1978	1977
Type	Estate	Estate	Estate	Smallholders	
Method	Accounts	Accounts	Accounts	Survey[a]	Case[b] studies
Yield level (kg made tea)	1000	2000	1976	1368	n.a.
Labour input (man-hours)					
Field operations					
Plucking	1380	3064	3040	1603	1170
Weeding	328	148	112	214	—
Pruning	56	68	176	189	131
Fertilizer application	39	29	32	35	10
Other work	64	—	—	576[c]	195[c]
Total	1867	3309	3360	2617	1506
Estate overheads	145	602	560	n.a.	n.a.
Factory	236	290	560	n.a.	n.a.
Total	2248	4201	4480	n.a.	n.a.

[a] Survey of Kenya Tea Development Authority farmers, with a bias for larger farmers. [b] Daily measurements of labour inputs in a few case study farms. [c] Transport from the field to collection centres. *Sources:* Sri Lanka: Department of Census and Statistics (1976); Kenya (1) and (2): Estate Records (1978, personal communication); Kenya (3): Lugogo: (1978, personal communication); Kenya (4): Swoboda: (1978, personal communication).

(a) *Tea estates.* Large-scale tea production in estates is generally characterized by the following farm-management features:

1. Tea production has developed over time into one of the most intensive types of tropical land use. In the examples in Table 8.11 the yields are between 1000 and 2000 kg of made tea per hectare, and the gross return varies between $1400 and $2000. The increase of yield over time has been more conspicuous than with most other tree crops. Improved planting and plucking techniques in conjunction with some fertilizer increased average yields in north-east India within 40 years from 625 to 1075 kg ha^{-1}. In the Kenya Highlands yields improved from about 1000 kg in the 1960s to a plantation average of about 2000 kg in the 1970s, mainly due to fertilizers and herbicides. Fields with high yielding clones, in a good tea climate, planted at high densities and with high applications of fertilizer (320 kg N ha^{-1}) yield in Sri Lanka 3000–4000 kg of made tea per hectare. Recently introduced pegging techniques for newly planted tea also contribute to an earlier yield.

2. A tea plantation is permanent and yield variations are minor. It is common to find tea plantations being used continuously for 50 years, and in Japan plantations are still operated that are over 150 years old. Replanting

becomes necessary when there are too many vacancies. A general problem of the tea industry is the prevalence of old plantations which require rehabilitation. Tea permits monoculture and tea may be planted after tea. Disease problems related to monoculture (root rot, red spider) are, however, gradually coming up. Intensive tea production is a soil-conserving type of land use. The hydrology effect is close to that of a natural forest. However, low-yielding tea with many vacancies usually implies heavy erosion damage.

3. The main costs in tea production are those of labour. Roughly 80 per cent of the costs of establishing a plantation and 52 per cent of the operating costs of an estate are for wages and salaries. Labour inputs into field operations vary between 2000 and 5000 hours ha^{-1}, and most of this is for plucking (see Table 8.11). 4000 shoots have to be picked for 1 kg of made tea. Total labour demands, including overheads and factory, hover around 2 ME ha^{-1} of tea, and much of this is female labour employed in plucking. Tea estates are very sensitive to wage levels, and success in economic terms depends on the availability of much low-cost labour, since operations cannot easily be mechanized. Plucking shears with which the harvesting performance per day can be increased from 10–15 kg of green leaf to 100–150 kg are used in Russia, Japan, and Taiwan, but only occasionally in Assam or Sri Lanka, because they reduce the quality of the leaf. The need to pluck at regular intervals and to process the leaf shortly thereafter imposes a rigid timing of operations. Thus, tea estates can be disrupted much more easily by strikes than can sisal or rubber estates.

The labour economy of tea estates has been drastically changed by the interacting effects of fertilizers and herbicides. Fertilizers roughly doubled yields in Kenya and increased the labour requirements for plucking accordingly. Herbicides, however, reduced the weeding input to insignificant proportions. The over-all effect is a slight increase in total labour per hectare and a change from weeding to plucking operations.

4. Tea production involves high capital investments. In Kenya, for instance, in 1977 mature tea was traded for about $6300 per hectare. To this the costs of the factory have to be added.

(b) *Smallholder tea.* Tea used to be a typical estate crop for several reasons: first because the green leaf must be properly plucked and processed within a few hours, and this is difficult to organize with numerous smallholdings, and secondly because of the importance of improved techniques in vegetative propagation, fertilizer use, plucking, and pruning techniques. Earlier efforts with smallholder tea in Sri Lanka failed. Nevertheless smallholder tea has become an important and successful industry in Kenya and more than half of the tea produced in that country now originates from smallholder-owned factories. Smallholder tea has also been introduced in other countries, including Tanzania, Uganda, and Ruanda.

The key to the striking initial success in Kenya was the very thorough organization, especially the close supervision of planting, husbandry, plucking (two leaves and a bud only), and collection of green leaf. Tea is grown in smallholdings that include numerous other activities like maize, potatoes, and milk production with grade cattle.

The welfare effect of well-organized smallholder tea is substantial, mainly because of the widespread employment and income effect in small farm communities. Experience in Kenya indicates, however, that a number of specific problems of smallholder tea production have to be considered:

1. There are much higher transport costs than on estates. The leaf has to be transported to collection points, and a dense road system has to be maintained from crop revenues.

2. Tea has to compete with other crops, and smallholders tend to neglect the tea when returns per hour of work are lower. Estates tend to operate their tea at fairly constant levels of intensity irrespective of price, while smallholders tend to be highly price-elastic in their input pattern. Tea estates are therefore a more stable element in the economy of the country.

3. Smallholder production in Kenya has not yet reaped the benefits from the use of fertilizers and herbicides. Yields are at roughly half the level of the estates, and the weeding problem is a major one.

4. Smallholder tea is less soil-conserving than most estate tea. There is less fertilizer use and less vegetation cover. The prunings are used for firewood, and vacancies are frequent.

5. Experience in Kenya indicates that two types of smallholder tea are viable. In one the tea hectarage per farm is so small that the work can easily be done with family labour only (about 0·25 ha). Such farms would usually combine tea with subsistence food and some dairy cows. In the other type tea is grown by small entrepreneurs who employ hired labour, and who usually tend towards a small-scale tea monoculture, often supplemented by some dairy activities. Clearly most of the expansion is with the latter group of suppliers.

Most of the problems of smallholder tea are not necessarily connected with small-scale operations, but are due to a lag in the adoption of innovations. Estates are quicker in the adoption of innovations, and with smallholders the welfare effect is higher. So the best solution to the production task from a national point of view is apparently the combination of both modes of production.

8.6. Estates and smallholdings with tree crops

Crops like cacao, rubber, oil-palms, and coconuts are tree crops proper. They differ from shrub crops dealt with before in that they need more years to reach the bearing age, their vegetation cycles are longer, and they present a less intensive form of land use. High costs for establishing plantations and for harvesting and low costs for plantation maintenance are the most obvious characteristics of this group.

8.6.1. *Cacao holdings*

The size of the cacao estate is not determined by the size of the processing unit as is the case with tea. Cacao-processing facilities can be designed to deal with the expected production, and it is feasible to process the product in small amounts. Cacao is therefore produced in holdings of widely varying size. Estate production predominates in Ecuador, Brazil, Costa Rica, and New Guinea. Smallholder production is prevalent in West Africa.

There are few crops where the gap between performance in research stations and in commercial farms are as wide as with cacao. Most of the estate cacao in Latin America is produced with very low inputs and low yields, as shown by the example from Ecuador (Table 8.14). Here cacao is a forest crop, maintained under shade trees, without fertilizers and pesticides. The slashing of undergrowth, harvesting, and processing are the only operations. Recent increases in cacao prices induced, however, new planting with high chemical and high labour inputs, which yield 1.0 t ha^{-1} dry beans and more. Most African cacao is produced in smallholdings.

Guillot's case study (1978, personal communication) of cacao farms at Boutazab shows in an exemplary way what can be expected once a 'rich' cash crop is added to traditional, relatively static subsistence-oriented food crop production under the condition of ample fertile land in a humid tropical lowland (see Table 8.12).

1. The traditional elements of the farming system remain largely unchanged: several plots are planted with food crops, each on a specific soil type and each with a specific crop association of maize, manioc, bananas, cucumbers, and sweet potatoes. Food crop production is geared to subsistence needs, and it is supplemented by rather important and time-demanding hunting and fishing activities which provide the animal protein.

2. The establishment of cacao plantations is a by-product of the food production. Cacao seedlings are planted into the cleared land. They are weeded with the weeding of the food crops, and they take over after them. The tendency is for more cacao to be planted than can be properly weeded, and this for two reasons: (a) planters tend to overestimate the labour capacity of their families, and they are not yet accustomed to the employment of hired labour, even though some Pygmy labour is available; (b) planting cacao establishes a claim to the land, and cacao is planted with preference close to a road or on particularly fertile land which thus becomes the individual property of the farmer. Also plantings occur with preference along the road so as to claim the hinterland for the planter (see Fig. 8.7). The critical time for the planter is the weeding of the cacao in the initial

SYSTEMS WITH PERENNIAL CROPS

TABLE 8.12

Farm-management data of cacao holdings

Country	Ecuador	Nigeria	Congo
Location	Quevedo	Alade (Yoruba)	Boutazab
Rainfall (mm)	2200	1592	1600
Year	1978	1963–5	1976
Type	Estate	Smallholding	Smallholding
System	Cacao–beef	Cacao–yams	Cacao–manioc
Method	Model	Sample	Sample
Number of holdings	1	43	14
Persons per household	n.a.	8·90	7·00
Labour force (ME)	35	3·80	3·00
Size of holding (ha)	680	7·77	17·50
Tree crops	380	2·73	4·76
Thereof cacao	300	1·93	4·76
Arable crops	—	1·10	2·52
Pasture	220	—	—
Bush and bush fallow	80	3·94	10·22
Cropping index of arable land	—	27	20
Cattle (numbers)	337	—[a]	—[a]
Yields			
Traditional cacao (t dry beans)	0·30	0·173	0·80
New cacao (t dry beans)	1·30	—	—
Economic return ($ per holding)			
Gross return	47 793[b]	615[c]	872
Purchased inputs	1063	9	3
Income	46 730	606	869
Wages for hired labour	29 415	62	1
Family farm income	—	544	868
Off-farm income	—	32	5
Total family income	—	576	873
Net return before taxes, rent and interest	17 315	—	—
Sales as per cent of gross return	100	75	66
Productivity			
Gross return ($ ha⁻¹ total land)	70	79	50
Gross return ($ ME⁻¹)	1366	162	291
Income ($ ha⁻¹ utilized land)	78	158	120
Income ($ ME⁻¹)	1335	159	290
Labour input (man-hours)	63 536	5517	2930
of which hired (per cent)	100	37	1
Man-hours per ME	1815	921	977
Income per man-hour ($)	0·74	0·11	0·30

[a] Some goats and sheep. [b] 31 per cent thereof from cattle. [c] Net of seed and sprays.
Sources: Ecuador: (personal information); Nigeria: Upton (1967*b*); Congo: Guillot (1978, personal communication).

years. Once the cacao has reached 3–4 years it usually provides a dense shade. However the maintenance weeding effort is too high for many. A significant part of the planted cacao disappears again under bush regrowth.

3. The size of the cacao plantation is primarily a function of the labour capacity of the family and of the willingness to invest labour into a crop which only produces after 6–7 years of growth. Such willingness varies widely, and we find therefore wide differences between families in cacao hectarage, cacao production, and cacao income. A few small entrepreneurs move ahead of the others and gain a dominant position in their community, employing Pygmy labour. The Pygmies, with a cultural background of hunting and gathering, are increasingly involved as hired labour with low productivity and low wages.

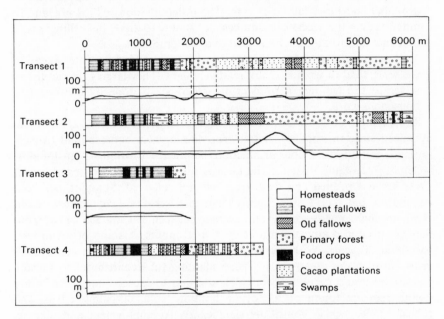

FIG. 8.7. Land use with growing distance from the road at Sembé, Congo (cacao and shifting cultivation). (From Guillot 1977, p. 164.)

4. Cacao farmers are rapidly involved into the cash economy. Some income is received from non-farm activities, and those who have had a spell of non-farm employment seem to be particularly active innovators. Food is bought to supplement on-farm production. Female labour which used to work for subsistence food only is gradually involved in work with the cash enterprise.

5. Cacao enforces sedentary living. Families tend to settle and have their houses close to the road. Land use is more intensive along the road than in the hinterland. Fig. 8.7 shows the land-use patterns which appear: (i) Close to the road are plots with food crops and short-term fallows; (ii) Next in

the transect are cacao plots mainly; (iii) Primary forest dominates with increasing distance from the road. Shifting cultivation gradually moves away from the use of virgin land towards a rotation cycle of two years of cropping, followed by three to thirty years of fallowing. In the average seven years of fallow are found, which indicates that the long-term tree fallow has been replaced with the medium-term bush fallow. Also the fallow pattern becomes more diversified, because land close to the house and the cacao plot is preferred for cropping.

6. A wide differentiation of income and the emergence of a class of commercial farmers can be observed, and this even though ample land is available and hired labour is rare. The differentiation is thus not due to differences in the access to resources, but simply to drive, the willingness to innovate and to invest labour into tree plantations.

7. It is remarkable that in this very early stage of land-use intensification the work load of women is still relatively light. They largely restrict themselves to the tending of food crops (clearance is done by men) and to household work.

The example of Yoruba cacao holdings in Nigeria (Table 8.12) shows an advanced stage in smallholder tree-crop development under humid tropical lowland conditions. Food production is practised within a bush-fallow system. A substantial part of the farm is under extensively cropped cacao. Trees are not pruned and dead trees are not removed. Spraying has been introduced, but crop losses, mainly through Black Pod (*Phytophtora palmivora*), are high. Yields are low and average between 300 and 500 kg ha^{-1}, but they are produced almost without purchased inputs. A stand of cacao lasts for about 30 years. With increasing age more and more bush and grasses grow between the cacao trees. Cacao planting thus contributes to turning forest land into a derived savanna. In addition to cacao a number of other useful trees are found on a Yoruba holding: cola, oil-palms, citrus, and robusta coffee, which contribute significantly to employment and income. The labour economy is no longer dominated by clearing and harvesting. Zuckerman (1973) found that 72 per cent of the total labour input is for weeding (Flinn 1974, p. 7). The system is no longer balanced. The area sufficiently fertile for cacao is gradually declining while arable subsistence cropping is expanding. The system could be turned into a stable one by intensive tree-growing techniques, much mineral fertilizer, and valley-bottom development, but development in this direction is not yet evident. In the early 1970s the system was very much in the process of producing derived savannas. But (as is shown by Table 8.12) it nevertheless produces a relatively high level of productive employment per family.

Dualistic holdings generally exist, with the men responsible for the tree crops and the clearance work and the women raising the food crops. The arable branch of the holding, organized by the women, tends to continue alongside the tree crops with practically no modification as long as this is possible.

8.6.2. Rubber holdings

Rubber is grown extensively by both large and small producers, although the objectives and the techniques of production differ. Production in estates is geared to the principle of maximizing financial returns, while smallholder rubber is part of an integrated farming system, with rubber serving as a supplementary source of income.

(a) *Rubber estates in Malaysia*. Rubber estates are mainly situated in Malaysia, Sumatra, and Sri Lanka. Tables 8.13 and 8.14 provide some

TABLE 8.13

Examples of input–output relations of mature rubber

Country	Sri Lanka	Sri Lanka	Sri Lanka	Indonesia
Location	Dartonfield	n.a.	n.a.	Sumatra
Year	1977	1977	1977	1975
Type of production	Estate (Crepe)	Estate (Crepe)	Smallholder (Ribbed smoked sheets)	Smallholder (Sale of latex)
Method	Accounts	Accounts	Accounts	Model
Labour input (man-hours)				
Weeding	243	60	60	24[a]
Fertilizer application	30	30	30	32
Plant protection	10	20	10	48
Roads, drains, etc.	43	30	10	8
Tapping	1304	1345	1064	800
Total	1630	1485	1174	912
Yield (kg)	1600	1345	1064	1772
Economic analysis ($)				
Gross return	1972	1659	819	648[b]
Material inputs				
Fertilizer	31	44	36	72
Pesticides	9	54	4	5
Herbicides	—	—	—	10
Tools, transport	44	30	30	17
Total	84	128	70	104
Manufacture	174	303	78	— [b]
Overheads	447	439	25	—
Income	1183	661	576	440
Wages	453	402	296[c]	137[c]
Net return before taxes, interest, and land rent	730	259	280	303
Income per hour	0·73	0·45	0·49	0·48

[a] Use herbicides. [b] Sale of Latex. [c] Field only. *Sources:* Sri Lanka (1): Peries (1978, personal communication); Sri Lanka (2) and (3): Chandrasiri (1978, personal communication); Indonesia: Ptsipef (1975).

TABLE 8.14

Farm-management data of various types of estate with permanent crops

	Sugar cane Kenya 1976 Project plan	Cacao Malaysia 1971–2 Accounts	Cacao Brazil 1975 Accounts	Rubber Sri Lanka 1969 Accounts	Rubber Ivory Coast 1971 Project plan	Oil-palm Malaysia 1969 Accounts	Coconut Sri Lanka 1969 Accounts
Crop	Sugar cane	Cacao	Cacao	Rubber	Rubber	Oil-palm	Coconut
Country	Kenya	Malaysia	Brazil	Sri Lanka	Ivory Coast	Malaysia	Sri Lanka
Year	1976	1971–2	1975	1969	1971	1969	1969
Method	Project plan	Accounts	Accounts	Accounts	Project plan	Accounts	Accounts
Number of estates	1	1	1	n.a.	1	1	1
Area of mature stands (ha)	12 300 (2600 estate + 9700 outgrowers)	324	33	303	10 000	2131	200
Yield (ha^{-1} and year)	54 t cane	0·539 t dry beans	0·45 t dry beans	1·006 t R.S.S.	2500 t dry rubber	21·0 t fruit bunches	7300 nuts
Estate capital ($ ha^{-1})							
Plantation	n.a.	n.a.	n.a.	n.a.	2016	n.a.	n.a.
Processing	n.a.	n.a.	n.a.	n.a.	312	n.a.	n.a.
Total	4262	n.a.	1141	n.a.	2328	2046	n.a.
Economic return ($ ha^{-1})							
Gross return	1695	230	478[a]	389	700	740	165
Operating costs	1192	200	186	253	283	448	96
Plantation maintenance (per cent)	68	54	9	11	14	28	50
Harvesting (per cent)			20	42	47	22	17
Processing (per cent)	14	10	9	14	16	14	—[b]
Maintenance of buildings (per cent)	11	36	n.a.	12	11	4	16
Salaries for management (per cent)			44	9	12		
Social charges (per cent)	n.a.		18	11	n.a.	32	17
Other overheads (per cent)	7		n.a.	1	n.a.		
Net return before depreciation, interest, and land rent	503	30	292	136	417	292	69
Wages and salaries as percentage of operating costs (per cent)	33	n.a.	77	67	68	n.a.	n.a.

[a] Prices in 1975 were much higher than in 1971–2. [b] No processing, sale of nuts to processing factory. *Sources*: Sugar-cane: (personal information); cacao: (1) Wood (1974, pp. 61–3), (2) Knight (1976, pp. 248 and 256); Rubber: (1) Sri Lanka, Ministry of Agriculture (1975), (2) Ivory Coast (1972*a*); Oil-palm: Little and Tipping (1972, pp. 18, 22); coconut: Sri Lanka, Ministry of Agriculture (1975).

information about the economics of estate production. The main farm-management characteristics are as follows:

1. Rubber production, although yielding much more than rain-fed arable farming under similar circumstances, is less intensive than, for instance, tea or coffee production. The established estate depicted in Table 8.14 yields 1000 kg ha^{-1}, valued at \$389. With traditional varieties it is about 7 years from planting to tapping, which is a significantly longer unproductive period than with the permanent crops dealt with so far.

2. Soil fertility is well preserved in rubber plantations, which have the appearance of a forest in temperate climates.

3. Manual labour is the main item of cost. Salaries and wages amount to between 67 per cent and 74 per cent of the total costs (see Table 8.16): 42 per cent of all costs (before depreciation and interest) are for harvesting, mainly tapping. There is no pronounced peak in labour demand. Tapping, the most important cost item, is carried out regularly through most of the year, with fixed assignments for each worker. This makes it possible to work with an all-the-year-round division of labour, one group being engaged in tapping, another in weeding, and so on.

4. High costs in plantation establishment in relation to the returns per hectare make depreciation one of the most important cost items. Thus rubber production requires a lot of capital per unit of output compared with other perennial crops.

5. One of the important advantages of estate rubber is the ease of intro-duction of innovations. New tapping techniques and new clones radically changed the economic prospects of rubber production. New clones yield—under the conditions of estate production—2000 kg ha^{-1} of dry rubber and more. New clones are, however, likely to be seriously affected by various diseases, and great care is required to prevent the loss of mature trees. High yields are largely the result of budding techniques. In some areas crown-budding is employed to combine a sturdy root system provided by one clone with the stem of a clone which yields much latex and a budded crown from a clone which is known for being resistant to wind damage (Sumatra) or fungus diseases (Amazonas Basin).

Economies of scale are very pronounced with increased rubber produc-tion per hectare (see Fig. 8.8). With tea the increased yields mean higher plucking costs. With rubber, however, costs of tapping remain almost constant irrespective of the yield. Moreover, planting can take place earlier. Plantations traditionally required 7 years before tapping could begin, but with the new planting material tapping may start after $5\frac{1}{2}$ years.

(b) *Rubber smallholdings.* Compared with the tea or coffee bush, which has to be manured, weeded, pruned, and plucked regularly to keep it in proper productive condition, the rubber tree is remarkably undemanding. Processing

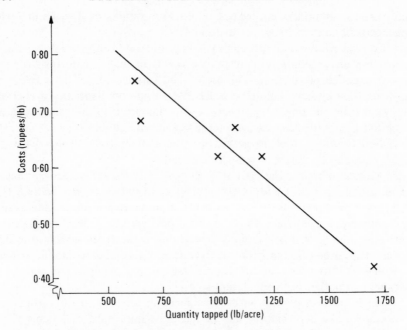

F<small>IG</small>. 8.8. Relation between yield per acre and costs per pound in seven rubber estates in Sri Lanka (from Hansen 1969).

is not tied to the producing farm. Rolling and smoking can be undertaken by neighbours or larger processing units. Rubber therefore attracted smallholders in Sri Lanka, Sumatra, Malaysia, and Nigeria to enter into commercial tree-crop production.

Rubber estates generally present a picture of orderliness, with trees grown in neat rows and the undergrowth kept down by constant cutting. Traditional smallholdings usually have a different appearance. The situation in Malaysia is depicted by Courtenay (1965, p. 124) as follows:

> . . . the valley floors are occupied by rice fields, irrigated by channels led along the hill foot from the central stream. Above . . . these canals run the main footpaths, just beyond which are located the Malay houses under the shade of coconut-palms and fruit trees. Behind and above the houses, the hillsides are planted with rubber, which may often be tapped by the rice farmer's wife or children.
>
> Smallholder rubber is often interplanted with fruit, coconuts and arable crops. The trees do not always form continuous stands but may be in clumps separated by other vegetation. Beyond the environs of the smallholder's house, the mixed stand of rubber and other crops usually gives way to a pure stand of rubber in the midst of tall undergrowth.

Smallholder rubber plots in Malaysia average between 1 ha and 3 ha. In contrast with estate planting (300 trees per hectare), final-stand planting on smallholdings is usually very dense, with 500–900 trees per hectare, the reason being that smallholders aim at high yields per hectare, disregarding higher labour input, while estates try to economize on labour by aiming at high yields per tree. Rubber tapping is not necessarily a regular task, and this makes rubber such an attractive crop for smallholders, since during the peak rice-planting and rice-harvesting seasons little labour need be devoted to rubber. Smallholders prefer to tap when time permits or when cash is needed, although irregular tapping may lower average returns. In some cases, smallholders tap only during periods when schools are closed—weekends and vacations. Another advantage of rubber for smallholders lies in the possible employment of farm family labour at low costs.

Smallholders use unselected seedlings usually. They do not receive maximum yields because of poor cleaning, lack of manuring, irregular tapping, overtapping when cash is needed, and often wasteful tapping techniques. They are, however, low-cost producers, and so are competitive with estates, and returns per hour of work compare favourably with arable crops (see Table 8.15). Whereas estates are severely hit by falling prices, smallholders may

TABLE 8.15

Examples of input–output relations in tree-crop production
(per hectare of mature crop)

Crop	Cacao	Rubber	Coconut	Cashew	Mango
Country	Cameroun	Thailand	Malaysia	Kenya	Kenya
Year	1971	1972	1968	1977	1977
Type	Estate	Small-holding	Small-holding	Small-holding	Small-holding
Method	Accounts	Sample	Sample	Sample	Sample
Labour input (man-hours)					
Weeding	74	135	112	237	237
Plant protection	166	—	—	—	30
Pruning	13	—	—	74	—
Harvesting ⎫	203	99	78	222	133
Processing and marketing ⎭		13	55	30	n.a.
Total	456	247	245	563	400
Yield	567 kg	397 kg	3358 kg	449 kg	1348 kg
	dry beans	dry rubber	nuts	nuts	fruits
Gross return ($)	154	91	154	56	44
Material inputs					
Tools ⎫		5	1	1	1
Depreciation of kiln ⎬	33	—	4	—	—
Other inputs ⎭		10	4	—	—
Income before overheads	121	76	145	55	43
Wages	48	n.a.	n.a.	n.a.	n.a.
Net return before taxes, interest, and land rent	73	n.a.	n.a.	n.a.	n.a.
Income per hour of work	0·27	0·31	0·59	0·10	0·11

Sources: Cacao: Wood (1974, p. 61); rubber: Goss (1973); coconut: Selvadurai (1968); cashew and mango: Thorwart (1977, personal communication).

cultivate other crops more intensively, and in the meantime the rubber trees may rest for 'rejuvenation', to produce larger yields when rubber prices rise again.

Table 8.16 illustrates the farm-economic situation of two rubber-growing smallholdings near Colombo, Sri Lanka. Rubber is the major cash crop and the plantations are considered as a capital asset. In both cases, new plantings with improved plants have replaced the traditional mixed stands. Coconuts are primarily consumed in the household or sold locally. Rice provides the food for the household.

8.6.3. *Oil-palm holdings*

Until recently, most oil-palm production took place in West African smallholdings and was part of subsistence-food production systems. In the traditional situation the palm is closely related to a number of activities and produces such diverse products as food, soap, fuel, construction materials, medicine (from roots), and feed (Koby 1978). Oil-palms are part of both the fallow vegetation and the mixed cropping pattern during the crop years. Self-sown seedlings are allowed to grow among the arable crops, and they have to compete with bush during the fallow years, so establishment costs are close to nil. Palm densities are low to allow interculture. Seventy-four palms per hectare, yielding 2000 kg of fruit bunches, are typical figures for Ghana (Thornton 1973, p. 34). The palm yields a great number of products: cash, subsistence food, material for soap, thatching and building material, leaves as feed for goats, and wine (18–27 kg from a 10–15-year-old palm) (Thornton 1973, p. 35).

A genetic breakthrough changed the oil-palm from a low-yielding but strongly integrated crop into a high-yielding one which is far less integrated in the land-use system than the traditional palm. High-yielding varieties are grown as monocultures in estates or in commercialized smallholdings with distinct oil-palm fields. The genetic breakthrough changed several variables whose economic importance are as follows (see Fig. 8.9):

1. Seed germination and nursery techniques have been improved. Better plants can be supplied at lower costs. The use of polybags produces stronger plants, and fewer infillings are required in the field.

2. Cover-crop mixtures have been developed which, supported by some fertilizer, are effective in reducing erosion, shading the soil, creating a dense mulch cover, controlling weed growth, and fixing nitrogen.

3. The use of herbicides allows the effective elimination of undesired weeds, in particular of *Imperata cylindrica*. A plantation of oil-palms can effectively be established on land heavily infested with *Imperata* without prior ploughing.

4. High-yielding varieties are available and they are usually not susceptible to disease. Two types of oil are produced from the fruit: palm oil from the

pericarp and palm-kernel oil from the kernel. New varieties have much more fatty pericarp. With traditional varieties 13–14 per cent of the fresh-fruit bunch weight is oil, and with the high-yielding varieties 20–26 per cent. Traditional varieties yield 5–10 t ha^{-1} of fruit bunches compared with 10–20 t ha^{-1} with new varieties under West-African conditions. Oil-palms in Malaysia and Sumatra, under favourable rainfall conditions and with much mineral fertilizer, produce 30 t ha^{-1} of fruit bunches, i.e. 6·9 t ha^{-1} of pericarp oil and 0·9 t ha^{-1} of palm-kernel oil.

5. High-yielding varieties are early yielders. In West-Africa the palms start to produce 3–4 years after planting compared with 10 years for traditional varieties. In south-east Asia oil-palms produce $2\frac{1}{2}$–3 years after planting.

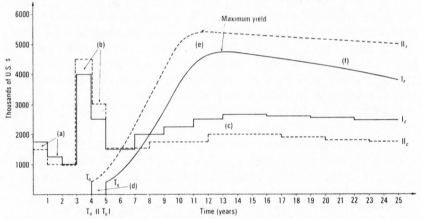

FIG. 8.9. The impact of technical innovations on the flow of costs and returns of an oil-palm estate at Owerri, Eastern Nigeria.
(From Walker (1976, pp. 37 and 67).)

——— I_c = Flow of costs of the Nigeria estate.
——— I_r = Flow of returns of the Nigeria estate (at $200 per tonne of oil and £213 per tonne of kernels).
- - - - - - II_c = Flow of costs of an improved estate based on innovations which are applied on very modern estates in Sumatra).
- - - - - - II_r = Flow of returns of an improved estate (based on innovations which are applied on very modern estates in Sumatra).
The possible impact of innovations:
(a) Savings due to (1) less costly clearing techniques, (2) improved germination techniques, (3) improved nursery techniques and (4) less weeding because of more effective cover crops.
(b) Increased factory costs which are caused by more effective extraction techniques.
(c) Savings in harvesting costs due to (1) low-stem varieties and (2) innovations in transport. The drawing also includes (3) innovations in weed control.
(d) Reduction of the time between planting and first yield.
(e) Higher returns due to (1) more fruit bunches, (2) heavier fruit bunches, (3) a higher fat content, and (4) a higher extraction efficiency.
(f) High-yielding varieties may be expected to show a slower decline in yields than un-improved ones, because the application of mineral fertilizer seems to have an important effect on the performance of older stands of high-yielding varieties.

6. High-yielding varieties are easier to harvest. Much of the produce is already available before the palm elongates and the height of mature palms is only a quarter of that of traditional varieties.

7. Oil-milling and the transport of fruit bushes to the mill have benefited from innovations.

Extensive areas where traditional oil-palm production takes place still exist, but increasingly both estates and smallholdings are planting new varieties.

(a) *Oil-palm estates*. Oil-palm estates were mainly established in Malaysia, Sumatra, and Zaire. More recently Ivory Coast, Benin, and Cameroun have entered large-scale production through para-statal bodies. Modern oil-palm cultivation lends itself to estate production. New varieties, and the corresponding husbandry techniques, can be applied *ad hoc* on large areas. The perennial nature of the oil-palm, its significant response to weed and bush control, its regular employment of the labour force on continuous harvesting, and the need to move the fruit to the mill as quickly as possible after harvesting in order to avoid loss in quality are factors that favour large-scale production. The estates are generally very large, varying between 500 and 2000 ha in Malaysia and 1000 and 4000 ha in West Africa.

The farm-management characteristics of oil-palm estates may be summarized as follows (see Table 8.15):

1. The initial capital investment is higher than with most other tree crops and is at about the same level as for tea and coffee. Establishment of the oil mill is the most important capital item; amortization charges are correspondingly high.

2. Gross returns are higher than with most other tree crops, but are lower than with shrub crops like coffee or tea. However, the natural risks of production (disease or other natural calamities) are low.

3. Oil-palm plantations offer an almost complete cover for the soil. The main cost item for modern plantation operation is mineral fertilizers.

4. The labour input is less than with tea, coffee, or rubber but higher than with coconut-palms. Most of the labour input is for harvesting. Estates usually operate with specialized labour units for harvesting, weeding, assisted pollination, and transport respectively. Rising wages—as they occur in Malaysia—induce the substitution of weeding by herbicides. The most effective way for increasing labour productivity is, however, the increase of yields per hectare.

5. Large estates organized by state firms, as is the case with oil-palm estates in West Africa, are usually burdened by high overheads and management costs.

(b) *Oil-palm smallholdings in Nigeria*. Most West African oil-palm production is still based on unimproved 'wild palm groves'. According to the analysis by Zeven (1967), there are three main types of oil-palm groves: secondary rainforest containing oil-palms, open palm bush with plant densities from a few palms per hectare to more than a hundred, and dense groves of palms. In addition there are here and there small- and medium-sized plantations established by small entrepreneurs.

Table 8.16 shows the farm-management data of such a small entrepreneur in eastern Nigeria. His plantation of 3·6 ha of mature palms yields 8·0 t of fresh fruit bunches per hectare, which is significantly less than the yield obtained by supervised contract growers in Sumatra (see Table 8.2). The produce is processed on the farm by traditional hand methods. The extraction efficiency is low, extraction rate (excluding the nut which is sold separately) amounts to no more than 8·1 per cent while modern processing factories obtain two and a half times as much.

Modern oil-palm production no longer has the close relations with the other farm activities that used to exist with traditional palm groves. Inter-culture is no longer practised. Newly established plantations carry a cover crop. The farming techniques on smallholder oil-palm plots are almost identical to those on large estates. Food cropping and cash cropping, which used to be integrated, have become distinct activities. A dualistic holding has developed with the modern plantation taking some of the land, while the rest shows the classical elements: the compound plot around the house with a lot of useful trees, the homestead field with interculture of maize, manioc, and yams under a light cover of wild oil-palms, and the outer field with a bush-fallowing system, mainly supplying manioc (see § 4.4.1).

Under traditional conditions the system is heading towards the competition for land between an extending plantation of modern oil-palms for cash and the requirements for fallow land to maintain yield levels of food-crops. Innovations offer new possibilities. In the long run most of the upland could preferably be planted with improved oil-palms, while food requirements could be covered by (1) the development of a multiple-cropping system in the valley bottoms, based on wet rice, and (2) by a multi-storey cropping system with zero tillage techniques on the compound and homestead plot (Lagemann 1975).

The problem of smallholder production is one of reorganization of farming, to take advantage of the genetic breakthrough, which is already exploited by the estates. An important effort to solve the problem has been made in Ivory Coast. Nucleus estates are increasingly surrounded by small-holders, up to 20 km away, who practise modern techniques of oil-palm production with the help of credit schemes and close technical supervision. The fruits are collected and processed by the mill of the estate. Initial experi-ence shows that plantation establishment in smallholdings is much cheaper than in the estates, but husbandry is not usually as good.

TABLE 8.16

Farm-management data of smallholdings with oil-palms, rubber, and coconut-palms

Country	Nigeria	Sri Lanka	Sri Lanka	Sri Lanka
Location	Okwe/ Umudike	Kaduwela	Kaduwela	Kandy
Year	1974	1969	1969	1969
Type of farming	Oil–palm– manioc	Rubber–rice	Rubber–rice	Coconut–rice
Method	Case study	Case study	Case study	Case study
Number of holdings	1	1	1	1
Persons per household	11·00	n.a.	n.a.	n.a.
Labour force (ME)[a]	4·10	2·70	2·00	2·70
Size of holding (ha)	6·48	2·42	1·18	2·20
Oil-palms	3·60	—	—	—
Rubber	—	1·22	0·61	—
Coconut-palms	—	0·80	0·20	1·40[b]
Total tree crops	3·60	2·02	0·81	1·40
Wet rice	—	0·40	0·30	0·80
Upland arable crops	0·50[c]	—	0·07	0·20
Fallow	2·38	—	—	—
Yields				
Oil-palms (fresh-fruit bunches ha^{-1})	8·0	—	—	—
Rubber (t sheet ha^{-1})	—	0·90	0·80	—
Coconuts (nuts ha^{-1})	—	5200	6100	3200
Rice (t ha^{-1})	—	2·00	1·80	4·10
Manioc (t ha^{-1})	13·0	—	—	—
Economic return ($ per holding)				
Gross return	1945	498	139	667
Tree-crops (per cent)	66	48	59	35
Arable crops (per cent)	34	52	41	65
Purchased inputs	10	43	12	107
Income	1935	455	127	560
Wages for hired labour	173	207	19	282
Family farm income	1762	248	108	278
Productivity				
Gross return ($ ha^{-1} crop land)	474	206	118	303
Gross return ($ ME^{-1})	474	184	69	247
Income ($ ha^{-1})	472	188	108	254
Income ($ ME^{-1})	472	169	63	207

[a] Including hired labour. [b] 159 coconut-palms, 70 fruit trees, 100 pepper vines, 200 plantains. [c] Occupied by 32 wild oil-palms and 99 other useful trees. *Sources:* Sri Lanka: personal enquiries; Nigeria: Lagemann (1975).

8.6.4. *Coconut-palm holdings*

Most commercial coconut production takes place in the numerous small-holdings of the Philippines, Malaysia, Indonesia, Sri Lanka, and the East African coast. Intermingled with the smallholdings are coconut estates.

(a) *Coconut estates in Sri Lanka.* Coconut estates in Sri Lanka vary in size from a few hectares to several hundred hectares. The characteristics of production are largely the same irrespective of size. Plant density is about 160 palms per hectare. Intercropping is rare. Some plantations are grazed by local cattle, but most managers prefer to farm without cattle. In the case study depicted in Table 8.14, each palm yields an average of 46 nuts a year, which makes 7300 nuts per hectare. About 5900 nuts are required to produce one tonne of copra. The nuts are picked 5 or 6 times a year, and the husk is separated from the nut in the field. The husks are used as mulching material. The kernel and the nutshell are separated at the farm. The flesh is dried in a copra kiln, the shells being used as fuel, after which the ash is scattered on the plantation. Besides wages, the fertilizer input is the main item of expenditure, amounting to 28 per cent of the total cost. Palms with a low yield, of less than 30 nuts a year, are interplanted with new palms, and chopped down as soon as the new palms begin to produce. There is a high degree of specialization. Permanent labourers are employed for weeding, transporting, and processing. Harvesting is done by outside task workers.

The essential characteristics of coconut-palm estates can be summarized as follows:

1. The intensity of land use in coconut-palm estates is lower than with other tree-crops. In the case study in Table 8.14, gross returns amount to $165 per hectare.

2. In addition, the lapse of time between planting and the first and the full crop is particularly long. Traditional varieties take 8–10 years before the first crop is picked. The full yield can only be expected after 15 years. The economic cropping time then may well extend to over 50 years.

3. The plantation requires comparatively few workers. The average figure for Sri Lanka is 20 ME per 100 ha of coconut-palms, which is lower than is noted in the case study (Lim 1965, p. 121). Labour costs are nevertheless the most important expense, because few other inputs are required to run a coconut-palm estate. There are no peaks in labour demand. The main work, harvesting and preparing the nuts, occupies about 60 per cent of the total labour and is spread over the whole year.

When estates establish plantations, they have to reckon with high costs, amounting, according to Piggot (1964), to $750–2000 per hectare, largely because of the long lapse of time between planting and the first harvest when the plantation has to be maintained without yield. The capital input is consequently high in relation to output and employment in mature plantations, but the day-to-day running of the estate requires little capital. The replacement of old palms is relatively inexpensive compared with the costs of new plantations, because young palms can be interplanted in old stands to replace them gradually and without a break in production. Even the processing facilities, copra kilns, do not require a high financial input. The size of kiln may vary according to estate production, so that estate size does not depend on the capacity of the processing facilities.

4. Estates usually carry pure stands, the soil being kept free from bush by slashing or cultivation. Coconut plantations in high-rainfall areas, however, lend themselves to mixed cropping or intergrazing, and this need not reduce nut yields provided the interplanted crops do not compete for water and nutrients. Thus interculture has to be restricted to seasons with high rainfall, and the crops have to be fertilized. Cashew, mangoes, bananas, pineapples, and various arable crops, like manioc or rice, may be found intermingled with coconut-palms. Grazing under coconut-palms is extensively practised in Sri Lanka and in the south Pacific. Fox and Cumberland (1962) report from an estate in Samoa that *Mimosa pudica* as a fodder crop under palms may carry $2\frac{1}{2}$ livestock units per hectare. The suitability of coconut-palms for interculture is based on two characteristics of the plant: (i) the way leaves are oriented permits a relatively high proportion of the solar radiation to reach the ground; and (ii) most of the roots are in the 30–120-cm soil layer. In pure stands the greater part of the surface soil is not effectively utilized (Nelliat, Bavappa, and Nair 1974, p. 263). Tall coconut-palms are particularly suited to interculture. Tall varieties are, however, more sensitive to disease, and their displacement by Malayan Dwarfs is a possible threat to the development of multi-storey cropping systems (Smith 1970, p. 594).

5. Considerable technical advances have been achieved in growing coconut-palms, as in the case of most other tree-crops. New varieties, which yield after as little as 6 years, and which produce 2000–3000 kg of copra per hectare of mature palms, are now available. Fremond wrote in 1968 about Ivory Coast: 'We know that yields of definitely even more than 3000 kg of copra per hectare, with nut-producing after 4 years, perhaps even less, are henceforth possible.'

(b) *Coconut-palm smallholdings*. Most coconut-palms are grown in smallholdings, because (1) the production process requires hand labour only, and (2) the palm lends itself to interculture. Smith describes the prevalent crop combination in Jamaica as follows:

> Traditional 'Jamaica Tall' coconuts have been established between the rows in banana fields and spaced very wide apart (about 100 palms per hectare). This has afforded a minor amount of competition to the bananas, which remain the farmer's prime consideration. The young coconuts suffer from banana shade, and the onset of bearing is delayed by up to 2 years. When coconuts come into bearing, the bananas are removed from the field and a natural sward is allowed to develop and is grazed by cattle. Some farmers plant an improved pasture of Guinea grass or Pangola grass, both of which do well under the light shade of the coconuts. On better soils bananas are again introduced as an undercrop when the coconuts are sufficiently tall [1970, p. 594].

The uses of the coconut are many and diverse, including copra, fresh nuts, fibre, fuel, leaves for roofing, and sap for beer. The dependence on large-scale processing facilities is not very pronounced. The main advantage of small-holder production arises from interculture in the early stages of plant development, which makes it much cheaper to establish stands of coconut-palms than in estates, where the young stand must be weeded for 8–10 years before harvesting begins.

Table 8.16 shows the farm-management data of smallholdings, three of which include coconut production. In all coconuts are not a major cash crop but are primarily grown for household consumption, while rubber and rice are mainly grown for cash. Another striking feature is the preference for non-farm work. In two cases, family members are employed outside the farm, and hired labour is employed for the farm work. All three major crops grown in the holding—coconut-palms, rubber, and rice—lend themselves to task work, which can be carried out without much supervision by the farmer. Here, as in most traditional smallholdings with coconut-palms, harvesting is done by specialized hired labour. In some parts of Sumatra and Malaysia the nuts are harvested by trained monkeys (*Macacus namestrima*) kept by contractors, who receive 10 per cent of the harvested crop.

8.7. Problems of farming systems with perennial crops

Both major types of exploitation, smallholdings as well as estates, show certain problems which may be summarized as follows.

8.7.1. *Problems of smallholder farming with perennial crops*

Perennial crops require a high input of labour and relatively few machine inputs, which would indicate that they could be grown more economically in smallholdings than in estates, where labour is more expensive. Experience shows further that tropical smallholders are usually very skilful in choosing optimum crops, crop sites, and crop mixtures. They are, however, inherently weak in husbandry by comparison with the performance of estates. Land, labour, and capital would frequently bring much higher returns if perennial crops were properly taken care of. Lack of proper husbandry is often accompanied by mixed cropping and interculture, although it is certainly not a necessary feature of these practices. Farmers in Taiwan show convincingly that very high levels of husbandry can be attained with interculture. However, few smallholders in the tropics cultivate their perennial crops at similar high standards. As a rule they can hardly be induced to cut down old trees. Weeding, bush clearing, and mulching are neglected in many cases. Pruning, which is not heavy work and offers much latitude in timing, is carried out negligently if at all, and this in spite of the fact that seasonal underemployment prevails.

Estates tend to maintain husbandry levels, irrespective of the ups and downs in producer prices, while smallholders tend to be highly price-elastic in their inputs. With falling producer prices small-farm plantations are neglected and with increasing prices they are re-established. The list of short-comings in husbandry could easily be expanded, but it is much more difficult to explain why many of the smallholders are poor in husbandry. There is certainly no single reason, but rather a host of possible explanations. Those most frequently put forward by the farmers themselves include the following:

1. The production of seedlings is usually a specialized task that can hardly be carried out efficiently in smallholdings. Optimum smallholder development consequently depends on specialized nursery producers supplying the seedlings and on the necessary loans.

2. Smallholders almost invariably say that husbandry falls short, even of their own standards, because of peak demands in labour and lack of the necessary cash for equipment.

3. Some difficulties are due to the length of the vegetation cycle of perennial crops. A smallholder usually inherits or acquires a mixed stand of trees and shrubs of varying age that are not planted in rows. Replanting in rows in one part of the plot after the other is difficult to organize. Infilling is a much more convenient way, but perpetuates a shrub and tree garden where yield-increasing innovations are difficult to apply.

4. Other losses in yield are due to the relatively short time-horizon of smallholders. Practices that reduce the total yield of the perennial crops but bring about early returns, like heavy interculture in young plantations and early tapping of rubber, are often preferred, because smallholders more or less consciously reckon with higher discount rates than do estate owners.

5. In many cases, husbandry is poor because the returns from an additional hour of work are not considered worth the additional effort. Most small-holders are still inhibited by the idea that their soil is not good, that the climate is adverse, that improved practices would not work, and that, consequently, increased efforts are not worthwhile. Because of this attitude, husbandry is often poor, and poor husbandry results in low returns per hour of work, which again leads to even poorer husbandry.

Although these and many other arguments may be valid, they do not afford a complete explanation of the lack of proper husbandry. If labour were indeed the main limiting factor, it should express itself in high marginal returns and the existence of a significant relationship between the labour available and the gross returns. This is not usually the case, however, The shortage-of-labour argument often offers no rational explanation, because a small but properly cultivated area probably yields higher returns per hour of work than do larger, neglected fields. The small but significant number of better farmers who work more intelligently shows that with less effort per unit of output, but with greater care, more money can be earned. Moreover, those smallholders who are not willing to take on additional work can usually employ seasonal labour for fairly low wages.

It would seem that the reason for poor husbandry is primarily a lag in cultural adaptation. Smallholders in many parts of the tropics, in particular in Africa, experienced within a few decades the change from shifting cultivation with temporary homes to sedentary farming with perennial crops. Shifting cultivators know how to adjust to changing natural conditions, and so in this respect smallholders are very knowledgeable. But they are less well prepared for the tasks of transforming natural conditions by conscious human effort and of improving yields by the proper maintenance of the plants—something that farmers with centuries of land shortage or decades of cash cropping behind them have learned how to handle. The quality of the husbandry found in especially worthwhile cash activities, and in areas that have had a land shortage for a very long time, indicates the starting points for better husbandry:

1. There must be a very strong incentive in the form of a high cash return per hour of work; or
2. There must be a serious, long-term land shortage, which leaves no choice but to take care of plants, soils, and animals; or
3. There must be production under close supervision.

The fact that many plantations are poorly husbanded may be considered an indication that, as a rule, neither the pressure nor the incentive has passed a certain threshold value required to get the smallholders on the way to proper husbandry.

The fact that peasant production can still compete with estates, despite obviously poor cultivation and husbandry techniques, is explained by the low costs of production and the lower susceptibility to risks. Investment in an estate consists of the cost of:

1. The factory.
2. The establishment of the plantations.
3. Workers' houses, a hospital, roads, etc.
4. Bridging the initial period when there is no yield.

The advantage of peasant production is that the financial outlay is low apart from building the factory, or can be raised by the farmers themselves. Smallholdings have no production overheads. The recurrent cost schedule is different, since up to a certain level most of the labour that is employed is either unpaid family labour or receives far less than the rates laid down by unions that apply on estates. Instead of having to pay back interest, a family is prepared to forgo the full income from the land until permanent crops produce yields. But it is not just a question of different demands on cash, or of the availability of labour that can scarcely be employed anywhere else; the whole situation is different as far as costs are concerned. A large-scale estate must bear the costs of clearance work, whilst in the family holding the land is in any case cleared to grow subsistence crops. It costs little extra to plant seeds or to set plants of a future perennial crop in the cleared land. The mixture of perennial crops with arable crops in the early years bridges the period when there is no harvest. A plantation that competes with arable crops grown on the same land certainly does not produce the same yield as a pure crop, but it is comparatively cheap and simple to set up.† These combined benefits give the smallholder a distinct cost advantage.

8.7.2. *Problems of the estate economy*

The farm-management advantages of growing perennial crops on a large scale do not lie chiefly in the labour economy. The use of labour-saving equipment and transport is important only with perennial field-crops like sisal or sugar-cane. The competitiveness of estates growing shrub or tree crops in relation to the farmer's cheap production methods is based mainly on:

1. The rapid and consistent use of technical advances in plant production.
2. The more efficient organization of delivery of the crop to the processing factory.
3. The more efficient processing of the product, and
4. The better access to markets and capital.

Most managers of large-scale holdings, particularly the employees of large plantation companies, take advantage of modern technology. They harvest large quantities per hectare and the product is of considerably better quality. The processing of the crop is more effective and results in a quality product that can be sold at higher prices. Moreover, by-products and residues are used more efficiently. All told, a return or income per hectare and per worker is obtained that is often higher than that of a smallholder. Estates, therefore, produce a high, taxable income. In addition, the proportion of the income charged as tax is much higher than in the case of smallholders.

On the other hand, there are a number of disadvantages.

1. The estate economy is exposed to high risks and uncertainties. Estates have to meet high interest charges on borrowed capital. The production is usually devoted to only one crop, and a shift in emphasis from one crop to another takes many years. The return from estates fluctuates widely because of natural hazards, for example wind damage with bananas, and because of changing prices on the world market. In some countries, the private-estate economy is threatened by nationalization, which acts as a barrier to investment. Semi-commercialized smallholdings are much better protected from risk: subsistence is guaranteed by cropping plants for immediate consumption, the proportion of over-all costs paid in wages is low, and there is normally no political danger.

2. Estates with perennial crops depend on an ample supply of low cost labour. The fact that the tropical countries can supply low cost labour is

---

† Thus in Kenya the cost of establishing a large-scale coffee estate before the first yield was \$750 per hectare whereas the research of MacArthur (1966) in neighbouring small-holdings shows a cost of \$437 per hectare (including the wages for workers belonging to the family).

one of the reasons why large-scale estates with perennial crops were established there, and explains their power to compete with products from temperate climates or products from the chemical industries. Rising wages, which are spreading with the organization of unions, hit large-scale estates with perennial crops very hard, because operations can be mechanized only with great difficulty or at unfavourable rates of substitution. Small-holdings operate with family labour and local seasonal workers, who are prepared to accept wages disproportionately lower than those of estate workers.

As risks in the marketing and political sectors increase, and as wages rise, smallholdings are becoming more competitive with large-scale producers. This is true more of perennial crops than of arable crops, which are more amenable to motorization and mechanization. Naturally this general statement needs to be qualified in individual cases. The various perennial crops and advances in agricultural technology differ in their suitability for large- or small-scale production.

Experience indicates that estate production is more competitive than small-holder production in a few fairly specific conditions.

1. When the production process is technically demanding. Where this is the case, the difference in yield between estates and smallholdings is likely to be particularly great, and this is especially the case where cultural change among the smallholders is slow.

2. If the produce of the land is bulky, requiring high transport costs. Here, the processing factory should preferably be located on the holding, with the fields as close as possible.

3. Where the produce is perishable and has to be delivered for processing or sale in large quantities of even quality at definite times.

8.8. Development paths of farming systems with perennial crops

There is a great deal of scope almost everywhere in the tropics for increase in the plantation area, in particular to replace shifting and fallow systems. Increases in area depend only partly on capital, labour, and technical information. Plantation establishment—both in smallholdings and in estates —is largely a function of prices and price expectations. However, perhaps even more important than the scope for extension in area are the numerous possibilities of increasing output per hectare. The following considerations are among the most important.

8.8.1. *The impact of technical innovations*

Organized agricultural research in the tropics began with perennial field crops like sugar-cane, progressed to arable crops and shrub crops, and was concerned only at a relatively late stage with tree crops. Growers are now in a position to reap the benefits of a great number of breakthroughs, particularly with tree crops. With traditional varieties, 30 per cent or less of the shrubs or trees produce 70 per cent and more of the crop. The development of high-yielding varieties and clones has significantly changed the picture. With various tree crops a doubling of yields is within reach. In addition, numerous other agronomic innovations are in the process of being introduced. They are summarized by Webster and Wilson (1967, p. 217) as follows:

(a) The use of improved planting material, selected to give high yields of good quality produce, but possessing good secondary characteristics.
(b) Cheap and practicable methods of propagating the improved planting material on a large scale.
(c) Efficient transplanting and early care of the young trees, in order to obtain a full stand of plants that grow vigorously and reach the productive stage as quickly as possible.
(d) Good horticulture practice in respect of such matters as spacing, shade management, windbreaks, pruning, etc.
(e) Methods of soil management, including soil conservation, cover crops, weed control, mulching, and manuring, which will maintain soil fertility and sustain high yields.
(f) Control of pests and diseases.
(g) Improved methods for the exploitation of the crop, such as modern techniques of tapping and yield-stimulation in rubber, or the use of synthetic growth substances to induce early and uniform fruiting in pineapples.

In recent years the introduction of arboricides and herbicides has greatly reduced labour costs and improved bush and weed control. The relevant economic aspect of these innovations is that rising wages can be absorbed by higher yields per hectare. In many cases the quantity of product required for the processing unit may in future be produced more economically on a fraction of the original land. The remaining land then becomes available for other uses.

8.8.2. *From mixed cropping to pure stands and interculture*

Yield-increasing innovations are best applied in pure stands, where they may be adapted to the special requirements of the crop. Mixed cropping is a handicap to mechanization also. We may assume, therefore, that mixed cropping is likely to decline. However, where land is scarce but where labour, capital and technical knowledge are relatively plentiful, various forms of interculture will probably be extended and intensified. This trend is seen clearly in Taiwan for example. Also there is the tendency to benefit from crop interactions by the establishment of modern forms of multi-storey cropping which combine high outputs with high efficiencies in the use of support energy.

8.8.3. *Towards a more pronounced spatial differentiation of land use*

The expansion of tree-crop production in smallholder farming implies changes in the spatial pattern of land use within farms and between regions. Tree-cash-crop production tends to expand on forest-type land while annual food-crop production tends to move into savanna-type land or valley bottoms (Cu-Konu 1978).

8.8.4. *From holdings with one cash crop towards diversified commercial production*

The tendency towards pure stands in order to apply effectively innovations specifically suited to one plant need not necessarily result in monoculture. This may be true of some perennial crops, but is certainly not true of all. In high-potential areas with high population densities, it is much more likely that production development will lead to even more diversified holdings. The introduction of intensive animal husbandry is of particular importance in this connection.

The development of Kikuyu agriculture, as depicted in Table 8.17, is a case in point. Originally, shifting cultivation with millet, maize, and beans prevailed. Increasing population pressure led to intensified fallow systems and permanent farming, reduced available grazing, and resulted in a process of involution. The efforts of the Swynnerton Plan brought a sweeping change in the farming systems. Coffee and other cash crops were introduced. Commercial milk production with grade cattle began, and some leys were established. Tea production, hybrid maize, potatoes, and the application of mineral fertilizer spread in the mid-1960s. The holdings became commercialized while diversification increased. In the mid-1970s grade cattle and hybrid maize are continuing to expand, but population growth and subsistence demands are tending to increase more rapidly than cash cropping. This and the rise in costs of mineral fertilizers in relation to product prices is reducing the tendency towards more diversified and commercialized holdings. There is the danger that a very promising evolution may turn again to involution.

8.8.5. *Increasing importance of perennial food-, fruit-, and fodder-crops*

Production of perennial tropical crops used to be considered as an activity mainly carried out on estates for export purposes. The facts are different. Tropical smallholders always relied heavily on perennials for covering subsistence requirements. Bananas are one of the staple foods of tropical rural families. With increasing population densities and with the encroachment of cropping on sloping, erodable land, it might well be expected that cooking bananas will gain in importance. Industrialization and urbanization also favour tree-crop development. Firstly there is the market demand for citrus,

TABLE 8.17

Evolution of the farming system in the Kikuyu districts of Kenya

Year	Type of farming	Cropping pattern	Livestock economy	Land rights
About 1860	Shifting cultivation	Maize, beans, mixed cropping	Ample grazing, Zebu cattle, goats	Ample land, rights of land use
About 1920	Fallow system	Maize, beans, sweet potatoes, mixed cropping	Limited grazing, Zebu cattle, goats	Limited land, rights of land use, communal grazing
About 1950	Permanent cultivation	Maize, beans, sweet potatoes, banana, wattle	Roadside grazing, mainly goats, some Zebu cattle	Rights of land use, most communal grazing turned into individually cropped plots
About 1960	Permanent crops + permanent cultivation + some ley farming	Coffee, maize, beans, sweet potatoes, banana, potatoes, vegetables, leys	Roadside grazing + some grade cattle	Private property rights; land can be leased and mortgaged
About 1978	Permanent crops + permanent cultivation + dairying	Coffee, tea, beans, hybrid maize, sweet potatoes, application of mineral fertilizer and manure	Grade cattle in zero-grazing systems, poultry, pigs	Private property rights; land can be leased and mortgaged; an active land market

This table is nothing more than a rough guide to the developments which took place and which are expected in the Kikuyu areas. See Ruthenberg (1966, p. 217).

mangoes, coconuts, papayas, etc. Secondly there is the tendency for suburban families and for part-time farmers to establish extended tree gardens.

There are more reasons to assume that perennial crops will gain in importance. They supply much productive employment, environmental amenities, and security. They reduce leaching and erosion; they favour intercropping and multi-storey physiognomies that seem to be very efficient converters of solar energy into dry matter and usable products. They produce reliable yields, which is particularly important in arid and semi-arid climates. Perennials seem to be much more efficient converters of mineral fertilizer into dry matter and food.[†]

[†] Experiments with oil-palms, for instance, show returns of 22 to 80 kg of oil per kg of nitrogen (Walker 1976, p. 30). With wet rice the response usually varies between 10 to 20 kg of paddy per kg of nitrogen. Nitrogen applied to oil-palms consequently produces 4 to 30 times more calories in the product than nitrogen applied to rice.

Modern production techniques with perennials mostly fall into the class of 'entropy-efficient' innovations that economize on the use of exhaustible resources (Randall 1975, p. 804).

8.8.6. *Integration of estate and smallholder production*

Of particular interest are the tendencies to combine in 'nucleus-estate' projects the advantages of estate production—the rapid application of innovations in producing and processing—with the advantages of smallholder production—the low labour costs in plantation development and the small-holder's preference for investment in tree or shrub crops. Schemes of this nature show a promising start with oil-palms (Malaysia and Ivory Coast) and tea (Kenya). The idea is that commercial estates (often public enterprises) are surrounded by a ring of smallholders, who obtain credit, advice, material inputs, machinery, and markets through the scheme management and who are subject to close supervision in their husbandry practices. Nucleus estates, combined with supervised smallholder production, seems to be a major alternative to the dualistic estate–smallholder economy that still prevails in extensive parts of the tropics. It is likely to become more popular because:

1. The capital costs are lower than with state-financed estate development.

2. Production is less sensitive to rising wages and labour unrest.

3. Projects of this nature are in line with the political philosophy of a great number of countries.

9. Grazing Systems†

9.1. Definition and geographical distribution of grazing systems

MOST of the livestock in the tropics is kept by arable cultivators, and in some of their systems livestock densities are very high, much higher than in grazing systems under similar natural conditions. Most additional animal production in the tropics will in future have to come from fertilizer-based land-use intensification in sedentary smallholdings. However, vast stretches of low-potential or scarcely populated land in Latin America and in Africa are utilized by pastoralists and ranchers who operate extensive grazing systems.

1. In total nomadism, the animal owners do not have a permanent place of residence. They do not practise regular cultivation, and their families move with their herds.

2. In semi-nomadism, the stock owners have a permanent place of residence, which is kept for several years. In the vicinity of the residence, supplementary cultivation is carried out. Semi-nomads usually travel for long periods of time with their herds to distant grazing areas. The husbandry practices of partial nomads, i.e. of farmers in permanent settlements who live mainly by arable farming but who own cattle grazing on the communal land, are similar (see § 4.3.4).

3. In nomadism the same land and water resources may be used by different pastoral units. In ranching, livestock, land, and water are within one organizational unit. Stationary animal husbandry organized according to commercial principles is practised on large stretches of land.

Arable cultivators increasingly master their environment and change it according to their objectives and, in irrigation systems for instance, production takes place in a largely man-made environment. Pastoralists and ranchers have to adapt their stock to the water resources and the fodder produced by a natural vegetation of grasses, shrubs, and trees. Climate, soil, vegetation, and animals form an integrated system, which the stockman, if he wishes to operate profitably, can influence only indirectly through such matters as the way he keeps his animals, the composition of the herds, animal density, times of grazing, and other considerations. Such considerations are especially

† This chapter only deals with extensive grazing systems. Some examples of intensive dairy systems are presented in Chapter 5.

important in dry regions, where the rainfall is not sufficient for arable cropping or tree-crops, e.g. in the marginal zones of the Sahara and the semi-arid regions of Australia and East or southern Africa. Nomadism and ranching in the humid savannas, on land that is also suitable for arable farming, are practised principally in areas with low population densities, as in Madagascar, for example, or in various parts of Latin America.

Commercial ranches are almost always very large. The establishment of extensive holdings with permanent stock-keeping was considerably helped by technical advances in fencing, water supply, and vehicle transport, and holdings of this kind are found principally in America, Australia, and various

TABLE 9.1

Types of indigenous stock-keeping in the tropics

Rainfall (mm per year)	Predominant type of farming	Animal mainly kept
Under 50	Only occasional nomadic stock-keeping	Camels
50–200	Nomadism with long migrations	Camels
200–400	All types of nomadism, transhumance, and supplementary arable farming	Cattle, goats, sheep
400–600	Semi-nomadism, transhumance, and partial nomadism, with stronger emphasis on arable farming	Cattle, goats, sheep
600–1000	Transhumance and partial nomadism—any semi-nomadism is mostly the result of ethnic tradition	Cattle
More than 1000	Partial nomadism and permanent stock-keeping—any semi-nomadism is mostly the result of ethnic tradition	Cattle

parts of Africa. The common aim of ranchers is the conversion of naturally grown herbage into marketable produce, and this objective is achieved by employing small numbers of people for large herds of livestock. In pastoral land use on the other hand, large numbers of livestock are kept by numerous households, and the aim is above all to transform the energy stored in grasses, herbs, and shrubs in the tribal area into a regular supply of food.

Rainfall is the primary factor determining indigenous ways of using grassland (see Table 9.1). Where there is less than 50 mm average rainfall per year, which means in effect no rain at all in many years, we find only occasional visits by nomadic herds, but in the semi-desert and the dry steppes, with between 50 and 200 mm of rain, there is usually systematic nomadic grazing with camels, such as is found, for example, in the Sahara and northern Kenya. Where there is a rainfall of 200–1000 mm, semi-nomadism and partial nomadism are of much greater economic importance, and we find this kind of economy chiefly in the African savannas. Where there is a higher rainfall, permanent stock-keeping is a better economic proposition and semi-nomadism where the rainfall is more than 1000 mm is a matter of tribal tradition rather than the only practical form of land utilization that conditions will allow.

Animal densities in grazing economies fluctuate greatly according to place and type of farming. In areas with a rainfall of 50–100 mm, 50 ha and more are needed for one livestock unit (the equivalent of one cow or seven sheep). In regions with 200–400 mm, the requirement is often 10–15 ha, and where there are 400–600 mm, 6–12 ha are unnecessary. Livestock densities increase with increasing rainfall, and a bimodal distribution of rainfall allows an area to carry much more cattle than a unimodal one. Grass grows abundantly in tropical grazing areas where there is a high rainfall of 1000–2000 mm a year. Where such high rainfall coincides with an even distribution of rain, such as occurs in some areas of Central and South America, animal densities of up to four livestock units per hectare are found.†

9.2. Total nomadism

9.2.1. *General characteristics of total nomadism*

Total nomadism is usually carried on under decidedly marginal production conditions, or where it is not worth trying to overcome natural difficulties by applying cultivation techniques. The basic principle lies in as much adaptation as possible to natural conditions. From the point of view of agricultural evolution, nomadic stock-keeping belongs to the same pre-machine category of land use as shifting cultivation, but this by no means prevents the type of adaptation being an extremely complex and carefully calculated system (Schinkel 1970).

The nomad's animals have to keep moving because water and fodder are not sufficient for them to be kept permanently in one place. A consequence of the animals' movement is the movement of the household, since the people live directly off animal products, particularly milk, and they are thus obliged to stay near the herd. This permanent migration precludes regular arable farming. The nomads' production is not tied to one particular place, since it comprises only animals and the implements necessary for animal-keeping.

† Water ranching with buffaloes is reported from Indonesia. During the rainy season the Barito River and its tributaries flood an alluvial plain of about 600 000 ha. Large parts of this area become covered by water hyacinth and a variety of grasses which are nutritious fodder for ruminants. Buffalo owners erect platforms of hardwood logs which are above the rainy season water table. Animals are coralled at night on the platforms. At sunrise they descend, and graze while swimming or wading (Groenewold 1974, p. 289).

The pastures grazed by the animals are either freely available to all the members of the tribe or the group, or the nomads have to rent them.

The nomads who practise *horizontal migration*, as in the Sahara or Arabia, chiefly keep large animals (camels) and trek with their animals to where rain falls and fodder grows. On the other hand, there are nomads who practise *vertical migration*. They are rare in the tropics, although typical of nomadism in the subtropics of the northern hemisphere, where people who possess principally sheep and goats migrate between different altitudes that temporarily carry fodder. Their routes are mostly shorter than those of the nomads of the plains, since the seasonal growth of fodder at various altitudes helps to ensure an even supply.

Nomadic groups operate chiefly in areas with little access to a market, and aim at diversified production in order to cover personal consumption as fully as possible. They have only one line of production, namely stock-keeping, with the result that this activity has to perform a disproportionate number of roles compared with that of any single activity in arable holdings or commercial stock-keeping. It has to provide:

1. Milk and meat for the household.
2. Wool and skins for the household.
3. Dung for fuel.
4. Mounts and pack-animals.
5. Cash products to pay for the purchase of crop foods, other plant products, and non-agricultural consumer goods.

These aims of stock-keeping must be fulfilled as regularly as possible, because storage presents much greater problems to a nomadic than to a stationary household. The natural conditions in a dry climate favour specialization in animals such as camels and sheep, which are capable of tolerating heat and trekking long distances. But the various needs of the household mean that the dominant type of animal is often supplemented by horses, mules, and donkeys for transport, and by goats, hens, and sometimes cattle for a more diversified food supply. Cattle are more demanding with respect to water and fodder and are not so mobile as camels or sheep, and for these reasons they are not characteristic of total nomads but of semi-nomads and the stationary population. The nomadic household depends especially on a regular supply of milk, which is the staple food. The importance of the continuity of food supplies in nomadic systems is so emphasized that live-animal products (milk and blood tapped from live animals) are the ones most utilized, with terminal products (meat and the blood of slaughtered animals) being mainly consumed only on special occasions (Dyson and Hudson 1969, p. 76). Thus animals like the camel, with a long lactation period, are preferred, for the female camel supplies about 2–4 kg of milk daily more than is needed by a calf during a 12–18-month lactation period (Capot-Rey 1962, p. 303; L. H. Brown 1963 p. 9; Schickele 1931, pp. 35 *et seq.*). In the hot season, camels need to be watered only every two or three days, and in the cool season only every 2–3 weeks.† They are remarkably tolerant of water with a high sodium and magnesium content. They rarely stray, and can defend themselves against small predators like jackals, and, therefore, when they are not needed near the camp for milk or as pack-animals they graze in herds of 30–40 at distances up to 80 km from the camp. Where there is no danger of theft, camels are left to themselves, and they are therefore able, without any use of labour, to eat fodder that cannot be reached by other types of domestic animal because of the long distance from water.

To ensure that animals, and particularly lactating animals, have a fairly regular supply of fodder, the nomadic household treks after the available vegetation, the camp being shifted as soon as the fodder in its vicinity is used up or becomes too dry. Usually the pastoral mode of life is characterized by seasonal migrations. Because of the uneven spatial distribution of watering places, large areas of rangeland can be used only during the short rainy season, when temporary pools are available. For the greater part of the year the herd must withdraw to areas with a moister climate or to valleys with off-season grazing and waterpoints. In very low rainfall areas these seasonal migrations may extend over several hundred kilometres. Mostly they extend to shorter distances (10–20 km) between rainy season upland grazing areas and dry-season lowland grazing areas (Barral 1977, p. 54). These seasonal migrations have to be distinguished from drought-induced migrations which may extend over long distances and often lead into territories unknown to the pastoralists. The nomad is by no means always on the move, but his mode of living and farming is characterized by a frequent change of camp, which in turn makes it practically impossible to crop fodder. There is in any case no permanent base at which to store fodder reserves, so that when and where fodder is scarce the shortage is overcome only because the nomad selects and breeds animals with greater ability to store fat, as portable reserves in the animal's body, and to travel long distances, so that areas with little fodder can be crossed.

The aim in breeding is not so much a high yield of meat, wool, or milk per animal as a high level of resistance to trekking, drought, heat, cold, disease, and periodic shortages of food. Advances in breeding geared towards increased output per animal and requiring improved animal feeding can scarcely be applied by nomads, who have to be prepared for a high production risk, particularly because migration leads frequently to the spread of animal diseases whose control is made more difficult by the need to travel. There are no stables or shelters to protect animals from extreme weather conditions,

† Webster and Wilson (1967, p. 337) write in this connection: 'During the northern winter, or rainy season, the moisture content of the trees and bushes browsed by the dromedary is quite sufficient to supply the total water requirement of the animal, and water may not be drunk at all during this period. In the dry season the vegetation is desiccated and extra water, over and above winter requirements, is required for heat regulation.'

theft, or predators, and stock losses during droughts with extreme shortage of fodder and water are particularly serious. Experience shows that most nomads can expect that at irregular intervals a large proportion of their animals will die.

The elements of nomadic enterprises are well summarized by Barral (1977):

1. Different types of livestock.
2. Rainy season grazing areas with (i) temporary and (ii) permanent waterpoints.
3. Dry season grazing areas with permanent waterpoints.
4. Distant migration areas in case of a major drought (certainly not all pastoralists have knowledge and access to such areas, and some, in case of drought, move not knowing where they will arrive, and whether they will be accepted by the local people).
5. Land for crop production by nomads (short-maturing millets) or access to sedentary cultivators for the exchange of animal products against crop products.

Because of the high production risks, the nomads try to accumulate herds that are as large as possible. The production aim is not financial gain, but rather the supply of household needs, provision against possible emergencies, and increase in social status, which rises in proportion to the number of animals owned. Nomads frequently have, in their livestock, assets of considerable value, but these cannot be regarded as economic capital, only as hoarded wealth. Livestock do not represent or indicate wealth: they are wealth. Moreover they are a means of social fulfilment and personal satisfaction, whereas comparable satisfaction cannot be obtained through money (Dyson and Hudson 1969). The society of the nomad is not organized on egalitarian principles, such as we often find among arable farmers, but is structured according to the number of animals owned. The possession of animals is by no means evenly distributed among the families of a tribe. Normally 10–30 per cent of the families own 60–90 per cent of the animals.†
On the other hand, the natural conditions of production make it possible for the poor to acquire animals and rise in the social scale. For the wealthy nomad there is little sense in migrating with very large herds. A frequent change of camp would be necessary, and to avoid this it is usual for the owner of a large number of livestock to copy the practice of arable farmers in adjoining areas and have his animals tended in groups of limited size by herdsmen, who are paid with a proportion of the young animals.

† In two Masai group ranches we found the following distribution: Olkarkar: the richer third of the members owned 76 per cent of all cattle and the poorer third 6 per cent. Kiboko: the richer third of all members owned 67 per cent of all cattle and the poorer third 9 per cent. These data probably underestimate the degree of concentration, because members with larger herds tend to send their sons with some cattle to other Masai ranching groups (Jahnke, Ruthenberg, and Thimm 1974, p. 40).

The labour input in nomadism is sometimes high in relation to the output. Herding and watering are the two major activities. Both tend to promote co-operation between families. The pooling of certain types of animal for specialized grazing and the joint effort in bringing water up from deep wells is easier if families work together. Labour requirements in nomadism occur mainly in the dry season, while those of arable cultivators occur mainly in the rainy season (Barral 1977, p. 63).

Nomadism is characterized by a great degree of interdependence between different pastoral units and between pastoralists and arable cultivators, and this the more so the lower the grazing potential. Different pastoral units interact in their common use of water and grazing resources. Herds meet at water places which facilitate the spread of diseases. Pastoralists depend on arable cultivators through the exchange of milk, meat, and butter against millet and sorghum, or the exchange of manure against food. In West Africa pastoralists are often the caretakers of livestock owned by arable cultivators.

9.2.2. *An example: camel nomadism of the Gabra, Northern Kenya (Torry 1973)*

The Gabra do not cultivate. By breeding four species of stock—camels, sheep, goats, and cattle—each adapted to a particular niche within the environment, they seem to have made the fullest use of a semi-desert, most of which receives less than 200 mm of rain per annum. The pastures are few and widely scattered, while the permanent water holes are fewer and fairly concentrated in space. Water and pasture conditions interact with the different needs of different animal species to impose a regime of multiple, mobile, and dispersed settlement. In the dry season the Gabra expand from main camps into satellite camps for camels and small stock. These camps move 60–130 km, making an average of twelve moves per season. Dry-season grazing is assured by the following differentiations:

1. Camel cows in milk and their calves graze around the main camp which is usually a larger unit consisting of about 17 homesteads with 150 persons (p. 103).

2. The rest of the camel herd grazes outlying areas in the hot plains where an important part of the feed originates from xerophytic shrubs and bushes. Their grazing radius extends up to 80 km from a permanent point of water. Camels supply meat, milk, some blood, and transport to the household. Their milk supply extends through the dry season, when hardly any other milk is available. Their drought-surviving capacity is unsurpassed by any other animal species the Gabra can keep. Camels are therefore essential for the livelihood of the Gabra. They are, however, slow in reproducing, they are sensitive to tick-borne diseases which are frequent in bush country. They are therefore supplemented by small stock and cattle.

3. Small stock graze more closely to water points and camps. Because of

TABLE 9.2

Management data of a nomadic livestock production situation: the camel nomadism of the Gabra in Kenya

Location	Gabra
Rainfall (mm)	100–300
Year	1970–1
Type of production	Camels, small stock
Method	Interview
Persons per unit	8·5 [a]
Labour force (ME)	6·7[b]
Land availability per household (ha)[c]	2202
Livestock per unit	
Camels	30
Cattle	15
Small stock	130
Camel herd composition	
Cows (per cent)	41
Bulls and mature steers (per cent)	3
Heifers (per cent) Immature steers (per cent)	34
Calves, female (per cent) male (per cent)	22
Performance indicators	
Hectares per livestock unit	37
Calving rate (per cent live births)	40
Mortality as percentage, year 0–1	n.a.
cows other	10
Offtake (per cent)	n.a.
Output (kg liveweight ha^{-1})	2·56
Output (kg milk ha^{-1})	1·85
Production per unit	
Subsistence (milk kg)	4065
Meat slaughtered animals (kg)[d]	2761
Animals sold	3 sheep/goats
Hides sold (number)	3
Economic return ($ per unit)	
Gross return	509[e]
Purchased inputs[f]	—
Family income	509
Sales in percentage of gross returns	2
Productivity	
Gross return ($ ha^{-1})	0·23
Gross return ($ ME^{-1})	76
Labour input	20 100[g]
Income per hour $	0·03
MJ per man-hour	4·65

[a] Usually two households acting as a unit. [b] All persons above 7. [c] Total areas divided by the number of households. [d] The Gabra consume slaughtered animals only. [e] Not including blood and collected foods. [f] Amounts are expected to be negligible. [g] Estimate for a dry year, when herding and watering requirements are particularly high. *Source:* Torry (1973).

their ability to feed on shrubs and bushes and their greater drought
resistance, small stock can utilize areas which are not open to cattle. Small
stock supply some milk and meat in small units. Cash income is mainly
derived from selling small stock.

4. Cattle-keeping is a minor affair with the Gabra and largely restricted to
a few bush areas with higher rainfall.

Table 9.2 gives the costs and returns of a modal Gabra homestead. The
low output per hectare and animal reflects the marginal conditions of the
area. The monetary return per homestead is not low by East African standards, but it has to be considered that living on meat mainly is a rather
expensive way of securing subsistence. The possibility of improving living
standards by eating less meat, selling more animals, and buying food-crop
foods (practices which seem to have spread amongst the Kenya Masai) is not
yet realized by the Gabra.

The striking aspect about Torry's analysis is the high labour input. The
Gabra normally work 9 hours per day and about 3000 hours per annum, only
to secure subsistence. Children above the age of 7 years are fully employed in
the production process. Two to three families (of about three to four persons)
group themselves into homestead units to economize in herding and watering.
Conspicuously high are the labour requirements of watering which absorb
roughly half of the total working time. Camel watering is toilsome. It implies
walking over long distances, a time-consuming organization of water use at
the well, and in particular working in a 'human chain' to bring the water from
a deep well to a trough. Labour hours per labourer are higher with the Gabra
than with any other system described in this book, and the return per hour
of work is the lowest.

9.2.3. *Problems of total nomadism*

Nomadism allows the use of areas of land that could not be used by a
permanent population without motor transport and a system of ranching.
Compared with the rancher, who must invest large sums in tracks, vehicles,
fencing, and watering places, the nomad operates at a low level of expenditure
and produces cheap animal products. Total nomadism represents therefore
an agricultural system that is adapted to a pre-technical situation, and that
might still be considered the best economic choice where price conditions
and natural conditions do not justify investment in ranching. Total nomadism,
however, is not open to the opportunities of technical progress that are
available to ranchers. It is technically difficult and expensive to keep a check
on the health of wandering herds, and cross-breeding with more productive
types causes more harm than good if the fodder basis is not modified. Because
of the marginal nature of the nomads' territories, it is scarcely worth their
while to try to improve the feed situation by cropping fodder, sowing more
productive grasses, and rotating the pasture.

Static production techniques in total nomadism contrast, however, with economic and social changes in and around the area where they live. This presents the nomads with a number of problems, which include the following:

1. Diminishing grazing areas. The expansion of arable cropping leaves the least productive zones to the pastoralists. Reliable dry-season grazing areas in valley bottoms are increasingly occupied by arable cultivators. The feeding value of crop residues is often substantial. According to Toutain there is no major decline in feed output due to arable cropping on former pastoral land (1978, p. 7). The problem is primarily with the wide variation in the output of straw and other crop residues, according to rainfall. The variation of dry matter supply over the years is much less with natural vegetations.

2. Growing competition from livestock kept by arable cultivators in traditional dry season grazing areas which have been turned into crop areas.

3. Loss of income because the motor vehicle is competing with traditional animal transport—the use of draught cattle and tractors reduces the sale of draught camels, and in many places prices for draught camels have dropped to a fraction of their former level.

4. Loss of income because urban traders are competing with the nomads in their traditional trading function.

5. Loss of the political control that they were traditionally able to exert over entire regions.

6. The attempts that development administrators are increasingly making to control the incidence and limit the degree of movement of the human population. Public authorities rarely take the side of the pastoralists.

The nomads have had to accept a considerable reduction in status, employment, and income at a time when new requirements have established themselves, and the consumption of tea, sugar, and alcohol has become common. Moreover, the population of nomadic tribes has increased rapidly in recent times. Measured in terms of traditional methods of land use, their territories are proving to be more and more 'relatively overpopulated'. The Sahel drought in 1972–4 brought the problem into the open.

Much evidence about the impact of droughts on pastoralists (nomads and semi-nomads) and their grazing areas has been collected and may be summarized as follows:

1. The immediate reaction to drought is a fuller exploitation of the available resources. More distant grazing grounds are frequented, and trees are lopped and bushes chopped down.

2. Commercial ranchers tend to sell stock in case of drought, and with pastoralists the sale of mature male animals has also been observed (Bernus 1974a). Generally, however, sales remain minor in the early stages of the drought. They are a desperate last step for those without any hope of migration. Cattle prices drop rapidly during the drought (Barral 1977,

p. 103). West African pastoralists prefer therefore to migrate with their herds into unknown territory, even in the knowledge that the likelihood of heavy losses is high (Santoir 1977). The expected loss through sales at low prices is usually deemed higher than the expected loss through migration.

3. In years of drought migration patterns change and much longer distances are involved. The chances of herd survival are best for those who depart early (Barral 1977, p. 100). In West Africa pastoralists have to pass the crop belt which developed between their traditional grazing areas and the grazing in the Sudan zone further south. This involves not only long distances but high losses in the savannas through feeding problems because Sahel cattle are not accustomed to the low quality of dry-season grazing in the Sudan zone (Barral 1977, p. 95), and through tick-borne diseases and trypanosomiasis.

4. Droughts change herd composition. They primarily reduce the cattle population. Camels and goats have an amazing survival capacity (Bernus 1977, p. 214). With cattle the losses are primarily amongst the young stock and the old. Cows in the age class 4–8 show a higher survival rate than younger and older animals (Santoir 1977, p. 48).

5. Droughts reduce forage production for a number of subsequent years and they change forage composition. Toutain's preliminary estimates after the Sahel drought indicate that total forage production in 1974 was 20 to 25 per cent below the 1955 level (1978, p. 4). Droughts drastically reduce tree and shrub populations (Boudet 1978). Perennial grasses are more affected than annuals (Granier 1975, p. 225). The seasonality of fodder supply becomes more pronounced.

6. Droughts tend to promote ongoing trends in cultural change. They quicken the change in diet from animal to plant products. They induce pastoral people to take up arable farming, in particular irrigation farming.

7. Droughts seem to affect different people differently. The Sahel drought inflicted heavier losses on some ethnic groups (Mourides) than on others (Fulani) (Santoir 1977). Livestock losses rarely destroy the herds of total communities (Barral 1977, p. 97). It is much more the case that some families move too late or are worse hit by disease (Bernus 1977). Data from Campbell indicate that the relatively poorer Masai families lost relatively more cattle in the 1975–6 drought than the wealthier ones (1978, p. 12).

Most conspicuous, however, is the tendency of those pastoralists who remain with sufficient livestock to return to their traditional pattern of life and animal husbandry (Santoir 1977, p. 49).

9.2.4. *Development paths of total nomadism*

Since nomadism is found mainly in areas where ranching is not a paying proposition, and since these particular people have little desire for change, it is usually advisable to leave them undisturbed, i.e. in the anteroom of economic development. It is particularly difficult and expensive to make nomadic tribes stationary and to integrate them into a modern economy. The natural habitat does not favour the transfer from nomadism to semi-nomadism or partial nomadism. Where herds are concentrated near the camp the surrounding pastures are overgrazed, and parasites and deficiency diseases increase. It costs $6000–15 000 per family to make them stationary within the framework of irrigation projects. The nomads' attainments in arable farming are mostly poor, unless every operation can be closely supervised, as in the Gezira scheme in the Sudan, but as a rule there is insufficient state control and state authority. Nomads are particularly reluctant to farm small areas intensively. Experience has shown that nomads can be induced to become stationary much more easily through mechanized grain cropping than through irrigation farming. Where this is not possible, towns and industry offer more incentives to become stationary than does crop production. Where employment in non-agricultural pursuits and in arable farming is insufficient to absorb the nomad population, it is probably economically and socially wise to let them continue their traditional way of life instead of wasting funds trying to improve a land-use system that will anyway gradually disappear in the course of economic development.

9.3. Semi-nomadism

9.3.1. *General characteristics of semi-nomadism*

Total nomadism, as depicted in the last section, is restricted to decidedly marginal climatic zones. Its economic importance, measured in terms of both subsistence and marketed production, is very slight, whereas semi-nomadism and partial nomadism, which are practised particularly in the African savannas, are of greater significance. We find stock-keeping, with little or no supplementary arable farming, among the semi-nomads of West Africa (Fulbe, Peul), Madagascar (Sakkalave), or East Africa (Masai, Turkana, Pakot, etc.). The principal animal kept is the cow, but there are often quite large herds of goats and sheep. Partial nomadism is even more important and characterizes stock-keeping among those African tribes that practise a considerable amount of arable farming.†

The organization of stock-keeping in the savannas is fairly uniform. The fodder basis consists of open bush, untended pastures, fallow, and harvest residues. The herdsmen usually have expert knowledge of their pastureland. As Allan (1967, p. 4) writes:

They know the feed value of the different grazing-and-browse species, which they usually distinguish by specific names; and they recognize ecological associations,

† Partial nomadism is further considered in § 4.3.4.

or pasture types, and can assess their value and stock-carrying capacity at different times of year. Masai herdsmen in Kenya and Tanganyika pointed out to me various species and associations that they regarded as good for supplying mineral deficiencies, for conditioning animals, for improving the potency of bulls, and for making milk and beef.

They often show great skill in adapting to the growth patterns of different fodder plants in the rainy and dry seasons and to the different feed values of shrubs, trees, and grasses. They have exact knowledge of the position, quality, and capacity of watering places, but they traditionally operate under technical conditions and with economic aims that result in low output and make the introduction of technical advances difficult or even impossible.

The livestock have to go long distances to watering places and to pastures, for at night only a kraal can protect them from wild animals and thieves. This prevents them from taking advantage of valuable night-grazing time. During the day, the herds often graze close together, since this provides better protection against wild animals, but they are consequently almost continuously on the move. During the rainy season there is plenty of fodder, but in the dry season there is little. Annual grass and bush fires reduce the reserves of fodder for the dry season and help to impoverish plant growth and the soil. On the other hand, burning helps to control growing bush, to remove old grasses, to stimulate grass growth in the dry season, and to eradicate parasites, particularly ticks. Firing is therefore a part of the system that can scarcely be dispensed with. Most of the extensive tropical grasslands of the present day are man-made, by the joint action of fire and grazing.

Pastoralists usually supplement their cattle herds with substantial numbers of small stock. Small stock supply meat, have better survival capacities in case of drought, and contribute to increased output. Different livestock species have different feeding habits and dietary requirements (browsers, grazers, and mixed feeders). A larger biomass and often a higher economic output can be attained by keeping goats and sheep in addition to cattle (Gwynne 1977, p. 57).

The cattle herds of pastoralists typically include 50–60 per cent of breeding females compared to the ranchers 20–5 per cent (Pratt and Gwynne 1977, p. 36) and this for a number of reasons:

1. Pastoralists live primarily on milk. Increasing population numbers lead to increasing cow populations. The need for milk results in a direct conflict between the young stock and the pastoralists' survival. High rates of calf mortality can often be explained by the fact that humans consume milk which is needed to raise calves properly.

2. Cattle herds with a high proportion of breeding females have a high capacity for increasing stock numbers after a drought. In this connection French (1967, p. 153) writes: 'The retention of aged beasts in the herd is deliberately based on the knowledge that such older animals have usually recovered from and developed some immunity to several diseases. Such

older cows may reproduce less frequently than younger animals, but they do provide an insurance against recurrent disaster.'

3. Pastoralists graze private cattle on common land. It is therefore, from the point of view of the individual family, fully rational to try to maximize the number of cows, even if this implies the progressive deterioration of the range. They produce milk for food and calves for herd growth at very low opportunity costs to the individual. The institutional setting of private cattle on common land leaves no alternative to the individual decision-maker who cannot influence the decisions of other herdsmen who have access to the same water places and grazing grounds. The tendency towards hoarding cattle, and in particular cows, and to overgraze is thus firmly anchored in the institutional setting.

The performance of pastoral livestock production varies widely. The calving rate usually lies between 50 and 80 per cent and calf mortality between 10 and 30 per cent. The number of weaned calves per 100 cows varies thus between 40 and 70 per cent (Dahl and Hjort 1976, p. 35). Calving is mostly spread over the year in order to provide for a regular supply of milk. Animals grow up alternately gaining weight in the rainy season and losing weight in the dry season, and five years are usually required for steers to reach maturity. There is neither the technical nor the economic basis for conserving fodder (hay or silage), and traditional rules for rotational grazing which existed in some grazing areas have usually not survived. Table 9.3 indicates the productivity of various types of livestock in low potential pastoral areas of East Africa. Milk yields per cow, above the requirements of the calf, are around 200 kg per year. The meat output of a herd of 100 animals which would require 500 to 1000 hectares of range land, is around 2 t which in terms of live-weight per hectare would amount to 4–8 kg only. The information on sheep and goats in Table 9.3 indicates that the output in terms of kg of protein and energy in relation to the live-weight of the herd is much higher than with cattle. Dahl and Hjort estimate that a family of two adults and four children would require a herd of 64 cattle or 28 camels or 130 sheep or 221 goats to cover household food requirements, if the family were to live on livestock products only (p. 140). This implies land requirements for subsistence only of 300–600 hectares per family.

Cattle in pastoral societies represent variable degrees of wealth, social status, and community influence. They are a man's legacy to his son. Those with few cattle aim at building larger herds, and those with large herds try to maintain them. Both have a clear-cut objective in life which goes much further than the arable farmer's working for subsistence and cash, and that is increasing herd size. Another important characteristic of semi-nomadic systems is their social mobility. The property distribution pattern often suddenly changes due to disease and drought. A relatively rich family may become very poor in a few weeks. Similarly, a relatively poor herdsman may

TABLE 9.3

Estimated output of nomadic production in East Africa

1. *Cattle:* Product estimate for a herd of 100 cattle
(20 calves; 41·8 males and females, 1–4 years old; 32 cows; 6·2 bulls)

	Amount of protein (kg)	Energy (kcal)	Total yield (kg)
Milk, 210 days lactation	244·8	5 497 065	6615
Blood from living animals	22·8	90 000	300
Blood from slaughter	4·2	18 480	56
Meat, normal slaughter (mean cdw 120 kg)	139·2	2 268 000	960
Meat, natural death and emergency slaughter	(max. 139·2)	(2 268 000)	(960)
Total, without natural death, etc.	411·0	7 873 545	—
Total, including natural death, etc.	550·2	10 141 545	—

2. *Camel:* Product estimate for a herd of 100 camels
(17 calves; 44 young males and heifers; 34 adult dams; 3 old oxen; 2 bulls)

	Amount of protein (kg)	Energy (kcal)	Total yield (kg)
Milk	918·34	17 374 000	1460
Blood	38·40	163 968	480
Meat	141·75	664 605	675
Hump fat	—	527 400	60
Total	1098·49	18 729 973	—

3. *Sheep:* Product estimate for a herd of 100 sheep
(29 female lambs, 0–18 months old; 39 breeding ewes; 21 male lambs, 0–12 months old; 8 male lambs, 12–18 months old; 3 rams)

	Amount of protein (kg)	Energy (kcal)	Total yield (kg)
Milk	130·7	2 340 000	1950
Meat (15-kg carcass weight)	37·5	1 513 125	375
Total	168·2	3 853 125	—

4. *Goats:* Product estimate for a herd of 100 goats
(32 female kids, 0–18 months old; 35 breeding female goats; 24 male kids, 0–12 months old; 8 male kids, 12–18 months old; 1 billy goat)

	Amount of protein (kg)	Energy (kcal)	Total yield (kg)
Milk	90·7	1 847 300	2450
Meat (27-kg carcass weight)	44·1	410 508	567
Total	134·8	2 257 808	—

Source: Dahl and Hjort (1976, pp. 157–219).

become a large herd owner by carefully husbanding his livestock. We therefore face a non-egalitarian society with a high degree of mobility. Herdsmen, of course, try to insure themselves. Cattle are exchanged to symbolize formal contracts of friendship and assistance. The transfer of cattle from the groom's family to the bride's is needed to validate a marriage (Dyson and Hudson

1969, p. 78). Herdsmen usually tend to hoard cattle in order to build a system of human bonds aimed at increasing individual security. This may be done in the following ways:

1. Some of the animals of relatives and acquaintances are kept in a family's own kraal and they give some of their animals to other people, so that in case of disease the losses for any one family are not total.

2. Herdsmen lend animals to a neighbour or relative who has lost his animals through disease or theft, and thus ensure his help in their own times of need. The tendency to dispose of animals by lending is encouraged by the fact that large herds entail a rapid consumption of grass and necessitate long treks.

3. The owners of large herds, who in any case do not want to keep them in one place, lend some animals to poorer members of the tribe, and in this way guarantee their allegiance.

4. If a man wants to marry, he has to give cattle to the bride's father. In poor families this amounts to one or two, and in rich families to ten cows and more. If the woman is treated badly by her husband, she can return to her father without his being obliged to return the cattle. Conversely, the husband can send his wife back to her father and demand back his cattle if she behaves badly or if she is infertile. Consequently, both parties have a material interest in the success of the marriage, but equally both are obliged to hoard animals for some part of their lives in order to meet their obligations if the case arises.

Thus when an assessment is made of animal populations, it is crucial to remember the social role of animals as well as the motive to maximize productivity. von Rotenhan (1968), for example, established in Sukumaland that 40 per cent of the Wasukuma's partially nomadic animals were tied as bride-prices of 10–20 head of cattle per marriage. Similar conditions prevail among semi-nomads.

The tendency to over-grazing is further encouraged by the fact that all members of the tribe have access to the communal pasture. People who acquire household surpluses from arable farming, or who work as seasonal labourers in the mining industry or on the plantations, are therefore inclined to hoard cattle in order to gain security as well as prestige. Permanent arable farmers tend to lend out their animals to semi-nomads like the Fulbe in West Africa. Thus animal-keeping and animal-owning are by no means identical.

9.3.2. *An example: semi-nomadism of the Bahima, Uganda*

The Bahima have long-horned Ankole cattle, which graze grasslands in north-east Uganda, 1000–1300 m above sea level and with a rainfall of 700–900 mm. Their economic position is better than that of many other pastoral tribes, because they have at their disposal land which receives plenty of rain and for the present is not over-grazed, largely due to the lack of watering places and a bimodal pattern of rainfall, which prevents arable farming. Milk is the main product, and the number of cows is deliberately kept high, which also allows the herd to expand rapidly when grazing conditions are favourable.

The animals are the private property of the families. The head of the family bequeaths the animals to his sons at his own discretion, but the pastureland is the communal property of the tribe, and every member of the tribe can graze as many animals as he likes. Watering places belong to whoever has dug them for as long as they are occupied; once a watering place is abandoned, anyone can use it. The herds of the Bahima are comparatively large, and on average a typical family of 10 people owns 33 animals, which are practically all cattle.

In general, several families live and migrate together. Temporary kraals and huts are built at seasonal watering places. Where plenty of water is permanently available, larger and more permanent kraals and huts are built, and are shifted about every 2 years. The Bahima have plenty of grass, so they do not move the kraals or construct temporary secondary kraals because they are in search of fodder but because they are:

1. Searching for water.
2. Trying to escape disease through movement.
3. Complying with the custom of abandoning a place where an adult has died and is buried.

Table 9.4 shows the farm-management data of the Bahima economy and compares them with the characteristics of the Karamoja economy and of smallholder ranching as it is in the process of being realized in the Ankole Ranching Scheme, on land where tsetse has been eradicated. Modern ranching absorbs far less people than traditional methods. Herd composition in traditional ranching includes more cows and calves and less steers, because of lower calving rates and higher mortality rates. Modern ranching has higher stocking rates and a much higher offtake, and it has to be considered that about one-third of the offtake of the traditional livestock economies is from fallen animals (mostly animals that were slaughtered shortly before dying). Modern ranching, however, produces no milk and requires high inputs. In the table, only those inputs directly involved with the small ranches are taken into account, not the overheads of the project. The modern ranching project's internal rate of return to the national economy amounts to not more than 9 per cent (Jahnke 1974, p. 113). Modern ranching in this case is more productive per unit of land and per unit of labour in terms of the gross return but not in terms of income. A comparison of the higher output with the lower labour absorption capacity of modern small-scale ranching clearly shows that the land would produce more livestock if fewer people lived by semi-nomadism.

TABLE 9.4

Management data of semi-nomadic livestock production and small-unit ranching in Uganda

Location	Nyabushozi	Karamoja	Ankole
Rainfall (mm)	750–1000	500–750	750–1000
Year	1969	1969	1969
Type of farming	Semi-nomadism		Small-unit ranching[a]
Method	Various surveys		Planning data for fully operating unit
Number of holdings	n.a.	n.a.	100
Persons per household	5	5	5
Labour force (ME)	2	2	4
Size of holding (ha)	—	—	1200
Land availability[b]	47	48	—
Livestock per household			
Livestock units (cattle mainly)	33	33	640
Herd composition as percentage of total cattle			
Cows (per cent)	44	46	40
Bulls (per cent)	2	4	1
Mature steers (per cent)	3	5	7
Heifers (per cent)	27	17	26
Immature steers (per cent)	7	12	18
Female calves (per cent)	10	9	4
Male calves (per cent)	7	7	4
Performance indicators			
Hectares per livestock unit	6·7	4·1	2
Calving rate, per cent[c]	60–70	60–70	75
Mortality as percentage, year 0–1	10	10	5
Cows	8	8	3
Other	5	5	3
Average weight at 3 years (kg)	250	250	400
Offtake (per cent)	12[d]	13[d]	18
Output (kg live-weight ha⁻¹)	5·3	9	56
Output (kg milk ha⁻¹)	12·4	20	—
Production per unit			
Subsistence, milk (kg)	2767	2733	—
Meat, slaughtered animals (kg)	191	183	—
Meat, fallen animals (kg)[d]	296	246	—
Animals sold for slaughter (number)	1·7	2·5	167
Hides sold (number)	4·3	2·56	—
Economic return ($ per household)[e]			
Gross return	327	351	15 230
Purchased inputs	—[f]	—[f]	6176[g]
Income	327	351	9054
Wages	—	—	840[h]
Family income	327	351	8214
Sales as percentage of gross return	32	37	100
Productivity			
Gross return ($ ha⁻¹)	7	7	13
Gross return ($ ME⁻¹)	163	175	381
Income ($ ha⁻¹)	7	7	8
Income ($ ME⁻¹)	163	175	226

[a] Expectations based on the initial performance of the scheme. [b] Total area divided by the number of households (Jahnke 1974, p. 288). [c] Live births, not calves alive after 1 year. [d] The meat of 80 per cent of the fallen animals is considered as being consumed. [e] Accounting values on a world market basis (Little-Mirrlees Method). Subsistence food production is valued at 750 shs (U.S. $105) per family of 5. [f] No information is available and the amounts are expected to be negligible. [g] Depreciation on ranch investments (without livestock) is included, overheads for the scheme are not included. [h] The ranch is expected to operate with one manager (the rancher), who receives family income, and three hired workers. *Source:* Jahnke (1974, pp. 96–117 and pp. 283–302).

9.3.3. *Problems of semi-nomadism*

Most types of semi-nomadism are economically wasteful. In comparison with large-scale production on ranches, the productivity per hectare, per man-equivalent, and per animal is usually low. Where plant growth has not yet suffered severely from over-grazing, animal densities are often high, since the stock-keepers know how to maximize the number of animals on a given area. But where long-term over-grazing has been practised, the animal numbers per 100 ha are already smaller than on ranches. In south-west Africa, for example, the animal densities in reservations are on average only 75 per cent of those on ranches operating under the same soil and climate conditions. However, lower productivity per animal is much more serious economically than the low animal densities. Whereas on ranches 20 per cent of the animals are normally sold every year, and these are high-quality animals, the semi-nomads and partial nomads of Africa slaughter and sell only 6–10 per cent of their stock, and these are poor-quality animals. Brown (1963) goes so far as to suggest that the efficiency in meat and milk production among the semi-nomads and partial nomads of Kenya is only one-third of what ranches achieve in the same conditions.

The low productivity of semi-nomadic stock-keeping cannot be attributed to unfavourable natural conditions, which is the case among the total nomads, but is caused primarily by the combination of individual animal ownership with communal grazing and pre-technical veterinary methods. The tendency to over-grazing and to neglect of those measures that could help to increase the efficiency of grazing is rooted in traditional institutions. Everyone tries his best to maximize his numbers of animals, and no one feels it incumbent upon himself to improve the land, since he has to share it with others. Communal bodies that undertake to organize the use of pastureland have not been able to survive, or even to establish themselves.

The institutional setting leads to the permanent over-grazing of a large part of the grassland, and a decline in the quality of the pasture sets in. Perennial grasses with high returns per unit area are displaced by annual grasses and bush. Areas prone to erosion are grazed without any consideration until the grass cover is totally removed. Brown (1963, p. 16) describes a district in Kenya as follows:

> Baringo District has reached an 'over-grazing end point' where most of the grass and the topsoil has already gone over large stretches of country and the ground is blanketed with thornbush, largely useless to man and beast alike, which cannot be eradicated without the expenditure of large sums of money . . . Land in a 35–40-inch rainfall area, once capable perhaps of supporting a stock-unit to 4–5 acres, is now scarcely capable of carrying a stock-unit to 20 acres . . . In the drier part of the district, with a 20–25-inch rainfall, land once capable of supporting a stock-unit to 10 acres is now scarcely capable of supporting one to 30–40 . . . Baringo is a case where the human population, in an attempt to maintain enough stock for their needs, have already to a large extent destroyed their own habitat.

A consequence of communal grazing and private cattle ownership is the so-called cattle cycle (see the example from Kenya, Fig. 9.1). The first stage is a relatively small number of animals, plenty of fodder, sufficient grazing in the dry season, and a high rate of reproduction. With increasing numbers of animals, the fodder supply becomes less adequate, and there is a particular shortage in the dry season. There is still sufficient grazing to cover the animals' needs in the dry season, and the stock can reproduce, but at more or less regular intervals there is an extremely dry year, when fodder growth and water are insufficient, causing some of the animals to die. In the following year the

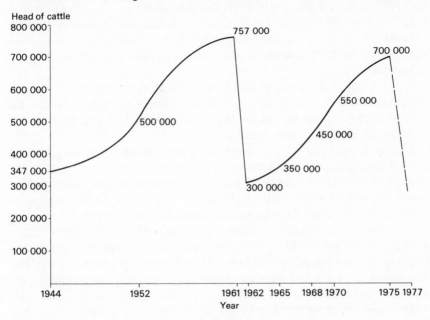

FIG. 9.1. Cattle numbers in Kajiado District, Kenya, 1944–77.
From Hoben (1976).

cycle begins again, with fewer animals, plenty of fodder, and a high rate of reproduction. In some parts of Tanzania, e.g. in the Masai area, this cycle seems to take about 10–15 years to complete. Another example of this, but with even more disastrous effects, was the Sahel drought (1972–4) which concluded a cycle of greater length, i.e. of several decades.

In partial nomadism the damage caused by the inefficient control of cattle-keeping goes beyond the loss of animals during the dry seasons. In arable farming with cash crops, a relatively large amount of money is earned, and it can be assumed that the more successful third of the farmers save consider-able sums. Some parts of these savings are invested in stock and thus are lost to the nation's economic capital. In addition there are such developments as, for example, the 'cotton cycle'. Cotton cropping brings in considerable amounts of cash, a part of which is saved and invested in stock. This results in a concentration of stock in cotton-growing regions, with the simultaneous extension of arable farming. The area of pasture per livestock unit decreases and erosion damage increases. The animal-feed situation becomes worse than usual. In extremely bad years the result is a loss of animals, which again stimulates the desire to buy more. Cotton cropping, the way to earn cash, is extended even further, and the process begins all over again.

In particular, semi-nomadic stock-keeping is characterized by the following malpractices, which also apply to partial nomadism:

1. Uncontrolled animal densities, over-grazing, and erosion damage in healthy habitats near settlements and at watering places.

2. Insufficient fodder reserves, and thus an inadequate fodder distribution over the year.

3. Long daily treks between the kraal, the pasture, and the watering place.

4. Low levels of calving and a high mortality rate among calves—the calves compete with people for the milk from the cow and they are not fed adequately in the first months.

5. Daytime grazing, often in the midday heat, instead of grazing at night or in the evening or morning.

6. Theft of stock is regarded by several pastoral tribes as an honourable activity, so that animals have to be driven into a kraal for the night.

New areas for cattle-keeping can be opened up by constructive watering places, since cattle should be watered twice daily if possible. Grass can be grazed for 4–6 km round the watering place, so the available pasture is a function of the distance between the kraal and the watering place (see Fig. 9.2). Since there are few watering places, over-grazing is the result in the catchment area. Digging new watering places alleviates these conditions only temporarily. With little control or management of grazing, the range

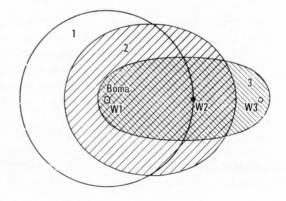

Fig. 9.2. The size of the grazing area, showing its dependence on the distance from the watering place to the kraal. (1) Watering place at the boma: large pasture area. (2) Watering place removed: smaller pasture area. (3) Watering place far removed: small pasture area. From Groeneveld (1968, p. 228).

deteriorates around the new watering points. The same mechanism applies to the improvement of veterinary services. They reduce calf mortality, increase the buildup of livestock numbers, and lead necessarily to over-grazing. Provision of water points and veterinary services under the condition of a fully stocked range are 'habitat deteriorating' services. No lasting improvement can be expected without a land-tenure reform which abolishes the implications of private cattle on common land.

9.3.4. *Development paths of semi-nomadism*

Animal production in arid and semi-arid Africa is undergoing rapid changes. Extensive grazing areas are increasingly occupied by arable farmers. New grazing areas are opened up by new wells, tsetse eradication, and chemotherapy. Husbandry techniques, herd composition pattern, marketing practices and preferences, and property distribution patterns all change (Bernus, Boutrais, and Pelissier 1974).

Pastoral societies have no self-regulating mechanism which would allow the adaptation of livestock populations to the carrying capacity of their land. The regulation occurs through disaster. No lasting improvement in efficiency and no ecologically stable system can be expected without two basic adjustments:

1. The dry range is relatively more overcrowded than most arable areas. Over-grazing is to a high degree the result of efforts to provide subsistence food. 'The root cause of overstocking is excessive human population . . .' (Pratt and Gwynne 1977, p. 39). Out-migration is essential for any lasting change in the area.

2. The institution of 'private cattle on common land' has to be replaced by land allocation to individuals or groups or to be controlled by a quota system whereby grazing rights are both allotted to the herdsmen and enforced.

Whereas the total nomads usually use areas that are not amenable to commercial methods of farming, the methods of the semi-nomads dissipate the very resources that would make technically modern management worth while. But before grassland can be managed in an organized fashion, institutional controls have to be established that provide incentives to introduce technical advances. In this respect, there are a number of possibilities.

(a) *Individual ranches.* On the subject of individual ranches, Webster and Wilson write (1967, p. 280):

In view of the great disadvantages of communal grazing, it is often suggested that a change to individually owned and operated holdings is desirable, in order to obtain good pasture management. However, over a great deal of the pastoral country individual holdings are impracticable for several reasons. First, seasonal migration of stock is often essential because large areas of pasture can only be used for part of the year. Secondly, provision of water to individual holdings is often impracticable or uneconomic in areas where water is scarce. Thirdly, individual holdings would often be unacceptable because few pastoral regions are uniform. Under individual tenure some people would get only poor grazing, whereas under the communal system all have equal access to good and poor areas. Fourthly, the fencing of pastures and water supplies on individual holdings would be expensive, and in many places the absence of fencing material would make it impracticable.

Small individual ranches are, however, a very interesting proposition where water points can be established at reasonable costs and where the distribution of grazing between upland grazing in the rainy season and valley grazing in the dry season can be organized within a 500- or 1000-ha holding. This should generally be possible in areas receiving more than 700 mm of rainfall annually. Small-scale ranching of this type may become a reasonable proposition for the use of areas claimed from tsetse or stocked with tsetse-tolerant breeds in the African savannas.

(b) *Group ranches*. Group ranching is a new concept developed in some Masai areas of Kenya. A large pastoral area is divided into units of between 15 000 and 40 000 ha, which promise to be viable ranch entities. Each group consists of about 60–300 herdsmen and their families and 3000–9000 head of cattle. The members of the group receive a title deed of the land, and a grazing committee is expected to allot grazing rights to every member and to organize common services, in particular dipping. Group ranches are an intermediate type of land reform. There are still private cattle on group land and there is still no centralized management of the group herd or the area. The establishment of groups is, however, an initial step in a process which is expected to lead in time to more centralized range management (Jahnke, Ruthenberg, and Thimm 1974, p. 37).

(c) *Co-operative or company ranches*. Another possibility is the establishment of co-operative or company ranches. These should be run by a manager and raise their own funds. The individuals concerned might receive shares according to their contributions. They might be employed by the ranch management where this is convenient. Actually, the name 'co-operative' can be misleading. The institutional set-up is often much closer to that of a company than of a co-operative.

(d) *Government ranches*. Wherever grassland areas are available that are not claimed by pastoralists and cannot be handed over to private enterprise, the establishment of government ranches is advisable. To obtain efficient management, they should be established and run by autonomous bodies as profit-making enterprises.

All types of development of the semi-nomadic economy imply the removal of a significant proportion of the people who participate in the pastoral way of life. There are too many herdsmen who try to live from their stock and who therefore attempt to retain, against advice, as many stock as they can for their own survival, thus continuing the destruction of their habitat. The relative degree of overpopulation is usually higher than in densely populated, high-potential areas.

A general tendency is to develop the untapped grazing potential of tsetse-infested savannas for cattle production and to change pastoralists from their nomadic way of life to sedentary grassland farmers (Nigeria, Ivory Coast).

In areas of semi-nomadism a great improvement could be achieved by a change in consumption pattern. Pastoralists traditionally live mainly on milk. The need for milk implies a great percentage of cows in the herd. If grains would be bought and consumed instead of milk—as is increasingly the case with the Masai in Kenya—then more steers could be raised and sold. The households would be better off economically and livestock numbers could more easily be adapted in case of drought. 'The ecologically unsound dependence on milk is at the root of the problem . . .' (Pratt and Gwynne 1977, p. 39).

9.4. Ranching

9.4.1. *General characteristics of ranching*

Ranching, which is the commercial alternative to the various types of nomadism, is carried out on large stretches of land. Most ranches cover several thousand hectares, sometimes more than 100 000 ha, and carry large, permanent herds of some 1000–10 000 animals. The term *ranch* is generally used to refer to properties with well-defined boundaries (fenced or unfenced), legally owned, or having a long-term lease, and with certain developments present that were effected by the owner or lessee.

In ranching, the number of animals and workers, the capital investment, and the return per hectare are very low. However, ranches have large stock numbers and a large fixed capital per man-equivalent invested in, for example, fences and pumps, and there is usually a high income per man-equivalent. Production is limited to stock-keeping and cropping of arable fodder is worth while only under favourable price conditions, which are rare in the tropics. In most cases, only one type of animal is kept in any considerable numbers, and the aim is a single product, like wool or animals for slaughter, and other products, like the sheep's meat or the cattle's milk, are relatively neglected. Ranches mostly occur in marginal zones where diversification is not commercially worth while.

In particularly dry, stony areas far from markets, sheep-keeping predominates (south-west Africa, Australia, southern Argentina). Cattle do not have the same mobility, and are more demanding with regard to the frequency of watering and the amount needed. Whereas sheep are kept in some cases where only 150 mm of rain fall annually, cattle-keeping usually requires more than 400 mm.

Ranching is by no means confined to arid and semi-arid areas (East Africa, South Africa, Australia). It is a traditional type of land use in extended semi-humid areas of Latin America. It has been practised for decades in semi-humid areas of Zaire (with tsetse-tolerant breeds) and it is gradually being introduced into the semi-humid lowlands of West Africa.

The main objectives of ranch management are summarized by Webster and Wilson (1967, p. 256) as follows:

1. To provide, as far as possible, a uniform and year-round supply of herbage for the maximum number of stock.

2. To utilize the herbage at a stage that combines good nutrient quality with high yield.

3. To maintain the pasture in its most productive condition, by encouraging its best grass types and by promoting as full a ground cover as possible. This will protect the soil from direct solar radiation and the beating action of rainfall, thus preventing run-off and erosion.

In evolving methods of management that will attempt to achieve these aims, ranch owners must give consideration to the following factors:

1. The influence of seasonal growth and of grazing on the maintenance of the sward.

2. The variation in the composition and feeding value of the herbage with stages in growth.

3. The value of certain trees and shrubs as browse plants.

4. The need for bush control.

There are considerable differences in the measures used by the ranchers to achieve these objectives. Ranches can be classified according to the following indicators of land-use intensity:

1. Whether pastures are fenced, or to what extent the fencing divides the area.

2. How far the watering points are from one another, or how much capital is invested in pumps, pipes, and troughs.

3. Whether and how the pastures are managed (control of bush, planned burning, weed control, sowing of more productive grasses, rotational grazing, etc.).

4. Whether and to what extent the supply and demand of fodder are balanced 'passively' by:

(i) calving or lambing at a time when there is plenty of fodder;

(ii) which type of production is chosen—on extensively managed ranches production is preferred that is less dependent on an even fodder supply than production on intensively managed ranches, so that rearing animals is preferred to fattening them or dairying, and wool is preferred to mutton;

(iii) fodder reserves 'on the stalk';†

(iv) balancing of fodder growth and animal numbers by selling when fodder becomes scarce, and buying when it is particularly abundant,

or are balanced 'actively' by:

(i) the establishment of improved pastures;

(ii) cropping of arable fodder on irrigated plots;

(iii) purchase of fodder.

From the point of view of intensity, we distinguish chiefly between the open-grazing system, the field or paddock system, and paddock systems with an actively balanced system of fodder supply. In the open-grazing system, the herds must be continually watched by herdsmen. The productivity per animal and per hectare is relatively low, because rotational grazing and separation of age groups, etc., are difficult to achieve without fencing. In the field or paddock system, the work of guarding the animals is replaced by investment in fencing and watering places. Market productivity is higher, chiefly owing to the better organization of the fodder economy. The third stage in ranch development, which is usually limited to ranches with irrigation facilities or in high-rainfall areas, is characterized by substantial investments in an actively balanced fodder supply and a correspondingly higher productivity per hectare.

Fig. 9.3 shows the principal stages in the development from open-grazing systems to paddock systems.

1. Stage I: Where a ranch has only one watering place, concentric grazing rings are formed. Zone (1) is bare ground and secondary bush round the water, with scarcely any fodder in the dry season. Zone (2) is an over-grazed area with predominantly weed growth and little fodder in the dry season. Zone (3) is over-grazed land with weeds and annual grasses. Zone (4) is a reasonably used area with good pasture and fodder reserves available according to the grazing technique. Zone (5) is scarcely used natural vegetation, since it is a long way from the water (4–8 km), and provides fodder reserves in the dry season.

2. Stage II. During the season when fodder is scarce, the fifth zone is a reserve. In so far as the dry season coincides with the cool part of the year, the animals can travel further and make use of the fodder growing in the outer zone. As several watering places are dug, the marginal zones of the ranch can be more easily developed. The distances covered by the animals

---

† Haymaking is rare in ranching systems for the following reasons (Webster and Wilson 1967, p. 269).

1. Much of the land is unsuitable for mowing owing to its topography, uneven surface, or the presence of rocks, ant-hills, trees, and bushes.

2. The low yields per hectare make haymaking uneconomic.

3. The grass reaches an appropriate stage for hay at the height of the rains. At this time the weather makes haymaking either impossible, or at least difficult and liable to heavy losses.

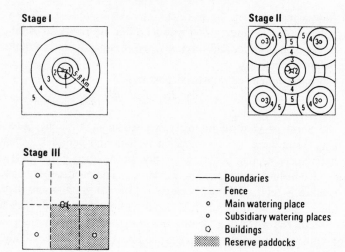

FIG. 9.3. Stages in ranch development. Stage I: no fencing, one water point, long daily migrations. (1) Bare soil, some weeds; (2) overgrazed area, vegetation mainly weeds; (3) over-grazed, some grass, much weed vegetation; (4) good pasture; (5) natural vegetation, scarcely grazed. Stage II: no fencing, several water points. (2)–(5) as in Stage I. Stage III: paddock technique. In the dry season the reserve paddocks are grazed. From Andreae (1966, p. 23).

become shorter. The result is the formation of vegetation zones round each watering place, with a better distribution of grazing and less over-grazing. 3. Stage III. Dividing the pasture into paddocks combined with fencing reduces the damage from over-grazing, facilitates separation of the animals into age groups, and allows reserve paddocks to be formed with hay on the stalk. The reserve paddocks are grazed in the dry season. Webster and Wilson (1967, p. 266) write:

A very simple system is that known as the 'three-herd–four-paddock' system. Each paddock is grazed continuously for three years and in the fourth year grazing is stopped just after the beginning of the rains, the accumulated grass is burnt off at the end of the dry season, and the paddock is then rested for a further one or two months at the beginning of the next rainy season.

The capital invested in progressing from Stage I to Stages II and III means that more cattle can be kept at a lower risk and will cause less damage to the pasture. It reduces the input of labour and shortens the time it takes to rear animals for slaughter from 4–5 years to 3–4 years.

The intensity of grassland use is reflected in the animal densities, which fluctuate between 2 and 50 livestock units per 100 ha, according to local conditions. The lower limit is where fodder and water are so scarce that permanent stock-keeping becomes impossible, and nomadism or hunting is the most economic activity. Ranches supporting enough fodder to keep

30–40 livestock units per 100 ha are mostly found in zones where the rainfall is sufficient for arable farming and where, if there were no lack of labour and markets, arable farming would be more competitive than ranching.

9.4.2. Examples

(a) *A Karakul ranch in Namibia (south-west Africa)*. Sheep ranches differ as to whether they aim to produce wool (as in Australia, Argentina), or skins (as in south-west Africa). Mutton is a by-product, if it is valued at all, on the large sheep ranches. The production of Karakul skins is suited to particularly dry areas. The majority of Karakul lambs are slaughtered immediately after birth, with the consequence that the ewe is less burdened and in a better state to survive the dry season.

The total area is divided up by fences enclosing between 5000 ha and 19 000 ha, and is then subdivided into 24 sections according to the position of the watering places. Eighteen sections are grazed in turn for 2 weeks at a time. Six sections, a quarter of the total area, are not grazed at all, but constitute a fodder reserve in case there is insufficient rainfall. Shepherding has been replaced by jackal-proof fences.

We can distinguish a long and a short lambing season. Two-thirds of the births occur between May and August, and one-third in January and February, and at both these times of year fodder is relatively plentiful. The sheep are shorn after lambing. The lambs are taken immediately after birth to the main yard and sorted to see whether they are suitable for breeding or whether they have good skins. Slaughtering takes place in the yard. Karakul pelts contribute about 50 per cent of the return of the ranch. The major cost items are depreciation on fixed investments, animal health, and wages. Most conspicuous is the low return per hectare, but the high return per labourer.

(b) *Cattle ranches in Kenya*. Cattle ranches, which are economically much more important than sheep ranches, may aim at different products; they can be divided into four groups:
1. Ranches with breeding herds, selling immatures.
2. Ranches with breeding herds and the fattening of steers.
3. Steer-fattening operations, buying immatures, and selling fattened steers.
4. Dairy ranches.

This division of labour between ranches for breeding and ranches for fattening is characteristic of countries with advanced economies based on labour division, like the United States of America or Argentina. It is rare in the tropics. Fattening ranches are only occasionally found, where nomads, semi-nomads, or arable farmers sell their unfattened animals, as for example in East Africa or Madagascar. They are of some importance in those areas of Latin America (northern Colombia, Sao Paulo, Brazil) where markets with substantial purchasing power are close to fodder-producing areas, where fodder growth at the various altitudes is conducive to the division of labour, and where immatures can be obtained in sufficient numbers at a relatively low price. Dairy ranching is also tied to the proximity of urban consumption centres. It is of increasing importance in, for example, Latin America, South Africa, and Kenya.

The data in Table 9.5 illustrate the characteristics of ranches in Kenya. The Taita data apply to newly developed, low-altitude land which is not cleared and not fenced. The area is not completely free from tsetse flies, and health risks are high. Under tropical lowland conditions the calving rate, the stocking rate, and the output per hectare are much lower than under the tropical highland conditions of Laikipia. Most of the capital in ranching—except land—is invested in livestock and water facilities. The operating costs are mainly animal health, and they are low in relation to total capital investment and the related capital costs. Ranching employs few labourers, and wage costs remain moderate in relation to total operating costs. One herdsman is usually required for 150 cows or 250 steers which implies that 12–15 labourers are employed for 1000 head of cattle (Pratt and Gwynne 1977, pp. 195–202).

(c) *Ranches in humid and semi-humid areas.* Most tropical ranching occurs in humid and semi-humid areas of Latin America where forested land has been turned into extensive grazing. The example from Costa Rica in Table 9.5 shows the return of ranching under exceedingly favourable natural conditions. The ranch operates on freely draining, highly fertile, alluvial soils of volcanic origin. Rainfall is ample and provides for 12 humid months. The animal health problems in the area are minor (mainly screw worms). The ranch is stocked with Brahman–Charollais crosses, and steers, surplus heifers, and cull cows are the output. About a third of the ranch area is under planted grasses. The major economic problem is bush control. Herbicides and labour for pasture maintenance are the major costs. Also stocking rates are kept relatively low—given the potential of the area—in order to maintain a vigorous grass vegetation which helps to reduce bush growth. Cattle actually graze only the tops of the grass vegetation. The productivity of the land is clearly much higher than with arid-zone ranching, but very low when compared with the potential. Ranching is economic only as long as land is ample and markets for perennial crops (bananas, cacao, etc.) are not attractive.

Table 9.6 gives the farm-management characteristics of ranching in other parts of Latin America. Most of these ranches operate in a semi-humid climate with a high potential for plant production. In several cases—for instance in the Llanos—soils are poor and extensive grazing is the only feasible way of using the land. In most cases, however, ranching is practised because land is ample. Ranches in semi-humid climates show the same

TABLE 9.5

Management data of ranches in Namibia, Kenya, and Costa Rica

	Namibia	Kenya	Kenya	Costa Rica
Country				
Location	Haribes	Taita[a]	Laikipia[b]	Siguirres
Rainfall (mm)	185	500	500	4000
Altitude (m)	n.a.	500	1500	10
Year	1969–74	1977	1977	1978
Type of ranch	Karakul sheep	Cattle-breeding	Cattle-breeding	Cattle-breeding
Method	Accounts	Accounts	Accounts	Accounts
Labour force (ME)	40	102	17	90
Size (ha)	72 226	42 400	7165	3500
Livestock (number)				
Cows	—	1981	718	2200
Bulls	—	84	33	110
Heifers	—	1588	284	1500
Steers	66	1226	189	1500
Calves	—	1174	450	1540
Sheep	11 920	—	—	—
Goats	651	—	—	—
Total	12 637	6053	1674	6850
Ranch capital ($)[c]				
Land	1 364 692	662 800	315 000	n.a.
Water facilities	n.a.	23 890	73 470	n.a.
Buildings, fences	55 160	34 700	15 470	n.a.
Vehicles	16 762	29 650	8990	n.a.
Livestock	259 022	425 220	108 460	n.a.
Performance indicators				
Calving rate (per cent)	—	80	91	70
Mortality (per cent)[d]	n.a.	1	1	1
Hectares per stock unit[e]	40	8·70	5·85	0·67
Output (liveweight ha⁻¹)	—	8·95	43·81	186·1
Lambs per 100 mothers	100	—	—	—
Economic return ($)				
Gross return	189 478	193 510	132 310[f]	458 082
Purchased inputs				
Livestock purchases	n.a.	62 300	—	5000
Livestock (health and veterinary)	36 142	19 400	11 610	24 250
Vehicles and machinery	11 553	13 590	8080	15 000
Maintenance and depreciation	45 479	18 080	17 390	42 000
Administration (without salaries)	11 469	36 780	14 050	20 000
Income	84 835	43 360	81 180	351 832
Wages	45 555	29 910	9300	135 000
Salaries	11 735	6060	11 890	18 000
Taxes	10 227	n.a.	n.a.	n.a.
Net return before interest and land rent	17 318	7390	59 990	198 832
Productivity				
Gross return ($ ha⁻¹)	3	5	18	131
Gross return ($ ME⁻¹)	4737	1897	7782	5090
Income ($ ha⁻¹)	1	1	11	101
Income ($ ME⁻¹)	2121	425	4775	3909

[a] Newly established African company ranches in the first good rainfall year after several years of drought.
[b] Established high-altitude ranch area. [c] Without working capital. [d] After weaning. [e] One head of cattle
= one stock unit, calves excluded. [f] Net of livestock purchases. *Sources:* Namibia: Spitzner (1975, personal
communication). Kenya: Wales (1978, personal communication). Costa Rica: Personal enquiries.

characteristics as ranches under more arid conditions: large size of holdings,
low labour inputs per hectare, high investments in livestock, low productivities per hectare, but high returns per labourer.

Return on capital and factors contributing t

Country	Paraguay		Brazil	Brazil	Bolivia	
Location	Eastern Chaco		Pantanal	Northern Rio Grande do Sul	Beni	
Rainfall (mm)	1150	1150	1300	1500	1650	1650
Method	Model	Case study	Case study	Case study	Model	Case study
Management	Average	Good	Average	Good	Average	Good
Size (ha grazed)	20 000	20 000	10 000	1000	4000	7500
Workers per ranch	14	14	10	3	4	3
Workers (ME per 100 ha)	0·07	0·07	1	0·3	0·1	0·11
Livestock (LU per ranch)[a]	4500	6700	2300	375	450	1600
Ranch capital ($)	459 000	592 000	249 000	110 000	43 000	128 000
Land (per cent)	43	34	52	68	26	16
Cattle (per cent)	46	34	37	21	53	64
Others (per cent)	11	32	11	11	21	20
On-farm prices[b]						
Cattle ($ per kg liveweight)	0·14		0·13	0·17	0·19	
Labour ($ per ME and year)[c]	460		390	480	484	
Land ($ ha⁻¹)	10		13	75	2·80	
Yield						
Take-off percentage[d]	17	20	13	22	15	18
Production (kg liveweight per ha)	18	30	11	36	7	19
Production (kg liveweight per LU)	80	90	48	96	62	89
Gross return ($ per ranch)	50 400	83 800	14 250	6100	5000	26 600
Costs of production[e]	18 900	26 800	10 800	4280	4050	14 200
Net return	31 500	57 000	3450	1820	950	12 400
Productivity						
LU per 100 ha	22	33	23	37	11	21
Gross return ($ ha⁻¹)	2·52	4·18	1·42	6·10	1·25	3·54
Gross return ($ ME⁻¹)	3600	5980	1450	2030	1250	3220
Income ($ ha⁻¹)	1·89	3·17	0·73	3·26	0·72	2·17
Income ($ ME⁻¹)	2720	4530	735	1086	722	2035
Net return ($ ha⁻¹)	1·57	2·85	0·34	1·82	0·24	1·65
Return on capital (per cent)	6·8	9·7	1·4	1·7	2·2	9·7

[a] LU = 1 adult cow held for 1 year. Other stock is computed proportionally to its liveweight and time held. Since the average weight of adult cows generally differs between areas and levels of management, the figures are not necessarily proportional to liveweight carried. [b] Average for all animals sold. Thus, prices presented reflect differences not only to general level of beef prices but in quality as well. [c] Unskilled labour. Includes wages in cash, wages in kind at on-farm

it in South America's beef-cattle industry in 1968

Colombia Costa, breeding zone		Colombia Llanos		Venezuela Apure (western Llanos)		Venezuela Eastern Llanos[f]	
1600 Case study Average	1600 Model Good	1500 Case study Average	1500 Model Top	1700 Model Average	1700 Case study Good	1300 Case study Average	1300 Case study Good
400	400	3000	3000	20 000	20 000	7000	7000
9	11	3·4	4·8	21	25	7	17
2·25	2·73	0·11	0·16	0·11	0·13	0·1	0·24
300	600	820	1000	5330	5330	860	1450
70 000	109 000	96 000	134 000	520 000	573 000	297 000	396 000
41·5	27	49·5	36	26	24	51	39
40·5	53	44·5	47	68	62	38	48
18	20	6	17	6	14	11	13
0·30		0·17		0·30		0·35	
470		350		740		890	
75		16		6·70		22	
21	23	11	23	8	14	21	19
61	147	11	35	8	13	10	22
83	98	40	105	29	48	78	106
9350	17 500	5430	18 500	47 800	70 500	29 200	62 800
5760	8400	2590	8510	19 150	32 500	10 300	28 400
3590	9100	2840	9990	28 650	38 000	18 900	34 400
75	150	27	33	27	27	12	21
23·40	43·80	1·81	6·16	2·39	3·52	4·17	8·98
1040	1590	1610	3860	2280	2820	4170	3690
19·80	36·00	1·21	3·89	2·18	2·82	3·59	7·07
884	1308	1187	2430	2080	2280	3585	2920
8·98	22·75	0·95	3·33	1·43	1·90	2·70	4·91
3·8	8·4	3·0	7·5	5·3	6·6	6·3	9·1

prices, and social security expenses compulsory for the employer. [d] Constant herd-sized assumed. [e] Includes wages for family labour and depreciation; excludes interests paid. In the case of family-sized units (4 or fewer ME), no allowance has been made for an administrator or manager. [f] Milking is common practice. The figures for beef production (10 kg milk = 1 kg beef liveweight), yield in terms of production, and income take milk into account. *Source:* von Owen (1969).

9.4.3. *Problems of ranching*

Ranching is far from being a system that can be run particularly easily or lucratively, since it is characterized by:

1. High fixed costs and very high initial investments.
2. A long-term commitment to a particular type of production.
3. High-risk production because of the constant threat of drought.
4. High costs for personnel in relation to the return from the ranch.

The high fixed costs stem less from the value of the land than from the investments necessary to develop and utilize natural fodder. The costs of animals are the primary outlay. To build up a ranch requires either a high level of buying, or slow stocking with animals born on the ranch, and at least ten years of low return have to be bridged. Another costly item is the procuring of water. Grassland is of scarcely any value for cattle-keeping unless there are watering places within 4–5 km; rotational grazing depends on watering places in every paddock. The watering places should have a reserve capacity of 50 per cent, so that sufficient water is left even in dry years to maintain the rotation. There is also the investment in fences and the cost of their maintenance, for ranching without fencing is often not a viable activity. Fences reduce labour, facilitate the control of animal densities, permit the separation of sexes and age groups, simplify veterinary care and breeding, and make it possible for cattle to graze at night. On the other hand, they are expensive in relation to the return from the fenced-off areas, especially as wide use is normally made of long-lasting iron or wooden posts which have to be transported from far away and impregnated against termites.

The result of investment is a long-term commitment to a particular type of production. The selected area is usually suitable only for either sheep or cattle, and it is virtually impossible to adapt production if economic conditions change.

The three related factors of high initial investment, specialization in a single type of production, and operation in marginal areas inevitably entail a high level of market risk. Ranchers must also consider the unreliability of the rainfall. There is a danger in planning the stocking rate on the fodder capacity of average years, when a few consecutive dry years can result in loss of animals or hurried sales. Furthermore, ranches with a single product are at the mercy of price fluctuations.

The high expenditure on personnel is a further feature of large ranches. Admittedly these costs are extremely low per hectare, but they are high in relation to the return of the ranch. A large part of the expenditure pays for the services of the ranch manager, especially where diseases are particularly hazardous, as in sheep-keeping in the tropics generally, or in cattle-keeping in East Africa, where east-coast fever, the tsetse fly, and other dangers are present. Ranch managers with some veterinary knowledge are needed, and they can demand a correspondingly high salary, which can be covered only by a large herd. Since the fodder capacity of pastures is small, it is necessary to set up ranches on a very large scale.

The management problems of large-scale stock-keeping in the tropics and subtropics are therefore not dissimilar to those of plantations. High initial costs and the long-term commitment to one type of production make the early stages particularly difficult, and only if there are stable economic and social conditions is there any incentive to invest in undertakings that will show a market profit only after one or two decades.

9.4.4. *Development paths of ranching*

In few areas of agriculture is large-scale production so superior to small-scale production as in the case of stock-keeping in dry tropical regions. Not only the net return on labour and land, but also the gross return is far above what can be obtained in traditional semi-nomadic stock-keeping. The economies of scale are not mainly due to mechanization. Much more important are benefits arising from access to agricultural and veterinary progress, from proper stocking policies, which justify the high fixed costs of investment in improvements like watering places, and from high levels of management. Large-scale ranching is thus in a far better position to reap the benefits of technical progress that become available in grazing and animal husbandry. The general tendencies that technical progress in ranching seems to favour may be summarized as follows:

1. The main prospect for increasing the output of livestock products lies in the improved management of existing breeds, in particular in better grassland management, rotational grazing, improved disease control, etc. Innovations are much more easily applied in large ranches than in partially- or semi-nomadic herding.

2. The improvement of the basic genetic types of local cattle is bound to be a long and complex process. However, upgrading is of increasing importance. The usual objective is to obtain earlier maturing breeds.

3. Ranching in arid regions is decidedly marginal and the scope for the intensification of land use is narrow because of the lack of moisture. In extensive areas of the semi-arid and semi-humid tropics, ranching could be supplemented by artificial pastures (irrigated or unirrigated) which supply dry-season grazing or silage and hay for feeding in the dry season. These techniques provide much scope for increased production, provided the price relations between beef and mineral fertilizer are sufficiently favourable to justify investments in artificial pastures.

4. Extensive areas in Africa suited for livestock are not yet utilized because of the prevalence of animal disease, in particular of *Trypanosomiasis*. The improvement of tsetse-tolerant breeds (N'Dama, Baoulé, West African shorthorns, etc.) and innovations in tsetse-eradication and chemotherapy

make it increasingly possible for these areas to be developed for ranching.

Much more favourable are the possibilities for livestock production on ranches in higher rainfall areas. Usually stocking rates decline rapidly a few years after forest clearance. The mobilized nutrients are leached and production continues at a low-level equilibrium situation, as indicated by the farm-management data in Table 9.6. In a great number of Latin American ranches, however, land use is intensified. Most important is the establishment of supplementary artificial pastures, increasingly supported by selective use of mineral fertilizers. Where markets are available (as for instance in southern Brazil) the trend is towards dairy ranching, and in humid areas (Amazonas Basin) buffalo ranching for meat and milk production may become an interesting proposition.

It should not be overlooked, however, that ranching is often a transitory phase in land use. Population growth and economic development usually bring an expansion of arable cropping at the expense of grazing systems. Innovations in animal health, animal husbandry, and fodder production usually favour more intensive systems. In the long run most ranches in semi-humid and much of the semi-arid tropics are likely to give way to crop farming or intensive grazing systems.

10. Tendencies in the development of tropical farming systems

THE various farming systems described in this book are all subject to change. Generally there are high rates of population growth. More technical innovations are becoming available and applicable. The growth of urban and rural purchasing power, government interventions and changing export patterns alter price relations. These developments are accompanied by institutional, cultural, and social change. The combined impact of all these changes usually implies a tendency towards land-use intensification.

10.1. Trends in systems

Fig. 10.1 shows in diagrammatic form the evolutionary paths of land-use intensification that seem to prevail in the four major climate zones of the tropics. Each line in the figure shows a possible evolutionary path, its relative importance being indicated by the thickness of the arrow.

In humid climates the main change is from shifting systems to irrigation farming, mainly with rice, and to the growing of perennial crops. These changes occur either directly from shifting systems or via the stage of fallow systems. Suitable conditions for permanent upland farming do not usually exist, while satisfactory ley systems for these areas are still in the early stages of experimental development. In semi-humid climates the change is mainly from shifting systems to fallow systems and—with increasing population densities—to permanent upland, wet rice and irrigation systems. Regulated ley farming which is widely practised in the subtropics has been successfully introduced into the semi-humid savannas of the tropics only in exceptional cases. In the semi-arid zone, the tendency is to change from shifting cultivation to unregulated ley farming and finally to permanent farming, with irrigation spreading wherever water is available. In high-altitude areas the evolutionary pattern is for the intensification of arable cropping developing rapidly from shifting cultivation via unregulated ley farming to permanent cropping, with the planting of associated perennial food and cash crops wherever feasible. Intensive grazing systems with dairy cattle or zero-

1. Humid climates

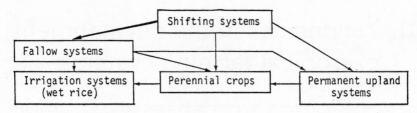

2. Semi-humid climates

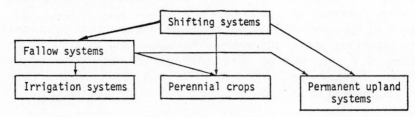

3. Semi-arid climates

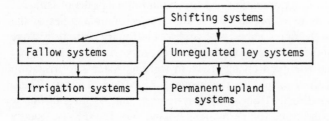

4. High altitudes

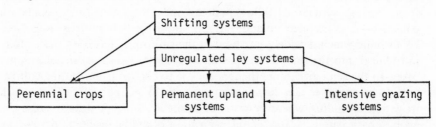

Fig. 10.1. General tendencies in the evolutionary development of tropical farming systems. The main tendencies shown are valid for indigenous smallholders only. The establishment of large farms and estates is usually connected with the *ad hoc* adoption of an advanced system of farming.

grazing systems tend to develop in tropical highlands close to major consumption centres.

The change in systems is accompanied by a change in the amount of dry matter produced and the percentage of it which is edible. The estimates in Table 10.1 provide an idea of the orders of magnitude under the climatic and edaphic conditions of Kakamega, Kenya:†

TABLE 10.1

Dry-matter production in the process of land-use intensification (an estimate of the orders of magnitude under the conditions of Kakamega, Kenya: rainfall 1926 mm, altitude 1553 m, in t DM ha^{-1}, year^{-1})

Farming system	Total above-ground dry-matter production	Above-ground dry matter in fallow vegetation and weeds	Above-ground dry-matter of crops	Edible dry matter
Natural forest	30	30·0	0	0
Shifting systems[a]	22	19·9	2·1	0·32
Fallow systems[b]	15	10·7	4·3	0·64
Permanent arable farming				
One maize crop interplanted with beans, traditional technique[c]	12	5·1	6·9	1·04
One maize crop not interplanted, modern technique[d]	15	4·3	10·7	3·20
Two maize crops, not interplanted, modern technique[e]	25	3·7	21·3	6·40
Permanent crop[f]				
Sugar-cane	31	0	31·0	7·70

Assumptions: [a] 8 fallow and 2 crop years in a ten-year rotation cycle. 1·5 t maize and 0·5 t beans ha^{-1} per crop year. 15 per cent of the above-ground dry matter in crops is edible dry matter (80 per cent of the harvested grains). The estimates for dry matter in fallows and weeds and in crop dry matter are averages over the 10-year rotation cycle. [b] 5 fallow and 5 crop years in a ten-year rotation cycle. 1·3 t maize and 0·3 t beans ha^{-1} per crop year. 15 per cent of the above-ground dry matter in crops is edible dry matter (80 per cent of the harvested grains). The estimates for dry matter in fallows and weeds and in crop dry matter are averages over the 10-year rotation cycle. [c] 1 t of maize and 0·3 t of beans ha^{-1} and year^{-1}. 15 per cent of the above-ground dry matter in crops is edible dry matter (80 per cent of the harvested grains). [d] 4 t of maize ha^{-1} and year^{-1}. 30 per cent of the total above-ground dry matter is edible dry matter (80 per cent of the harvested grains). [e] Two crops of 4 t of maize ha^{-1} and year^{-1}. 30 per cent of the total above-ground dry matter is edible dry matter (80 per cent of the harvested grains). [f] 70 t cane ha^{-1} and year^{-1}, 11 per cent sugar. The above-ground dry matter in the crop includes trash and stems.

† Some empirical evidence for the impact of changes in land use pattern on energy production in the Dominican Republic has been produced by Antonini, Ewel and Tupper 1975, p. 73 et seq.

1. Shifting cultivation, the first stage in land use, reduces total dry-matter production compared with the natural forest vegetation and produces an edible crop. Most of the dry matter is in the fallow vegetation which serves to regenerate soil fertility. Only a small fraction of the above-ground crop dry matter is edible. This may have been the type of land use which was practised at Kakamega several hundred years ago.

2. The change from shifting to fallow farming brings a further reduction in total dry-matter production. Crop yields ha^{-1} decline, but edible above-ground dry-matter output is significantly up because of the greater frequency in cropping. This probably was the prevalent type of land use about a hundred years ago.

3. Permanent arable farming with one crop per annum supplemented by intercropped beans, carried out with traditional varieties and without fertilizer, produces much more edible dry matter but implies an even lower level of total dry matter production. The tendency is towards depleted soils, low yields, and a very much reduced fallow-weed vegetation. Production occurs at a low-level equilibrium. This situation prevailed at Kakamega in the 1960s.

4. The introduction of hybrid maize and fertilizers usually eliminates the bean intercrop, but total dry-matter production is up, and edible dry matter very much so. A much greater proportion of the above-ground crop dry matter is edible. This applied for improved farms at Kakamega in 1978.

5. The introduction of double cropping which is perfectly feasible in the Kakamega area, would drastically increase total dry-matter production, while only a very small amount of dry matter is produced by weeds.

6. The growing of sugar-cane as a permanent crop which has recently been introduced into Kakamega District, increases total dry-matter production above the original level of the natural forest vegetation maintained by substantial support energies (mainly fertilizer). There is significantly more edible dry matter (sugar) and a great amount of dry matter becomes available as trash and bagasse.

The example shows a general trend: the change towards more intensive systems is arable upland cultivation without modern technology and its related support energies, increases edible dry matter until the low-level equilibrium situation is attained, while total dry-matter production is significantly down. The use of modern technology (high-yielding varieties, fertilizers, pesticides, etc.) and in particular the change towards multiple cropping or to a permanent crop brings total dry-matter production up to or even above the level of the natural vegetation, and produces much more edible dry matter at the same time.

10.2. Trends in cropping pattern

The general tendency towards land-use intensification usually involves four rather distinct but overlapping and interacting elements:

1. There is the trend towards high-yielding varieties which are not so much characterized by more above-ground dry-matter formation but by a much higher utilizable proportion of total dry-matter production. With traditional food-grain varieties only about 10–20 per cent of the total above-ground dry matter is edible, while high-yielding varieties attain 30–45 per cent (Snaydon and Elston 1976, p. 48). Most of the available high-yielding varieties concern the major marketed food grains. There is reason to assume that high-yielding varieties of minor food grains, grain-legumes, root crops, and fodder crops will become available and will be introduced into farming. Also there is reason to assume that varieties will be developed which improve yields on marginal land (deep-water rice, drought-resistant varieties, etc.).

2. Of no lesser importance is the tendency towards more intensive crops with higher outputs and higher input requirements per hectare. This may involve the change from millets to maize or from upland rice to lowland wet rice. Particularly important is the ongoing trend from grain crops to root crops such as manioc and sweet potatoes in high population density areas. This is often accompanied by the change from subsistence crops to cash crops either for the local market or for exports. Traditional small-holders showed a preference for securing subsistence food requirements first before expanding into cash crops. Increasingly, however, smallholders tend to expand cash crop production and to buy food, if this is advantageous. The need for more employment and the wish to increase incomes is behind the growing cash orientation in most cases. In many parts of the tropics the change towards more intensive crops is accompanied by a gradual transfer of the emphasis of production from upland farming to valley-bottom farming.

3. The third important element in crop intensification is the tendency towards a fuller use of the hydrologic growing season in rain-fed and irrigation farming by various types of multiple cropping. This seems to occur in distinct stages. Most traditional arable cropping systems in the tropics involve mixed cropping. With high-yielding varieties and the introduction of fertilizer the change is usually towards one stand of sole crops per year. With increasing population densities, however, the tendency is again towards various types of multiple cropping. In irrigation farming the trend is towards sequential farming and intensified nursery techniques. In upland farming the trend is towards permanent crops and mixed-row cropping and in land-short areas of humid climates towards multi-storey cropping patterns.

4. Land-use intensification changes the role of weeds and fallow vegetations. The spontaneous growth of wild vegetations is a necessary feature in shifting and fallow systems where they contribute to shading the soil, and regenerating soil fertility. Also in most systems with permanent cultivation of arable crops we find weeds developing under maturing crops which cover and protect the land during the seasonal fallow. Where, however, the hydrologic season is fully utilized by crops, wild vegetation has no useful role to play any more. Its function is taken over by crop vegetations.

10.3. Trends in animal production

In those extended parts of the low population density tropics which are not yet stocked with livestock, mainly in Africa and Latin America, but where vegetation is ample, as for instance the Guinea zone of western Africa or the Amazonas Basin, more land will be made available for livestock production. Innovations in animal health and animal nutrition are essential for the process.

Most additional livestock production, however, will have to come from already utilized land. Land-use intensification and the provision of purchased inputs change the role of animals in most farming systems. The provision of fertilizer reduces the dependence on manures and the provision of tractors on animal traction power. The availability of purchased feeds and of industrial by-products reduces the dependence on farm-produced feeds. The opportunity costs of arable grassland increase with land-use intensification, and more and more grassland is ploughed. Clearly there is the tendency for several types of animal production to move partly or mainly into specialized enterprises which are no longer closely related to crop production. This is obvious with poultry and fish. Pig production in the tropics shows similar trends. Such animal enterprises increasingly depend on processed root-crops and agricultural and industrial by-products. A major change in the price of energy would be required to change the trend towards this type of specialization.

With cattle the trend seems to be different. Traditional smallholder systems often include large numbers of cattle and buffaloes to provide traction power, milk, and fuel. The first step in modernization usually brings a drastic reduction in numbers. Farm-produced animal products are replaced by purchases. With increasing land-use intensification, however, more on-farm by-products become available. Mineral fertilizers widen the fodder basis and allow farmers to maintain animals on very small amounts of land. Fodder crops in the rotation and manures gain in importance as supplements to fertilizer in soil-fertility maintenance. Cattle, and in particular dairy cattle, thus contribute to higher crop yields, fuller employment, and higher incomes. Advanced stages in land-use intensification tend to increase the degree of complementarity between crop and cattle/dairy activities. This clearly is true only of areas with high population densities and ample provision of labour.

The trend goes from unimproved and unfenced pastures to fenced pastures; the establishment of artificial pastures; fodder conservation for the dry season; supplementary arable fodder production; and may lead, in areas with great land scarcity, from grazing to minimum or zero-grazing systems. The key parameter for future cattle production is apparently not land availability, but the price relation between fertilizer and cattle products. If fertilizer use on grasses remains uneconomic, as it still is in most parts of the tropics, then animal production will probably not remain competitive with arable crop production and will remain restricted to the use of non-arable land and of by-products of arable cropping. If fertilizer for grasses pays, then practically unlimited feed resources can be made available, probably at fairly constant costs per unit of output.

10.4. Trends in chemical inputs

Land-use intensification under tropical conditions seems to be even more dependent on chemical inputs than in temperate climates. The effective use of solar energy is tied to an ample supply of nutrients and to an effective control of pests and weeds. Machines can be replaced by hand labour in most cases, if price relations make this advantageous. However, fertilizers can be replaced by manures and composts in exceptional cases only. There are simply not enough nutrients in the available organic material to sustain a vigorous crop vegetation. Pesticides cannot be replaced by hand labour, and with herbicides the substitution rates are usually such that farmers are unlikely to forgo herbicides as long as they are obtainable. Most of the examples presented in this book indicate that high output modes of production are almost without exception tied to high chemical inputs, in particular of fertilizers. Tropical agriculture seems to be at the beginning of a period which is likely to bring a much wider use of chemicals.

There clearly is the danger that ecological damage may occur. It should be remembered, however, that hardly anything is as destructive in terms of maintaining a balanced environment as the expansion of impoverished smallholder farming producing unfertilized arable crops on depleted soils in a tropical setting. Destruction of forests, eradication of game, and serious erosion are the accompanying phenomena (Buringh, van Heemst, and Staring 1975). The use of chemicals tends to concentrate land use on the better soils, particularly irrigated areas. Marginal land, slopes, dry areas like the Sahel, and forest on acid soils could be left to the natural vegetation. The more humid and the warmer the climate, the more do chemicals encourage the trend towards a spatial concentration of production, and the more opportunity there is for untouched areas that preserve the environment.

There is also reason to assume that the application of very high inputs may be characteristic of an intermediate stage in land-use intensification only. This is particularly obvious with pesticides. In 1969 not more than about 12 per cent of the high-yielding IRRI rice varieties were resistant against two or three diseases or insects. In 1974 about 95 per cent were resistant against 5–6 diseases or insects (Herdt and Barker 1977, p. 4). The development of integrated plant protection systems combining breeding for disease resistance, chemical inputs, biological measures, and agronomy practices (such as rotations and interplanting) indicates that the trend is not necessarily towards higher and higher and more frequent applications of pesticides and herbicides. It is much more reasonable to consider the increased use of chemicals in tropical agriculture as indispensable to the change from soil mining to balanced systems of land use. Once these have been attained, chemically-efficient types of agriculture may prevail. The demand for nitrogen may be reduced by crops and crop production techniques with more symbiotic and non-symbiotic nitrogen formation in the field. More important may be changes in farming systems. Modern tropical upland farming with sole stands of annuals is highly expensive in terms of chemical inputs. The tendency towards multiple cropping and to perennials as major food and feed crops may prove to be fertilizer-saving, and the same applies to multi-storey cropping. The establishment of a vigorous vegetation of permanent crops usually requires chemicals, particularly fertilizer, but such systems, once established, are likely to be productive in terms of output and chemically efficient at the same time.

10.5. Trends in the labour economy

Land-use intensification in most farming systems involves higher labour inputs. The gradual process of the commercialization of the smallholder economy leads to more work hours per available family labour unit, a trend that is illustrated in a great number of examples in this book. Also land-use intensification tends to be accompanied by farm-size reduction and smaller farms employ more labourers per ha. Generally there is the tendency towards mechanization, but the form it takes varies widely, as for instance between the small-farm type mechanization on Taiwan and the large-scale mechanization in Brazil.

There is a general shift of emphasis within the labour economy of most types of production. Relatively less labour tends to be invested in cultivation, and more goes into harvesting. Land-use intensification increases the labour demand for weeding and crop maintenance operations, but this is often balanced by the labour-saving effect of herbicides.

The drier the area the greater the tendency to opt for labour-saving types of mechanization, while land-use intensification in semi-humid, humid, and irrigation areas is apparently related to rather high demands for additional labour. There is thus a trend towards a spatial concentration of employment.

Generally, however, there is reason to doubt whether the recognizable expansion paths of the tropical farming systems would lead to a sufficient volume of employment, given the likely rates of increase in the rural labour force of most countries. The various examples in this book indicate clearly that productive tropical farming is usually not feasible without substantial chemical inputs, but that it is certainly feasible with low labour inputs. Large-scale fully mechanized operations are possible with most crop and livestock activities. They seem to be economic from the large-scale farmers' point of view even at rather low wage levels, as is indicated by several examples from Latin America. Also smallholder farming, which commands a much greater part of the land resources in the tropics than large enterprises, tends towards increased labour productivities brought about by yield-increasing practices. Labour demand increases much more slowly than output. Most tropical cultivators perform not more than 800–1600 hours of agricultural work per annum, and with increasing labour demands in the process of land-use intensification the tendency is rarely towards a full use of the labour capacity of the family, but towards the employment of hired labour. A great part of the rural labour capacity is not yet employed, and there seems to be no major change in this respect.

Another disquieting aspect is the conflict which apparently exists between labour productivity and the protection of the environment, a problem that is greater in the tropics than in temperate climates. Increasing wages and a growing disutility of work with rising *per capita* incomes militates against the mixed- and multiple-cropping patterns which are so valuable from an ecological point of view. It is difficult to combine mechanization and diversified cropping patterns, which are so useful in maintaining soil fertility and in reducing erosion. Terraces are no longer maintained, except where high-yielding cash crops are grown. Forests are cut and burned for modes of livestock production with very low outputs per hectare, but attractive labour productivities. Many of the beneficial aspects of tropical agriculture, such as the rice terraces and the mixed- and multiple-cropping pattern, depend on low opportunity costs of labour and may give way to soil-mining modes of production if labour costs go up.

10.6. Summary

Two conflicting trends seem to dominate the change in tropical farming systems. Economic growth, the development of markets for inputs and outputs and the provision of technical innovations, tend to promote specialization and thus the development of more open farming systems which are more productive in terms of human objectives and are more dependent on the exchange of goods and services. Growing land scarcity, however, and the resulting trend towards land-use intensification, strengthen the interactions between crop enterprises, as is indicated by the tendency towards multiple cropping and between crop and livestock enterprises. The impact of these conflicting trends on actual farming varies widely according to soils, climate, and price relations. It is obvious, however, that with increasing demand for agricultural produce under the conditions of limited land, those systems will gain in relative advantage which are more efficient than others in the use of solar energy, and these are generally systems with substantial inputs of support energy. A drastic increase in the price of energy in relation to the prices for agricultural products would mean stagnation in agricultural growth, and thus hunger, in the tropics even more than in temperate climates.

11. Notes on methodology in cropping and farming system research

THE improvement of farming systems in terms of the multiple objectives farmers and national decision-makers may have, is to a large degree based on biological and technical innovations. Agricultural research which produces such innovations tends to move from less complex to more complex combinations. The design of agricultural research strategies and the adoption of innovations by the farmers is greatly enhanced if crop research generally is integrated with cropping system research, if livestock research is integrated with livestock system research and both with farming system research. The move from component research (i.e. crop or animal research) to sub-system or whole farm system research requires the development of appropriate methodologies and institutions.

11.1. Methods to identify and evaluate improved cropping systems
by Hubert G. Zandstra†

11.1.1. *Definitions*

1. Cropping system research. The analysis of existing cropping systems and their role in farming systems of different regions provides the basis for the major task of agricultural researchers: the development of improved production alternatives that are acceptable to farmers. To be acceptable, new technological components need to be identified and carefully combined to fit the prevailing production environment. This requires a holistic approach to agricultural research that is oriented towards a combination of crop enterprises encountered on or suitable for different land-types.

The research methodology described here was formulated by the Asian Cropping Systems Working Group in collaboration with the International Rice Research Institute (IRRI). It aims for a manageable research process that is particularly suited for small farms and that treats agricultural research as site-dependent (Cropping Systems Working Group 1975; Zandstra 1976*a*; Zandstra and Carangal 1977). It is conceptually based on the cropping systems as an open, reactive system. The research activities therefore focus on the description and classification of the environment, on the design of improved cropping systems, on their on-farm testing, and on methods for the formulation of production programmes. Whereas agronomic research increases the resource-use efficiency of a given crop, cropping systems

† Cropping Systems Program, The International Rice Research Institute, Los Baños Laguna, IRRI, P.O. Box 933, Manila, Philippines.

research in its quest for more efficient utilization of physical resources, considers the cropping pattern† as a variable.

Most farms, particularly small ones in developing countries, combine several production activities. In fact, the farm-household unit can best be considered a combination of production and consumption activities (Fig. 11.1). Farming systems research addresses itself to each of the farm's enterprises, and to the interrelation-

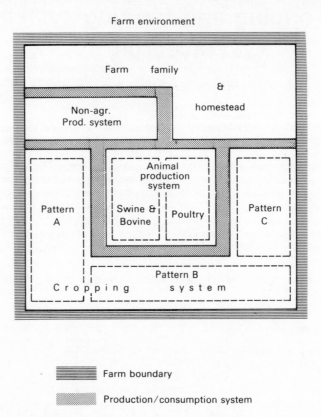

FIG. 11.1. Schematic presentation of small family farm-household system.

ships among these enterprises and between the farm and its environment. It employs information about the farm's various production and consumption systems and about the farm environment (physical, institutional, social, and economic) to increase the efficiency with which the farm utilizes its resources. Cropping systems research, on the other hand, is confined to the farm's crop-production enterprises. It takes into account relationships among the various crop-production activities, between the crop-production enterprises and other production or consumption activities on the farm or both, and between other

† By cropping pattern is meant the spatial and temporal combination of cultivars in any one plot. Generally, time is limited to one year, and a plot is defined as a contiguous area of land planted in homogenous manner during the defined period.

environmental factors (physical, institutional, social, and economic) and the farm's crop production enterprise. Strictly speaking, cropping systems research isolates the crop-production enterprises.

For research on rice-based cropping systems, this has required the development of an over-all framework as well as specific on-farm research methods. The following describes the methods for cropping systems research developed by several Asian agricultural research institutions and the International Rice Research Institute in the context of the Asian Cropping Systems Working Group.

2. Cropping systems. The productive base of a cropping system is plant growth, which is influenced by environment and management. Environmental conditions are factors that influence plant growth but are not subject to modification by management.† Plant growth and crop yield (y) can then be considered to be the result of two multidimensional vectors, the environment (E) and management (M), so that

$$y = f(M, E). \qquad (1)‡$$

In this relationship management and environment can be considered the endogenous and exogenous entities respectively of an open system (Baker and Curry 1976).

For the cropping systems researcher, management includes the type and arrangement of crops in time and space (the cropping pattern) and their management. It covers the choice of variety, and the methods of crop establishment, fertilization, pest management (weeds, insects, diseases), and harvest (component technology) for all crops in the pattern. The environment is composed of such land- and climate-related variables as available rainfall and irrigation, textural profile of the soil, phreatic level, soil toxicities, the topographic position of the field, use or non-use of bunding, daylength, solar radiation, and temperature, and of the availability of such resources as power, labour, and cash (Beek and Bennema 1972; Harwood 1974). The economic performance of the cropping systems depends, of course, on the economic environment (input and output prices).

To evaluate the relation $y = f(M, E)$, the cropping systems researcher focuses on the interaction between E and M, and seeks to determine how he should vary his cropping patterns to optimize returns for different production environments. From his understanding of the relation $y = f(M, E)$, he seeks to predict the best management vector M from information about the environment vector E.

Because $y = f(M, E)$ refers to a wide variety of crop-production environment, the cropping systems researcher must eventually come up with a statement about the effect of different management practices on cropping systems performance for a given environment. His recommendation must specify the management vector and the environment for which he recommends it. To do so he evaluates

$$y = f_i(M \mid E_i). \qquad (2)$$

That function describes the relation of the management vector M to the crop production vector y for a specific environment E_i. Operationally the transfer from eqn (1) to eqn (2) changes the environmental vector from a vector of variables to one of fixed constraints. Interactions that were in terms of E and M in eqn (1) become, therefore, terms in M only, in eqn (2).

† Note that this is a default relation: the set of environmental variables considered is a result of the researcher's decision about the extent to which he controls environment by management.

‡ This treatment was inspired by the treatment of soybean development used by Keller, Peterson, and Peterson (1973).

By evaluating eqn (2) for selected performance criteria (y may, for example, represent yearly returns per hectare to land and family labour, or yearly protein yield per millimeter of rain), the researcher can identify management vectors that result in high performance and recommend them for farmers' use. Similarly, by measuring the farmers' M and E, the researcher can specify the existing cropping system as a crop production process that is used to derive benefits in a given environment.

11.1.2. *The site-related cropping systems research method*

1. Selection of sites. The test sites are carefully selected. They should represent major agroclimatic zones, so that results have a good chance of being applicable to other areas with the same environment.

An important criterion for site selection is the estimated potential for crop intensification. The estimate is based on knowledge about the relationship between the environment and the crop-intensification potential of several agroclimatic zones. Undoubtedly, the extent to which the potential for crop intensification can be estimated depends on how well this relationship, $y = f(M, E)$, is understood and how well the environment is defined.

2. Site description. The first activity of the cropping systems researcher is to describe the existing cropping systems in a selected area. The researcher needs to identify the different production complexes of the region and to relate them to physical and economic differences in the environment. An example of environment classification based on environmental complexes (the production complex was dominantly rice–fallow) is that used in the IRRI-BPI (Bureau of Plant Industry, Philippines) site at Iloilo. There, soil texture and landscape position were used to classify the environment.

A useful framework within which to relate environmental factors to cropping-systems potentials follows Zandstra (1976c).

1. Environmental factors include physical resources (climate and land-related), economic resources (availability of land, labour, cash, power equipment, and materials) and socio-economic conditions, product prices, input costs, marketing costs, and customs reflecting preferences for certain foods or management practices.

2. The cropping systems researcher specifies the factors he wants to operate on, and those to consider invariant. The first set will be included in the management vector (subject to optimization), and the second set will be part of the environment vector of eqn (1).

3. In environmental classification, readily modifiable physical factors should be excluded: nitrogen and phosphorus fertility; easily-corrected, microelement deficiencies; and the normal incidence of pests. The relation of $y = f(M, E)$ is thus reduced to one in which standard crop-management practices in M are assumed to correct for variations in the readily modifiable factors in E. Those factors remaining in E are cropping pattern determinants and should be used for environmental classification.

4. A union of sites that have similar cropping pattern determinants is defined as an environmental complex or land type; a union of sites in which the relative performance of cropping patterns is substantially the same is defined as a production complex (Zandstra 1976b). A production complex is measured by cropping pattern performance and is, as such, an ecological unit. It may contain more than one environmental complex because there are various ways in which cropping pattern determinants can interact to produce a particular cropping

pattern performance. Rubel (1935) referred to this as the replaceability of factors. If the performance of cropping patterns is substantially different for any subset of sites within an environmental complex, one or more important determinants must have been overlooked in the description and specification of that complex. This provides the ability to test the adequacy of the environmental description method employed.

11.1.3. *Cropping systems design*

In terms of eqn (1), cropping systems design is the specification of the management vector M. The Cropping Systems Working Group (1976) defined it as a synthetic activity that employs the physical and socio-economic site characteristics obtained at the descriptive stage, together with knowledge of the effect of those characteristics on the performance of cropping patterns, to identify intensified patterns that are well adapted to the site.

The design activity (Fig. 11.2) is focused on a certain land type. A limited assembly of practices from the available component technology can be employed in design. The technology includes cultivars, tillage practices, planting methods, plant population considerations, knowledge of optimal spatial relations between intercrops, crop interactions, effects of crop combinations and cropping sequence on weeds, insects and diseases, water-management methods, and pest control methods (by hand, pesticides, crop resistance, or escape). The technology also includes accumulated knowledge about the performance of cultivars and about the

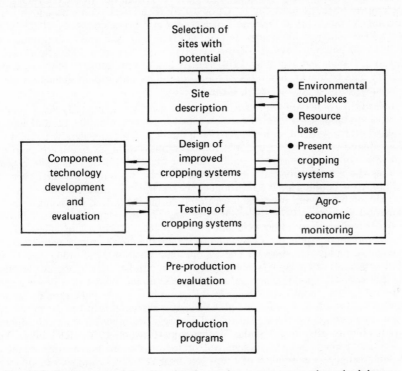

FIG. 11.2. Components of the site-related cropping systems research methodology.

management practices listed above, under the conditions specified in the environmental vector. Among those conditions are amount and distribution of rainfall and irrigation, landscape hydrology, drought, saturated soil, high precipitation and humidity during the crop establishment and harvest periods; temperature and daylength variations, extreme soil conditions, and predictable flooding, as identified for each land type during the descriptive stage.

Cropping systems programme specialists have gained considerable experience in the management of various crop-intensification techniques such as intercropping, relay cropping, sequential cropping, and ratoon cropping (Herrera and Harwood 1973; Baker and Norman 1975). Intercrops and relay crops have been found to use available light more efficiently. By choosing cultivars, planting times, and spatial arrangements, crops can be ordered with heights and densities that extend the time of full leaf spread, with maximum leaf area occurring for each component crop while solar energy is available to its canopy (Herrera and Harwood 1973; Sooksathan and Harwood 1976). Nutrient uptake and utilization have proved more efficient in corn–rice and corn–soybean intercrops than in those crops as monoculture (Suryatna and Harwood 1976). In addition, intercropping can sometimes provide a mechanism to reduce the effects of insects and diseases (such as corn-borer and downy mildew) on production (Suryatna and Harwood 1976). Canopy manipulation can also be used to reduce weed populations (Litsinger and Moody 1976). Intercropping can mitigate losses from damage to one crop through compensatory yield of other crops in the canopy. (For simulated canopy loss, see Liboon and Harwood 1976; for reduction of drought risks see description of the corn–sorghum intercrop used in El Salvador by Cutie 1975). The effects combine and become particularly important at low levels of fertilizer and pesticide inputs (IRRI 1976), so that well-designed intercrops show an advantage in land and cash productivity of 20 to 40 per cent over sole crops (Herrera and Harwood 1973; Syarifuddin, Suryatna, Ismail, and McIntosh 1974). Yearly labour requirements of intercropped patterns may be higher than those of monoculture, but the labour demand is better distributed throughout the year (Norman 1970).

Intensification of cropping systems for rain-fed or irrigated paddy rice primarily involves the addition of crops to the sequence. Where monthly rainfall is high (> 200 mm) for 4 to 5 months (IRRI 1974), the rice crop can generally be followed by an upland crop or intercrop. When 6 to 8 months of high rainfall are expected, a double-cropping pattern with paddy rice can be established. That pattern often requires the use of early-maturing varieties and dry seeding of the first crop. In addition to a double rice crop, drought-tolerant upland crops can follow rice to utilize available soil moisture and low rainfall during the tailend of the rainy season (Harwood and Price 1976; Herrera, Bantilan, Tinsley, Harwood, and Zandstra 1976). This discussion above does not take into account important effects of soil texture and topographic position, which substantially modify the cropping pattern potential in paddy-rice systems. The topographic position of a paddy determines whether farmers can accumulate water from other paddies for a rice crop; whether they can drain the paddy when needed, for good establishment of a direct-seeded crop (wet or dry); or whether they can shift to upland crops while rainfall has not completely subsided. Light-textured soils have shown much less potential for double-rice-cropping patterns, but they allow great flexibility for the establishment of upland crops after rice (Palada, Tinsley, and Harwood 1976; IRRI 1976). The quality of cropping systems design will improve as more information becomes available on the performance of crops and management techniques in different environments.

The process of cropping systems design (Fig. 11.3) by necessity employs certain performance criteria. Those criteria should include estimates of cropping pattern performance, the available resources, and a pattern's resource requirements. A difficulty arises in determining the resources available to the cropping pattern. The resources are most easily determined by substitutions; slack resources of the farming system are added to the resources used by the cropping pattern that is to be changed.

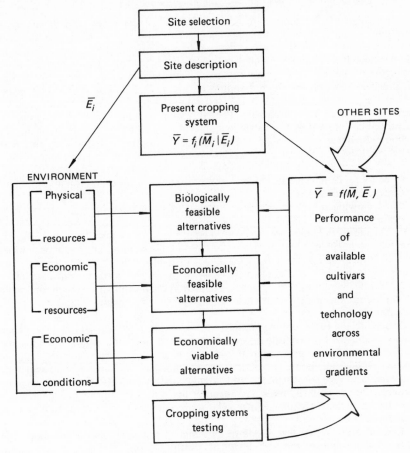

FIG. 11.3. Schematic presentation of the design of alternative cropping systems for a selected environment.

A more rigorous treatment (as a resource allocation problem) requires linear programming or similar routines for optimizing the total cropping system or still better, the complete farming system. That demands knowledge of the performance of all the component activities of the system as a function of resource allocation and costs, which goes far beyond an approximate estimate of cropping pattern performance.

The usefulness of expected pattern performance as a design criterion depends, of course, on the accuracy with which performance can be estimated prior to testing in farmers' fields. The estimate is generally obtained by extrapolating the measured performance of patterns or component crops from similar environments. As this estimate is improved and the knowledge of the inputs required becomes more precise, the performance criteria employed at design (*ex ante*) will more closely resemble those used after field testing of the cropping patterns (*ex post*).

11.1.4. *Design of the site-related research programme*

The design of the research programme for a site coincides with the design of cropping patterns for that site and should be completed at least one month in advance of the first seeding date at the site. In most programmes the yearly research programme is designed in workshop format in which all researchers at the site participate. Site researchers should be given prime responsibility for the presentation of previous research results and should be encouraged to contribute their insight about the existing farming systems, the potential for increased production, and farmers' reactions to alternatives. The workshop should draw on the support of advanced cropping systems researchers and subject matter specialists in the areas of economics, entomology, weed science, plant pathology, soil fertility, and plant breeding. Researchers should expect this workshop to take about 3 days and although the research programme for the site is designed before the cropping season starts, it may be useful to re-evaluate the research programme after each crop and make the necessary modifications.

The following steps are suggested for the design of the cropping patterns to be tested at the site:

1. Decide on the land types to be studied at the site and describe each of these as precisely as possible. The team need not conduct research on all land types in their area of operation; generally by using 2 to 4 of the most important (common) land types the team can cover the vast majority of production situations at the site.

2. Identify variables that in general constrain crop production, such as fertility problems, minor element deficiencies or toxicities or the common and dependable occurrence of crop pest.

3. Decide on the cropping patterns to be studied for each land type. These patterns should be carefully designed in accord with the physical and socio-economic conditions prevailing at the site. Farmer's cropping history, climate, product value, and potential market are all important factors to be considered.†
For each land-type the research team should probably limit itself to 3 or 4 cropping patterns. These patterns may be the same for different land-types. In fact it is desirable that the performance of one or more patterns can be compared between land-types.

4. Each cropping pattern needs to be assigned a management technology. Fig. 11.4 is an example of the complexity of a cropping pattern and the information required with respect to component technology. As the research team considers different alternatives it must evaluate the expected response and the cost involved for each alternative. After the design of the cropping pattern a simple cost-and-return analysis must be conducted. These factors should not be taken lightly as it has been estimated that to decide on varieties, tillage methods, planting methods, pest management, fertilizer additions, weed control methods, and harvest methods in addition to the timing of all operations, more than 30 decisions need to be made for a two-crop cropping pattern.

† See information required to design and test for economic criteria, page 36a to 36c, Fourth Cropping Systems Working Group Report.

During the first year the component technology chosen for the cropping patterns will depend primarily on information from the environmental description and previous research at the site and in similar sites. Over time more information on component technology will become available from research at the site and will increasingly form the basis for decision-making about the component technology levels to be used for the cropping patterns.

As the team discusses the component technology to be assigned to the cropping patterns, it will also identify areas of lack of information that need to be studied at the site. This may involve varietal screening, insect, weed, or disease control, the soil-fertility aspect, tillage methods, or the date of establishment of different crops. During the first year it is often useful to do time-of-planting trials for the important crops at the site over their potential range of planting dates. These trials should be monitored for the occurrence of insects and diseases.

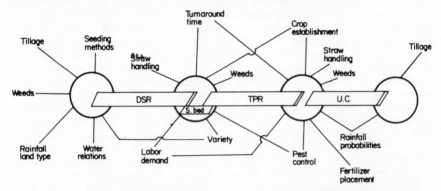

FIG. 11.4. To assign component technology to a pattern requires a careful selection from many alternatives. DSR=dry seeded rice; TPR=transplanted rice; UC=upland crops.

Component technology research is conditioned to the cropping pattern selected. It normally addresses only one crop of the pattern sequence and one or two variables, such as variety trials, tillage methods, and subsequent levels of weed control, or method and rate of nitrogen application. Component technology trials are often managed by the cropping system researchers. Where field × treatment interactions are considered important the number of fields should be larger than three and within field replication can be reduced to a minimum. *In such trials all component technology levels and management methods that are not being varied as part of the treatments must be the same as those for the same crop in the cropping pattern trials.* Limits to seeding dates that apply to that crop in the cropping pattern must be applied to the component technology trials. This is important as it will allow linking of the component technology research results to those of the cropping pattern trials.

11.1.5. *Cropping systems testing*

While cropping patterns and their management are being tested in farmers' fields, the assumptions made in the cropping systems research process, particularly those at the design stage, are to be verified. The assumptions are:
1. The proposed system is biologically suited to an important physical environmental complex of the site. Yields of crops in the pattern should therefore be adequate, and biological instability should not occur.

2. The system's requirement for economic resources, such as cash, labour, and power can be met.

3. The management of the specified pattern is optimal.

4. The system satisfies the selected economic performance criteria.

The first step of the testing process is to define satisfactory performance criteria (Fig. 11.5). To be useful in the context of site-related research these should not require complex computations. Nonetheless, the performance criteria must be conditioned by the factor costs prevalent in the site and present knowledge of farmers' decision making. Because of farmers' control over on-farm resources (land, farmer's time, family labour including exchange labour, water, daylight, and farm implements), the net returns to these resources provide a useful first estimate of the over-all benefit derived from a cropping systems by the farm enterprise. Further performance evaluation can be based on returns to material inputs and labour compared to their cost in the region; cash requirement compared to its availability; the required level of indebtedness compared to actual cash income of the farm; and risk as a function of yield variations (preferably by the subjective estimates of farmers) and levels of purchased input (Zandstra, Swanberg, and Zulberti 1975). Recent work on opportunity cost budgeting methods (Price and Barker 1977) has led to a relatively simple method for handling seasonal variations in labour wage rates. Another criterion to consider is the return to the factor most critically limiting to crop intensification: available water. The efficiency of the use of rainfall by cropping patterns can be used to evaluate this. In the Philippines, patterns tested in farmers' fields under upland and rain-fed bunded situations varied widely in rainfall-use efficiency, but the most efficient patterns reached 3 or 4 kg/mm rain (Table 11.1). These efficiencies are similar to those obtained by Rastogi (1974) and by Krantz and Kampen (1974); the indexes may provide a point of comparison for the efficiency of cropping systems with different rainfall regimes. Equally important performance criteria relate to biological stability. They include maintenance of soil fertility and the prevention of erosion, of buildup of pests, and of reduction in subsoil water availability.

The testing process (Fig. 11.5) requires more time and research personnel than the other activities described in the cropping systems research process (Fig. 11.2). The monitoring of patterns and the data-collection system must be both manageable and sufficiently rigorous to allow reliable estimates of the cropping patterns performance, its resource requirements, and the farmers' reactions to it. Identified management bottlenecks should be attacked preferably at the research station, but may at times require on-site studies. In addition, the Asian Cropping Systems Working Group has developed standard analytical and reporting techniques for the comparison of the economic performance of cropping pattern trials and existing farmers' patterns (Cropping Systems Working Group 1978). An example of such an analyses is provided for the 1976–7 results of cropping pattern trials of the deep-water-table land-type of the Manaoag, Pangasinan, Philippines site (Tables 11.2 and 11.3).

A major activity of cropping pattern testing is the fine tuning of the component technology. It is rare that the management identified at the design stage is adequate. For this reason, on-site research compares different varieties, planting methods, fertilizer regimes, and insect- and weed-management methods. A pattern's agronomic performance, its input requirements, and its optimal component technology allow an economic evaluation of its suitability according to the performance criteria established for that purpose. An important aspect of the testing methodology is the nature of on-farm cropping pattern testing. By on-farm testing is meant testing on farmers' fields of patterns that are managed by farmers (Harwood 1975).

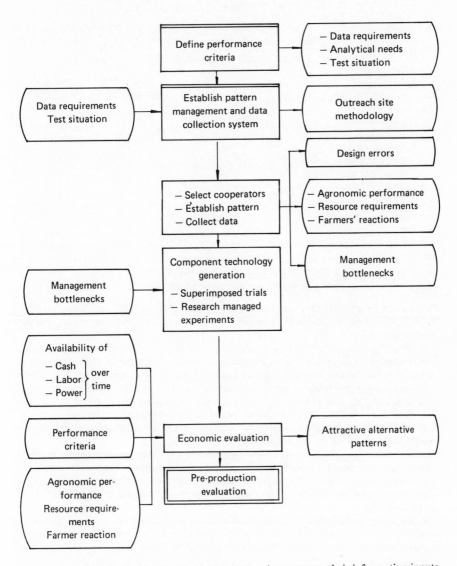

FIG. 11.5. Operations involved in the testing of cropping patterns, their information inputs, and results.

It can be argued that there are efficient research methods for testing the first and third assumptions listed above at research stations or under research management in farmers' fields. After an initial investment in measurement and surveys, the time and labour requirements of most operations can be estimated with sufficient accuracy to allow testing of the second and fourth assumptions. Why, then, insist

TABLE 11.1

Grain yields and net returns per millimetre of rain of 11 cropping patterns in a rain-fed, bunded rice-growing area, Iloilo, 1975[a]

Cropping pattern	No. tested	Total yield (kg grain/mm)	Returns[b] (US$/mm)
Rice	10	1·7	0·12
Rice–corn	8	3·3	0·15
Rice–sorghum	3	3·2	0·16
Rice–corn/groundnuts	2	2·7	0·50
Rice–corn/mungbeans	2	2·1	0·09
Rice–mungbeans	9	2·2	0·12
Rice–cowpeas	10	2·2	0·10
Rice–soybeans	6	2·0	0·07
Rice–groundnuts	6	2·1	0·34
Rice–rice	31	4·5	0·32
Rice–rice–pulses	13	4·7	0·29

[a] Rainfall during crop season ranged from 1882 to 2114 mm among locations. [b] Returns over variable costs, including family and exchange labour, but excluding cost of land.

on farmers' management and large plots (700–1000 m²) for cropping pattern testing? The reasons, gathered from IRRI's cropping systems programme, are the following:

1. Many management problems do not manifest themselves in small plots, because the researcher who has complete control over timing of operations often makes subtle modifications in pattern management to avoid problems. The site of research-managed trials is rarely selected at random within a defined environmental complex and is often determined with the experiment in mind.

2. Resource conflicts between the proposed cropping system and existing systems are difficult to measure in research-managed trials, because labour and power inputs are supplied by the researcher.

3. Farmers' modifications of cropping patterns and their management, particularly the timing of operation, are tell-tale indications of resource conflicts. Farmers' observations, although not easily interpreted, provide variable insights into the potential and the limitations of cropping systems tested under their management.

4. By using superimposed treatments that do not interfere with the farmer's operations, the component technology specified for a pattern can be more realistically evaluated under farmers' management than in research-managed trials.

These reasons all point to a need to expose the researcher to the farmers' reality and to arrive at an interactive method for identifying new cropping systems. Undoubtedly that requires a careful structuring of the test situation to which the farmer is exposed. Farmers' observations must be interpreted with caution, and the interpretations must be fed back to the farmers for verification.

The testing phase allows evaluation of the research team's ability to design improved cropping patterns on the basis of the environmental classification employed. It also allows an evaluation of the efficiency of the cropping pattern determinants as stratifying variables for design and future recommendations. In

TABLE 11.2

Performance of existing and experimental rice-cropping patterns, deep-water-table land-type, Manaoag, Pangasinan, Philippines (1976–7)

Class[a]	Cropping pattern Crops[b]			Observations (no.)	Crop yield (t/ha)			Rice equivalent yield[c] (t/ha)
	1	2	3		1	2	3	
1–F	TPR	—	—	6	1·89	—	—	1·89
2–F	TPR	Mung	—	27	1·04	0·23	—	1·96
2–BF	TPR	Mung	—	9	1·45	0·26	—	2·49
2–E	TPR	Mung	—	10	3·37	0·68	—	6·09
4–E	GC	TPR	Mung	3	13·40	3·59	0·33	7·59
5–E	TPR	Sorghum	—	4	3·31	1·82	—	5·13

[a] A different number is assigned to different crop combinations. Also, 'F' denotes farmers' present patterns; BF denotes the best 1/3 of the 'F' observations in terms of net income, and E denotes experimental patterns managed by farmers on their fields under researcher supervision. [b] TPR = transplanted rice; WSR = wet seeded rice; GC = green corn. [c] Based on the following price weights: rice, 1·0; green corn, 0·2; mung, 4·0; tomato 1·0; sorghum, 1·0.

TABLE 11.3

Costs and returns of existing and experimental rice-cropping patterns, deep-water-table land-type, Manaoag, Pangasinan, Philippines (1976–7)

Cropping pattern class	Obser-vation (no.)	Gross returns	Labour costs	Total material costs	Net returns	Return per		Year
						Man hour	Material cost	
1–F	6	2459	1075	217	1167	3·0	17·3	1
2–F	27	2321	1193	434	694	2·2	5·3	2
2–BF	9	3363	1462	449	1452	2·8	7·5	2
2–E	10	7768	2481	2093	3194	2·1	3·7	2
4–E	3	8501	2624	2781	3099	2·0	3·1	3
5–E	4	7017	2298	1604	3115	2·1	4·4	2

this manner the test results can lead to modifications in site description. Thirdly, testing cropping patterns in the farm setting provides important clues to techno-logical constraints to increased production (Fig. 11.5) such as lengthy turnaround times between crops; a lack of techniques for upland crop establishment in previously puddled rice fields under wet conditions; weed control in dry seeded rice; fertilization of zero tillage planted upland crops growing on residual moisture; and ratooning rice varieties and management of the ratoon crop (IRRI 1976; Zandstra and Price 1977).

11.1.6. *Institutional requirements of site-related cropping systems research*

In 1978, the site-related research method was applied by more than 40 research teams throughout South and Southeast Asia (Carangal 1977). Many of those teams receive advice and backup from regional or central research stations and university-based senior staff in national programmes. As the on-site research proceeds the capabilities required for the research model become clear for all levels.

(a) *Institutional requirements at the site.* The research team at the site is the instrument of cropping systems research. It is the contact point between the research structure and the on-farm reality it must address. The site team must, therefore, be able to identify different environmental complexes based on land-types, textural differences, irrigation and drainage characteristics, and slope of the fields.

The team must be trained in farm survey methods to determine the farm resource base and to identify the existing management practices and their relation to important environmental factors at the site. They must relate to the farmers and be trained in the interpretation of farmer's comments. In addition, the site team must be able to plan and execute experiments, analyse them, and interpret results. The site team must also be involved in the decisions made about the focus of their research. For these reasons, they need to participate in the definition of research priorities for the site and in the planning of the experiments and surveys. They must be encouraged to become a strong multidisciplinary unit that formulates hypotheses about the type of production technology required for the land-types in the site—hypotheses that are continuously tested against daily observations. The site team should be a dependable source of information about farm-level production tech-niques and the performance of technical innovations in the area covered by the site. It is particularly important that the site team consult with local extension and irrigation personnel. They can provide guidance in selection of farmer co-operators and provide details about the technological history of the site that are valuable to cropping systems researchers. Extension organizations should also be exposed to research plans and on-farm trials at an early stage.

(b) *Regional- and national-level support.* To operate the on-farm research at the site with B.Sc. and the occasional M.Sc. level staff, the team needs to be con-tinously supported and encouraged. Our experience is that the teams derive strong motivation from the realization that they are addressing the real, everyday problems of farmers; their solutions are immediately affecting the farmer-recipient group with whom they can identify. In addition to this motivation, the team need, however, to maintain contact with research institutions and recent research. They also need guidelines for environmental description, research design, farm surveys, and experimental designs.

This requires a group of specialists at the research centres with experience in site-related research in addition to the advanced training needed to advise research teams at sites. These groups can often be compared to researchers at existing regional or national experimental stations by encouraging multidisciplinary team discussions at these stations and by having them work with a number of site teams for support in research design, analysis, and interpretation. In addition to providing methodological and motivational back-up to the teams, the support group provides contacts with experts for consultations on specific problems, such as the identi-fication of rare pests, minor element deficiencies, or disease problems.

Up to this point, cropping systems research has been discussed in terms of operations research designed to incorporate available knowledge, processes, and materials (biological, physical, human, and institutional) into crop production methods suitable for identified environments with clearly defined farm resource availabilities and institutional support structures. Because of the operational nature of site-related research, the project depends completely on the technology available to them. This comes from national level experiment station and university research on one hand, and from the farmers in the region on the other hand. At the national level, there is a need for continued back-up by commodity- and discipline-oriented researchers to resolve bottlenecks to increased production identified at the farm level (Fig. 11.3). In addition, the national institutes need to continue the development of on-farm research methods that will improve on-site operations in environmental classification, in research on soil and crop management and plant protection methods, and in the economic evaluation of production alternatives. To achieve this, commodity- and discipline-oriented researchers should visit on-farm research sites and invite opinions about research needs and priorities.

(c) *Institutional constraints to cropping systems research.* A new production technique is often constrained by institutional characteristics because they were not designed to handle it. In the same way the change from strictly discipline- and commodity-oriented on-station research to interdisciplinary multiple-cropping-oriented research on farmers' fields, is constrained by the existence of research institutions and traditions that were not designed to cope with the requirements for multiple-cropping research.

The strong multidisciplinary nature of the site research teams require the participation of agronomists, soil scientists, agricultural production economists, and plant protection specialists. A similar, or still broader multidisciplinary requirement exists for advisory support at the regional or national level.

In most countries, the capability in soil and land research, soil fertility and crop improvement, farm management economics, climatic analysis, and irrigation and water management are found in different departments or institutions within the Ministry of Agriculture. Therefore institutions responsible for the generation of new production technologies—not a variety or a fertilization rate, but a completely specified and carefully tested sequence of crop and management activities—often need to acquire capabilities in disciplines not normally represented among their staff. In addition considerable training and management planning are needed to provide the operational and methodological support for multidisciplinary on-farm research. In some cases, existing institutions have combined their expertise to form site-related research teams for which staff of several institutions provide the expertise required. Such a model places heavy demands on site co-ordinators and may complicate the administrative structure. It has, however, the potential for strong disciplinary back-up and important feedback from on-farm research to policy makers.

11.2. Farming system research in the context of an agricultural research organization

by Michael P. Collinson†

Research about the farm system concerns the farm as a socio-economic unit. It is therefore more comprehensive and more complex than cropping system or livestock

† CIMMYT Economics Programme, POB 25171, Nairobi, Kenya.

system research, and more emphasis has to be put on socio-economic aspects. Farming system research comprises (i) the grouping of farms in meaningful classes, (ii) the description of the system, and this includes efforts to explain why things are as they are, (iii) prognosis about the likely expansion paths of the system, and (iv) normative work (optimal or satisfying solutions given certain objectives).

Farming system research has been a widely used tool for farm improvement and—at the development policy level—a basis for anticipating farmers' reactions to policy decisions. In recent times farming system research is increasingly employed in order to improve the relevancy of biological-science research in agricultural research organization.

Farming system work in the context of agricultural research organization is needed for two reasons:

1. Agricultural research is 'search for additional knowledge and for practices in an almost infinite space of not yet discovered knowledge on the behaviour of species, varieties and combinations of practices in any given agro-climatological zone' (Binswanger, Krantz, and Virmani 1976, p. 5). The number of possible and seemingly relevant lines of research and experiments is very large. Farming system research informs about the existing and improves prognosis. One of its major tasks is the testing of innovations in the system context. It thus contributes to the relevancy of biological-science research work.

2. Agricultural research is usually commodity- or discipline-oriented. However, solutions with the highest technical efficiency may not be optimal solutions in economic terms, and optimal solutions for a given crop may not be optimal in the system context. It is the task of farming system research to assess the worthwhileness of innovations in the context of the total farm unit.

Farming system research is thus a major tool to close the 'gap' between agricultural research and the performance in actual farms. This section concerns the organization of work. Let us assume a farming system research economist has recently been assigned to an agricultural research institute, where he is added to an interdisciplinary team of crop scientists. How should he organize his work?

11.2.1. *Objectives*

The Director of Agricultural Research has decided that the farming system research programme will have a dual role:

1. To focus the adaptive experimental work of the crop scientists on identified farmer problems and development possibilities.

2. To evaluate the stock of past experimental results and recommendations for relevance to farmers' situations in the region served by the centre.

Farming system work is highly location-specific. To provide appropriate technology and to cover larger numbers in order to make research economically viable, farmers have to be grouped into target populations. This should be classes (types) of farms with the same natural and socio-economic environment, similar resource endowments, constraints, and development opportunities. The term 'Recommendation Domain' is used to describe the grouping. In this case a framework of recommendation domains has been drawn up and regional planners have ranked them in priority order for a farming system programme on the basis of national and regional policy objectives.

11.2.2. *Strategy*

The farming system economist has drawn up his own programme strategy.

1. His priority will be to cover the recommendation domains of his region as rapidly as possible with descriptive surveys to give an initial understanding of the systems farmers operate.

2. In these descriptive surveys he will focus on the major crop or cropping system where some form of mixed cropping is practised within the farming system in each defined recommendation domain. Because agricultural experimental methodology is commodity- or resource-specific, the target crop or mixed-cropping system is predetermined by weighing three factors;

  (i) That which absorbs the highest proportion of resources or contributes the most to the identified resource problem. This is the enterprise likely to allow most fruitful manipulation of the farming system as a whole.

  (ii) That which centre-scientists believe has the greatest potential for improving system productivity in the natural and economic circumstances of the domain.

  (iii) That which is most closely related to special policy objectives; export potential, urban food supplies, local raw material supplies, and soil or timber conservation are examples.

With this enterprise focus the survey describes the farming system as it meets farmers' objectives and absorbs their resources. It details present production practices of the chosen enterprise and explains how far over-all system objectives and constraints dictate these current management methods. Evidence of the generality of objectives, constraints, and management practices across the target population in the domain is presented as frequency distributions. Relationships between objectives, constraints, and practices will be examined by cross-tabulations.

3. In addition to gain experience in quantative model building and data-gathering techniques for himself and his staff, the farming-system research economist decides to use a low-intensity data-collection technique to quantify parameters for a simple LP matrix as a representative farm model. His resources are limited, and he will only do this in a priority recommendation domain with a sufficiently simple farming system. He considers his first-year manpower and funds are adequate to complete investigation, planning, and advisory stages based on descriptive surveys in the three recommendation domains given highest priority. Objectives of the programme have thus been closely specified and analytical tools to satisfy the identified objectives.

11.2.3. *Preparation of investigations*

The sector structure of the location of research is made up of illiterate farmers, and demands a verbal interview investigation technique. The farming system research economist prepares a control chart to monitor the progress of his programme (see Fig. 11.6). The timing of the research will pivot around one or more stages in the sequence. Both the exploratory survey and the farmer survey are best carried out with the season's crops at an advanced stage of growth in the field. Greater insight results from direct observation of management methods in the field. At the same time the sequence should be complete in time to feed into the planning of the next season's experiments for the domain being researched.

Also part of the preparation of the investigation is the collation of information about the natural and socio-economic environment of the domain. The interaction of farmers' objectives with their circumstances results in their farming system. For example examination of the rainfall records show a bi-modal rainfall distribution giving two cropping seasons with one of the intervening dry seasons two months

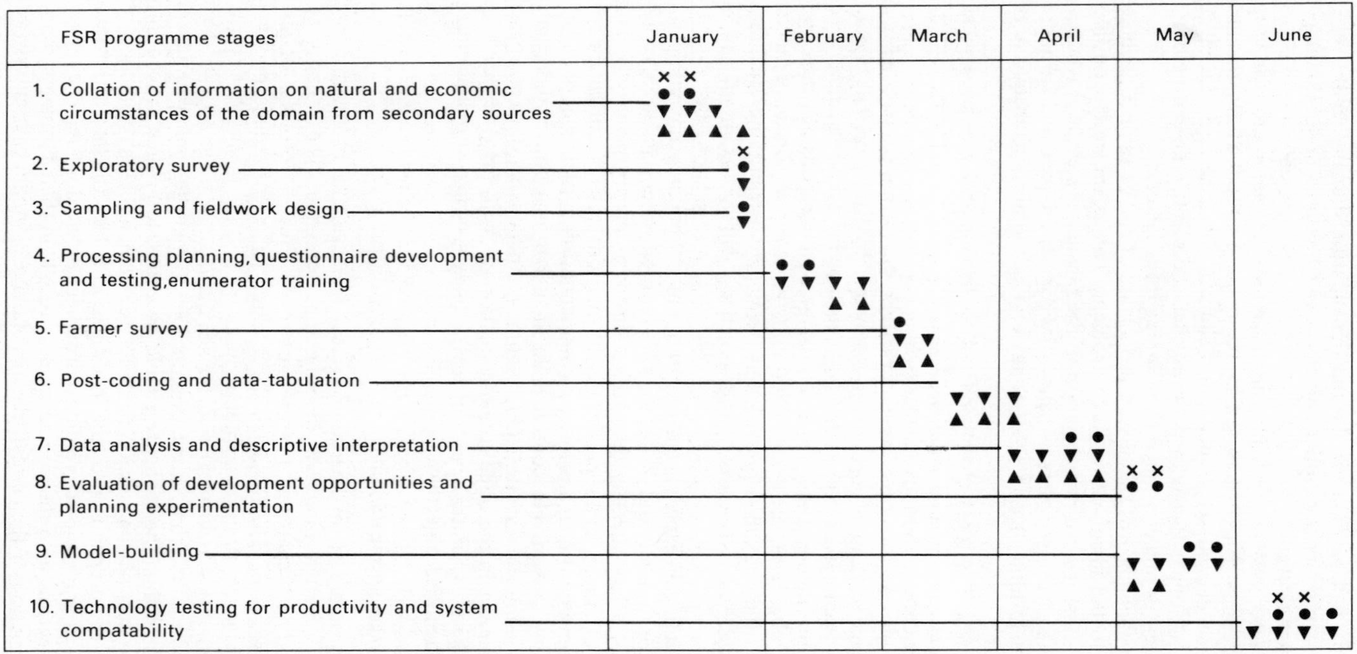

FSR programme stages	January	February	March	April	May	June
1. Collation of information on natural and economic circumstances of the domain from secondary sources	× × ● ● ▶ ▲ ▲ ▶ ▲ ▲					
2. Exploratory survey	◀ × ● ▶ ● ▶					
3. Sampling and fieldwork design	▶ ◀					
4. Processing planning, questionnaire development and testing, enumerator training		● ● ▶ ▶ ▲ ▲				
5. Farmer survey			● ▶ ◀ ▶ ◀			
6. Post-coding and data-tabulation			▶ ◀ ▶ ◀	▶ ◀		
7. Data analysis and descriptive interpretation				● ● ▶ ◀ ▶ ◀		
8. Evaluation of development opportunities and planning experimentation				▶ ◀ ▶ ◀	× ● × ● ▶ ◀	
9. Model-building					● ● ▶ ◀	
10. Technology testing for productivity and system compatability						× ● ▶ × ● ▶ × ● ▶ ▶

The chart shows involvement by four types of person represented by symbols

× Natural scientists   ● FSR economist   ▼ Supervisory field and clerical staff   ▲ Enumerators and clerks

Fig. 11.6. A control chart for the hypothetical farming system programme.

long. The researcher is forewarned that in a system with hand cultivation there may be problems in timely land preparation for the second season. Evaluation of the available data on local circumstances will focus later stages of the investigation onto critical areas of system management. Much of the available data on farmers' circumstances will have been collated while defining recommendation domains. Secondary information, particularly on local market prices and institutional channels, may be supplemented in the early part of the exploratory survey.

11.2.4. *The exploratory survey*

The exploratory survey is essential to the understanding of farming systems. It is used to describe how farmers in the domain are farming and to relate farmers' circumstances to the way they farm to understand why they farm that way. The farming system economist himself carries out a survey, with the participation of the centre agronomist. Each aspect of the farming system should be probed moving from the general descriptive to the more specific analytical level of evaluating the influence of objectives and circumstances on management. It is a recursive learning process over a two-week period and fifteen to twenty farmer interviews.

In addition to this central task of understanding the farming system the exploratory survey may have as many as eight supplementary objectives, all related to the effective implementation of the farmer survey.

1. To publicize the research and its work programme with government officials, local leaders, and farmers.
2. To help confirm the homogeneity of the Recommendation Domain.
3. To supplement secondary source information on farmers' circumstances.
4. To reconcile a sampling design with the availability of frames of sample units and a logistically feasible scheme of field work.
5. To evaluate farmer contact organizations with a view to using the best to help identify farmers and elicit farmer co-operation.
6. To describe attributes which in some areas are common to the whole population; examples are the land tenure system, preferred dishes, social obligations in cattle. This helps reduce the volume of the questionnaire in the formal farmer survey.
7. To identify and evaluate the measurement problems of parameters needed for quantative models. For example, where early plantings of maize take place on low-lying valley soils some distance from the main farms, visiting such plantings may double the enumeration time on each sample unit. A decision to visit requires an assessment of the importance of these plantings to the farmer, to the development of the farming system, and thus to the planning model.
8. To test respondent reaction to subjects to be covered by the farmer survey which may be sensitive or particularly complex.

The exploratory survey is used by the farming system economist to describe the farming system, and the management practices on the target crop or cropping system. Hypotheses are set up as to the seasons for present management practices on the target crop. Finally the economist and agronomist together consider possible improvements in management in the light of the apparent reasons for present practices. Hypotheses are set up on the likely attitudes of farmers to changes in each component of current practice. The description is verified as common through the target population and the hypotheses tested in the formal farmer survey.

11.2.5. *Sampling and fieldwork design*

Within the context of the single-visit data-collection technique already chosen for the farmer survey, sample design and fieldwork planning are complementary.

Two alternatives for fieldwork organization for single-visit surveys are:

1. Enumerators accommodated in sample clusters for the survey, with the cluster size planned to keep them busy for the whole field period and the field supervisor visiting regularly.

2. Enumerators transported by the supervisor to a sample cluster, its size designed to occupy the group of enumerators for one or perhaps two days.

Choice between these alternatives will depend on the relative availability or cost of accommodation and transport. Two-stage cluster sampling within the chosen domain helps to overcome the frequent problem of the lack of a proper sampling frame, by limiting pre-enumeration to the population of selected first stage units. It also eases the logistics in fieldwork.

11.2.6. *Processing plans, questionnaire development and testing, enumerator training*

The farming system economist has decided to visit sample farms only once and to pre-programme his visits. He anticipates it will take one hour on each sample unit to administer the questions to verify the pre-survey work aimed at understanding the farming system. He anticipates a further two hours to quantify the parameters essential to the planning model. Three-hour questioning of the respondent may jeopardize co-operation and reduce memory performance. He believes if he informs farmers in advance of the date and length of the visit, and breaks the interview into two by measuring field areas and plant stands in the middle without the assistance of the farmer, this will prevent the loss of co-operation or memory performance on the part of the respondent.

Prerequisites for questionnaire development are:

1. A detailed list of data requirements based on the verification needs of the exploratory survey and the parameters required for the chosen planning model.

2. A final decision on the method of collection for each data-set to be based on the evidence on measurement problems and possibilities provided from the exploratory survey.

3. A final decision on the means for data tabulation and processing; by hand or computer and if by computer with pre- or post-coding of questions.

Single-visit methods of data collection require the most care in questionnaire development and the most thorough pre-testing. The sequence and phrasing of questions are important and the space and layout for responses. With labour input and crop production as two key parameters in quantitative planning models and as variables with the awkward characteristics of frequency and irregularity of incidence the questionnaire sequences to capture these will be particularly important.

Where a permanent enumerator force is employed training will not be a recurrent problem. When new enumerators are to be trained it can usefully be combined with questionnaire testing. The essential step in testing is to take the draft questionnaire out amongst the target population and try it out. The farm system economist and the field supervisor will always be present to hear the enumerators administer the draft to selected respondents who are not included in the survey sample. They will be concentrating on four issues:

1. Is the question sequence followed the best for ensuring farmer co-operation?

2. Does the sequence give the minimum movement around the farm, without repetition?

3. Are the questions as phrased, and as interpreted by the enumerator where this is relevant, focusing the farmer's interest on the subject matter intended?

4. Is there adverse reaction from respondents to questions foreseen or unforeseen to be sensitive?

It is useful in the office for the clerical supervisor and clerks to go through the coding and tabulation procedures on the test questionnaire to ensure that processing will be a smooth routine.

11.2.7. *The farmer survey*

If preparation and planning have been effective, the farmer survey will be routine. The questionnaire will be administered to the sample farmers according to the field-work plan. With single-visit surveys in which enumerators are mobile, moving to different areas day by day, it will be imperative for the field supervisor to check questionnaires on completion and in the immediate area of the day's field-work. When enumerators must re-visit farms it should be on the same day whenever possible. In the hypothetical example a work-rate of one farmer each day will allow this; with higher work-rates, re-visiting may have to be on the following day. When the field team are changing areas daily and following a prepared farm-visit programme, re-visits on the following day play havoc with the logistics. It can be minimized by the use of a floating enumerator who is held for re-visits and may also be used to check the on-farm performance of the other enumerators.

11.2.8. *Post-coding and data-tabulation*

Again, with good preparation and planning, the post-coding of questionnaires, preparation of a codebook, and data-tabulation, either by hand or computer, will be routine. With small samples—up to 50 or 60 farmers—there is no substitute for hand-tabulation of the uncoded data to get a feel for the results and an understanding of the farming situation.

11.2.9. *Data analysis and descriptive interpretation*

The first stage of analysis and interpretation seeks to confirm, and perhaps to elaborate on, the understanding of the system gained in the exploratory survey. It provides the basis for planning the content of adaptive experiments to benefit domain farmers.

Frequency distributions are used to show the incidence of responses across the population:
1. Alternative methods used for management operations such as land preparation, planting, weeding, etc.
2. The timing of operations.
3. Farmers' views on the relative importance of resource constraints, and the timing of seasonal labour or capital scarcity.
4. Farmers' views on the frequency, timing, and relative importance of production hazards such as drought, waterlogging, pests and diseases, and price uncertainties.
5. Farmers' attitudes to possible changes in management practices.

Cross-tabulations are used to test hypotheses on the relationships between farmer objectives, farmer resource allocations, farmers' circumstances, and their management practices. For example: the exploratory survey gave the impression that sweet potato is a non-preferred staple grown as an insurance against a seasonal shortage of preferred grain staples and absorbing resources for land preparation and establishment at a time when maize, the target crop under research, needs weeding and top dressing. This hypothesis is tested at several levels:
1. By farmers' responses when asked the reasons for growing sweet potatoes.
2. By cross-tabulating the incidence and timing of scarcity of grain staples with the incidence and timing of sweet potato use.

3. By farmers' responses on the timing of the establishment of sweet potatoes.

In addition to testing the hypothesis, examination of the level of incidence of the growing and use of sweet potatoes will improve the understanding of its importance as an insurance strategy.

11.2.10. *The evaluation of development opportunities in the farming system and the planning of adaptive experimentation*

It should again be emphasized that the evaluation process described here is intuitive, based on an understanding of the farming system and farmers' circumstances in the domain under research. The farming system economist, from a holistic viewpoint similar to that of the farmer as the decision-maker, diagnoses the constraints on system productivity. He identifies two sets of development opportunities; those which involve overcoming existing constraints and those which improve productivity within these constraints. Overcoming constraints may have policy implications for prices and institutions and these are all research conclusions of immediate importance to regional and project planners for the area. For planning adaptive experimentation on the target crop the farming system economist reduces the opportunity sets to those involving the target crop or cropping system. The process of planning experimental content becomes a dialogue between the economist and the natural scientists concerned with experimentation on the target crop.

In addition to resource constraints on the whole system influencing crop management practices farmers' objectives may be compromising technically ideal management of the target crop in three ways:

1. With a subsistence crop the timing of the availability of the food may dictate both maturity length and planting time.

2. With a cash crop special market considerations, particular storage possibilities, and seasonal price variations, may dictate these same management practices.

3. With any crop risk aversion may dictate current management practices.

As a team the scientists discuss each management practice of the target crop. They review three aspects:

1. Present practice and the reasons for it.

2. Likely increments in productivity from changing the practice, including possible interaction with other changes.

3. Likely farmer willingness and ability to change the practice and limits to the extent of change.

For each practice a conclusion is required on whether farmers will be likely to change in view of the expected productivity increment from changes in the practice within the limits of farmers' flexibility, or within limits which can be widened by institutional or policy action. For reasons of risk, capital scarcity, and management complexity small farmers only change in small steps. Those practices which *a priori* hold hope of high productivity increments within the limits of farmer flexibility form the experimental variables for the domain-oriented adaptive research effort on the target crop or cropping system. Non-experimental variables and control plots in the experiments are held at farmer practice as described by the survey work to give a straightforward comparison of the productivity increments the changes can bring to the farmer.

11.2.11. *Model-building*

The quantative survey results on levels of resource availability and use over the agricultural year and over the range of enterprises are organized in the chosen LP matrix format. The matrix is a rough simulation of the existing farming system in the domain being researched. Warnings about the mindless manipulation of data can usefully be repeated. Such models can only improve the researcher's understanding of the reactions in the system to changes in management, the numbers turning up as results have no religious significance. Particularly useful are the reaction of three factors to the changes introduced:

1. Opportunity costs not always satisfactorily reflected through market values.
2. Farmers' objectives in subsistence production.
3. Farmers' objectives in risk avoidance.

Of course the whole of subsistence production can be seen as a risk aversion device, necessary because of the failure of the distributive sector to supply (preferred) foods reliably. The subsistence and risk-aversion activities present problems in modelling and in the interpretation of results. Understanding the objectives of the farmer, the circumstances in which he operates, and his farming system is imperative in balancing priority resource allocations to these activities with residual allocations through market prices. It is subsistence and risk-aversion activities which make system compatibility an equally important criterion in evaluation as improved productivity. The state of the arts in quantitative modelling does not permit the substitution of sophistication for understanding and intuition in these aspects.

11.2.12. *Technology testing for productivity improvement and system compatibility*

Technology testing through the representative farm model has two roles:
1. The one emphasized in the text; evaluating *past* research recommendations and experimental results in a systems context as a basis for the immediate selection of improved technology appropriate to the circumstances of domain farmers.
2. Evaluating *proposed* experimental treatments:
   (i) To improve understanding of the level of productivity increments necessary to have an impact on the system.
   (ii) To further assure researchers of the compatibility of their proposed treatments with farmers' objectives and resource endowments.

In both roles a further dialogue is required between the farming system economist and the natural scientists. Past experimental results forming the basis of current recommended practices are retrieved. Key treatments from these, and from proposed experiments, are built up into activities through discussion of the impact, practice by practice, on the level and timing of land, labour, and capital use. These synthetic activities show single-practice changes and multiple-practice changes representing the improvement packages characteristic of the content of area-based development efforts. Introduced into the model singly and as packages they allow interpretation of which practice combinations improve productivity and which are and are not compatible with the system. This interpretation allows the modification of current recommendations, dropping component practices which are not compatible with the system or do not improve its productivity. The evaluation produces a technological package appropriate to the circumstances of farmers in the domain being researched.

Appendix

Note (1) Key to vernacular or local terms

Arrowroot	*Maranta arundinacea*
Bambara nuts	*Voandzeia subterranea*
Berseem	*Trifolium alexandrinum*
Buckwheat	*Fagopyrum esculentum*
Bulrush millet	*Pennisetum americanum*
Cashew	*Anacardium occidentale*
Chick-pea	*Cicer arietinum*
Cowpea	*Vigna sinensis*
Elephant grass	*Pennisetum purpureum*
Ensete	*Ensete edula Horan*
Finger millet	*Eleusine coracana*
Green gram (Mung)	*Phaseolus aurens*
Guatemala grass	*Tripsacum laxum*
Guinea grass	*Panicum maximum*
Gum arabic	*Acacia arabica*
Kikuyu grass	*Pennisetum clandestinum*
Imperata grass	*Imperata cylindrica*
Lubia	*Dolichos lablab*
Mango	*Mangifera indica*
Matrimony vine	*Lycium halimifolium*
Nigerseed	*Quireta abyssinica*
Okra	*Hibiscus esculentus*
Pangola grass	*Digitaria decumbens*
Papaya, pawpaw	*Carica papaya*
Pearl millet	*Pennisetum glaucum*
Pigeon pea	*Cajanus cajan*
Phillipasara	*Phaseolus tribolus Ait.*
Pyrethrum	*Chrysanthemum cinerariaefolium*, and *coccineum*
Quiabo	*Hibiscus esculentus*
Safflower	*Carthamus tinctorius*
Simsim	*Sesamum* spp.
Star grass	*Cynodon plectostachyus*
Taro	*Colocasia esculenta*
Teff	*Eragrostis abyssinica*
Water hyacinth	*Eichornia crassipes*
Wattle	*Acacia decurrens*
Yam	*Dioscorea sativa*

Note (2) Choice of data

The data used in this book, with few exceptions, originate from sample surveys, case studies, or accounts, and so relate to existing farms which operate under commercial conditions. Wherever possible data have been chosen which refer to farms which produce most of the output, or which are of particular interest for the understanding of the system being considered, or its expected development path. The farm-management tables do not usually reflect the situation in which most of the people live. They are production-oriented.

Note (3) Use of man-hours

Most sources cite man-days but not man-hours. The hours per man-day in tropical farming vary between 2 hours and 10 hours, and the use of man-days is not very informative. Man-days have therefore been converted into man-hours, using information given either in sources or in local norms. Man-days given in Africa usually contain 5–6 hours of work and in Asia and Latin America 7–8 hours. The hours listed include going to and coming from the field. The tables always quote man-hours and no weighting coefficient is used to convert woman-hours into man-hours if this was not done in the publication cited. Man-hours per crop are without overhead labour inputs. Man-hours in the farm-management tables include field work and farm work at the homestead, unless a footnote says otherwise.

Note (4) Organization of tables

The table with farm-management data follow a fairly uniform pattern and are based on the following definitions and assumptions:

Persons per household: all persons who share the same kitchen.

Labour force: no effort is made to employ the same weighting coefficients for all systems. The data are taken from the sources cited. Where the source lists persons and not ME, as in several African surveys, then Collinson's (1962–5) coefficients are used: Male: $10–14 = 0.25$, $15–19 = 0.67$, $20–50 = 1.00$, over $50 = 0.67$. Female: $10–14 = 0.25$, $15–19 = 0.50$, $20–50 = 0.67$, over $50 = 0.50$. It is important to realize that the number of man-equivalents shows labour capacity according to norms which may be irrelevant under the cultural conditions of the people concerned.

Size of holding: all land claimed by the household members, used or unused. Most sources, however, cite the crop land only.

Farm capital: at market prices.

Yields: main products only.

Gross return: sales and household consumption. Changes in stocks, farm-produced inputs (manure, bullock labour, roughage) are not included. See footnotes for exceptions.

Purchased inputs: Seeds, purchased fertilizers, manures and pesticides, repairs, purchase of fuel, tools and depreciation of machines and farm buildings (not including farmhouse), custom services, livestock expenditure, veterinary services, insurances, and administrative overheads (not including salaries). Footnotes indicate exceptions.

Income: Value added on the farm. This figure is irrelevant to farmers but is relevant for the policy-makers and planners.

Farm family income: Return to the family before interest and loan repayment. MJ (megajoules) per man-hours: Estimate of energy production without by-products, garden produce, fishing, etc. Non-food outputs are neglected. In farms with large quantities of non-food output, no estimate of megajoules per man-hour are made.

The best data for the farm-management tables of this book would have been model farms derived from sample surveys or case studies of typical situations. Both types of data are very rare and in most examples averages derived from sample surveys have had to be used. Because of this, several examples show a more diversified production pattern than is actually the case on any individual holding.

The tables showing input-output relations of specific crops are differently organized. The gross return is given without by-products, because the value of by-products is available only in surveys from India. The material inputs include, however, the value of farm-produced inputs (seeds, bullock labour, manure, etc.).

Note (5) Exchange rates used (equivalents of US $1)

Bangladesh	1974	7·90	Taka
Bangladesh	1975–6	13·82	Taka
Brazil	1973–4	5·69	Cruzeiros
Brazil	1974–5	7·77	Cruzeiros
Brazil	1975–6	10·88	Cruzeiros
Brazil	1977	14·00–15·27	Cruzeiros
Brazil	1978	18·00	Cruzeiros
Cameroun	1974	232·00	CFA
Cameroun	1976–7	244·00	CFA
Central African Republic	1969	247·00	CFA
China (Mainland)	1977	1·00	Yen
Colombia	1976–8	36·00	Pesos
Costa Rica	1977–8	8·54	Colones
Ecuador	1974	24·87	Sucres
Ecuador	1977	36·00	Sucres
Ethiopia	1972–3	2·34	Ethiopian dollars
Ghana	1971	0·95	Cedi
Ghana	1974	0·87	Cedi
Guadeloupe	1969	4·94	French francs
India	1966–73	7·52	Rupees
Indonesia	1972	416·00	Rupees
Indonesia	1975	388·00	Rupees
Ivory Coast	1969	277·00	CFA
Ivory Coast	1970–6	240·00	CFA
Kenya	1962–8	8·33	Shillings
Kenya	1973–4	7.14	Shillings
Kenya	1977	8·00	Shillings
Madagascar	1968–74	250·00	CFA
Malawi	1971–2	0·72	Kwacha
Malaysia	1974	2·33	Malaysian dollars
Mali	1971	247·00	CFA
Mali	1977	500·00	CFA
Martinique	1970	5·55	French francs
Martinique	1976	4·77	French francs

Namibia	1969–74	1·36	Rand
Nigeria	1963–7	7·12	Shillings
Nigeria	1974	0·64	Naira
Pakistan	1971–2	4·76	Rupees
Philippines	1973	6·70	Pesos
Senegal	1971	240·00	CFA
Sierra Leone	1971–2	1·00	Le
Sri Lanka	1969	6·00	Rupees
Sri Lanka	1972–3	5·95	Rupees
Sri Lanka	1976	6·00	Rupees
Sudan	1972	0·40	Sudanese pounds
Sudan	1977	0·35	Sudanese pounds
Taiwan	1965	20·00	NT dollars
Taiwan	1973	35·00	NT dollars
Tanzania	1963–4	7·14	Shillings
Thailand	1956–74	20·00	Baht
Uganda	1968 and 1969	7·14	Shillings
Yemen	1971–2	4·54	Rial
Zambia	1969–70	0·72	Kwacha
Zambia	1972	0·78	Kwacha

Weighted average exchange rate of the US Dollar (against major foreign currencies (I) and consumer price index in the USA (II)):

1973 (March) = 100

Jahr	I	II	Jahr	I	II
1967	120	75	1973	100	100
1968	122	78	1974	102	111
1969	122	82	1975	94	121
1970	122	87	1976	105	128
1971	120	91	1977	105	136
1972	108	94	1978	95	n.a.

Source: (I) Federal Reserve Bulletin, Governors of the Federal Reserve System, Washington D.C., August 1978.

(II) US Department of Commerce (1977). Statistical Abstracts of the United States.

Note (6) Definitions for farm-system analysis

(a) *Farming system.* A farm system is a collection of distinct functional units, such as crop, livestock, processing, investment and marketing activities, which interact because of the joint use of inputs they receive from the environment, which deliver their outputs to the environment and which have the common objective of satisfying the farmer's (decision-maker's) aims. The definition of the borders of the system depends on the circumstances. Often it includes not only the farm (economic enterprise) but also the household (farm-household system).

(b) *Rotational system*

Bush fallowing: Intensive fallow systems with bush as the fallow vegetation.

Continuous cropping: One crop is grown after the other, without seasonal fallowing. Continuous cropping may be achieved by sequential cropping or relay-planting techniques.

Dry farming: Cultivation of cereals in rotation with 1 or 2 years of fallow in arid and semi-arid zones of the subtropics.

Fallow systems: Sequences of crop years and fallow years. Extensive fallow systems are shifting cultivation systems. Intensive fallow systems are (1) bush-fallow systems ($33 < R < 66$) and (2) grass-fallow systems ($33 < R < 66$).

Ley farming: A rotation of arable crops requiring annual cultivation and artificial pastures occupying the field for 2 years or longer.

Mixed farming: Farms wherein crop and livestock activities are integrated by growing fodder crops (on arable land or as artificial pastures) and the use of manure on the crop land.

Permanent cropping: Crop cultivation without fallow years ($R > 66$). It is compatible with seasonal fallowing (in the dry or cold season).

R value: Percentage of crop land actually cropped in a year. Frequency of cropping in a fallow cycles: $R = $ crop years times $100/$(crop years plus fallow years). The R value is identical with the cropping index.

Seasonal cropping: Cultivation of crops in one season only (mainly rainy season).

Semi-permanent cultivation: An intensive fallow system with more than one-third and less than two-thirds of the potential crop land cropped annually ($33 < R < 66$).

Shifting cultivation: Several crop years are followed by several fallow years and two-thirds or more of the potential crop land is fallowed annually. Shifting cultivation is an extensive fallow system ($R < 33$).

Unregulated ley farming: Intensive fallow systems with a grass fallow utilized by livestock.

(c) *Activities*

Arable crops: Short-, medium-, and long-term crops. Most of these crops require cultivation.

Biennial crops: Crops with a length of vegetation cycle of 18–30 months (manioc).

Crops: All useful plants on a farm, except pastures.

Enterprise: Any farm activity for which a separate economic analysis is feasible and meaningful: a single crop, a crop mixture, the raising of animals, etc.

Green-manure crops: Crops grown for their soil-improving qualities. They are neither food nor feed crops.

Long-term crops: Field occupation of 6–18 months.

Medium-term crops: Field occupation of 3–6 months.

Perennial crops: Crops (not including grasses and legumes in permanent pastures) occupying the field for more than 30 months.

Perennial field crops: Crops which require cultivation and which occupy the field for 3–12 years (sisal, often sugar-cane).

Short-term crops: Field occupation of 3 months and less.

Shrub-crops: Useful trees which are pruned in order to develop a shrub-like appearance (coffee, tea).

Tree-crops: Trees yielding fruits (or useful sap) and not primarily grown for timber.

Wet rice: Rice grown with artificial irrigation (water led to the field) and/or water impounded in the paddy field, and deep-water rice.

(d) *Cropping pattern*

Catch crop: A minor crop (in terms of output) is planted after the major crop to 'catch' remaining moisture. Catch-cropping is a subclass of multiple cropping.

Crop association: Mixture of arable crops in fallow systems (de Schlippe).

Crop mixtures: Any type of crop combination at a given point in time. It may be interplanting, mixed cropping, or relay-planting.

Cropping index: Number of crops per year on a given field times 100. It is identical with the *R* value. The cropping index is applicable for sequential cropping only.

Cropping pattern: The spatial and temporal combination of cultivars in any one plot (Zandstra 1976a, p. 2).

Cropping system: The cropping pattern utilized on a given farm and its interactions with livestock activities and farm resources (Harwood 1973).

Double cropping: Two crops are grown per year on one field one after the other. Double cropping is a subclass of multiple cropping and sequential cropping. The same concept applies to triple cropping, etc.

Intercropping: The growing of two or more crops in different but proximate rows. It has the following subclasses (Herrera and Harwood 1973):

crops of similar height and similar length of field occupation (barley + oats), crops of similar plant type but with different lengths of field occupation (6 months sorghum + 3 months millet),

tall crop with short crop below;

annual or biennial crops under a perennial crop (manioc below coconut palms)— this is the same as interculture;

short- or medium-term crops with taller long-term annual or biennial crops (soyabeans + sugar-cane, millets + manioc);

short-term with medium-term annual;

taller crop harvested first (3 months maize + 4 months sweet potatoes);

minor crop harvested first (2 months beans + 4 months maize).

Interculture: Arable crops grown below perennial crops.

Interplanting: All types of seeding or planting a crop into a growing stand.

Mixed cropping: Two or more crops are grown simultaneously in the same field at the same time, but not in row arrangement.

Mixed row cropping: The growing of two or more cultivars simultaneously in the same plot within a distinct row arrangement (Zandstra 1976a, p. 2).

Monoculture: Agronomic definition: a sole stand of one crop is repeatedly followed by the same crop; *economic definition:* only one crop is grown on the farm.

Multiple cropping: The growing of more than one annual crop on the same land in one year (Harwood 1974). Multiple cropping has the following subclasses: sequential cropping: ratooning; intercropping including relay-planting.

Multiple cropping index: The sum of the areas planted to different crops, harvested during a year, divided by the total area. The cropping index is 100 if the area is cultivated once during a year (Zandstra 1976a, p. 4).

Phased planting: A given crop (cotton, rice, maize) is not planted at the optimum time. Planting is distributed over several weeks usually to distribute risks and labour requirements (also called staggered planting).

Rain-fed farming: Arable cultivation relying on rainfall only.

Ratooning: The roots of a harvested crop produce a subsequent crop.

Relay-planting: The maturing annual or biennial crop is interplanted with seedlings or seeds of the following crop. Relay-planting is a subclass of intercropping as long as both crops are on the field.

Sequential cropping: One crop is planted after the other without in-between fallowing. Sequential cropping may take the form of double cropping, triple cropping, etc.

Sole crop: One crop grown alone in pure stands (also called single crop).

Upland crop: Crop grown on unirrigated land without impounding of water and not grown in flooded valley bottoms.

Bibliography

AEREBOE, F. (1919). *Allgemeine landwirtschaftliche Betriebslehre.* Parey, Berlin.
ALKALI, M. M. (1967). *Mixed farming in northern Nigeria.* Ministry of Agriculture, Kaduna.
ALLAN, W. (1967). *The African husbandman.* Oliver and Boyd, London.
ANDREAE, B. (1966). Weidewirtschaft im südlichen Afrika. *Geogr. Z.* **15** (special issue).
—— (1977). *Agrargeographie.* Walter de Gruyter, Berlin, New York.
ANGLADETTE, A. (1966). *Le riz.* Maisonneuve and Larosse, Paris.
—— DESCHAMPS, L. (1974). *Problèmes et perspectives de l'agriculture dans les pays tropicaux.* Maisonneuve and Larosse, Paris.
ANTONINI, G. A., EWEL, K. C., and TUPPER, H. M. (1975). *Population and energy. A system analysis of resource utilization in the Dominican Republic.* The University of Florida Press, Gainsville.
ARNHOLD, J. and LINDEMANN, K. H. (1974). Landwirtschaftliche Nutzungsformen und Probleme der Rinderhaltung auf Neurodungsflächen in immerfeuchten Regenwaldklimaten. *Tropenlandwirt* **75**, April and October, 1974.
ATTEMS, M. (1967). *Bauernbetriebe in tropischen Höhenlagen Ostafrikas.* (Afrika-Studien, no. 25). IFO Institut, München.
BAKER, E. F. I. (1974). *Research into intercropping aspects of farming systems in Nigeria* (mimeo). Ahmadu Bello University, Zaria, Nigeria.
—— NORMAN, D. W. (1975). Cropping systems in Northern Nigeria. In *International Rice Research Institute, Proceedings of the Cropping Systems Workshop,* pp. 234–361. IRRI, Los Baños, Philippines.
—— CURRY, R. B. (1976). Structure of agricultural simulators: a philosophical view. *Agricultural Systems* (1), pp. 201–18.
BANTA, G. R. (1972). Economic evaluation of multiple cropping systems. *IRRI seminar on multi-crop diversification in Taiwan, Taipei* (mimeo).
—— (1974). *Data collection and evaluation for multiple cropping* (mimeo). IRRI, The Philippines.
——HARWOOD, D. R. (1973). *Multiple cropping program of IRRI.* IRRI, The Philippines.
BARKER, R. and ANDEN, T. (1975). Factors influencing the use of modern rice technology in the study areas. In *Changes in rice farming in selected areas of Asia.* IRRI, The Philippines.
—— HERDT, R. W. (1978). *Rainfed lowland rice as a research priority—an economist's view.* IRRI, Los Baños, Philippines.
BARRAL, H. (1977). *Les populations nomades de l'Oudalan et leur espace pastoral* (mimeo). ORSTOM, Paris.
BARRAU, J. (1961). *Subsistence agriculture in Polynesia and Micronesia.* Honolulu.
BAUM, E. (1968). Land use in the Kilombero Valley. In *Smallholding farmer and*

smallholder development in Tanzania (ed. H. Ruthenberg), pp. 23–50. (Afrika-Studien, no. 24). IFO Institut, München.

BEEK, K. J. and BENNEMA, J. (1972). *Land evaluation for agricultural land use planning, an ecological method.* University of Agriculture, Wageningen, Netherlands.

BENNEH, G. (1970). The Huza strip farming system of the Krobe of Ghana. *Geographica polon*, **19**.

—— (1973). Small-scale farming systems in Ghana. *Africa*, **43**.

BENOIT, M. (1977). Mutation agraire dans l'Ouest de la Haute-Volta: le cas de Daboura (sous-préfecture de Nouna). *Cah. ORSTOM, sér.Sci.Hum.*, vol. XIV, no. 2.

BERNARD, A. (1978). *L'agriculture à la marge. L'exemple d'un cas extrême de la Sierra Equatorienne: Guangaje* (ms.). ORSTOM, Quito.

BERNUS, E. (1974a). L'évolution récente des relations entre éleveurs et agriculteurs en Afrique tropical: l'exemple du Sahel Nigerien. *Cah. ORSTOM*.

—— (1974b). *Les Illabahan (Niger). Atlas des Structures Agraires au Sud du Sahara*, no. 10. ORSTOM, Paris.

—— (1977). *Les tactiques des éleveurs face à la sécheresse: le cas du sud-ouest de l'Air, Niger* (mimeo). ORSTOM, Paris.

—— BOUTRAIS, J. and PELISSIER, P. (1974). Evolution et formes modernes de l'élevage dans les zones arides et tropicales. ORSTOM, *Sciences Humaines*, **11**, no. 2

BERTALANFFY, L. VON (1973). *General systems theory* (4th edn.). Brazillier, New York.

BIGOT, Y. (1974). *Revenues agricoles, diffusion des innovations techniques dans les unités expérimentales et conséquences immédiates de gestion individuelle et de politique agricole pour le Sud Sine-Saloum* (mimeo). Centre National de Recherche Agronomique de Bambey, IRAT.

BINSWANGER, H. P., JODHA, N. S., and BARAH, B. C. (1979). *The nature and significance of risk in the semi-arid tropics* (mimeo). ICRISAT, Hyderabad.

—— KRANTZ, B. A. and VIRMANI, S. M. (1976). *The role of the International Crop Research Institute for the Semi-arid Tropics in farming system research* (mimeo). ICRISAT, Hyderabad.

BLANCKENBURG, P. VON (1963). *Rubber farming in Benin area, preliminary report* (mimeo). Nigerian Institute of Social and Economic Research, Ibadan.

—— (1964). Afrikanische Bauernwirtschaft auf dem Wege in eine moderne Landwirtschaft, *Z. ausl. Landw.*, *Sonderh*. Frankfurt.

BONNEFOND, Ph. (1977). *Les exploitations bananières Ivoiriennes employant des tracteurs* (in preparation). ORSTOM, Paris.

BONNEMAISON, J. (1976). Tsarahonenana, des riziculteurs de montagne dans l'Ankaratra. *Atlas des Structures Agraires à Madagascar 3.* Mouton & Co. ORSTOM, Paris.

BOONMA, CH. and KLEMPIN, A. (1975). *Socio-economic conditions of farmers in the Pra Buddhabat self-help land settlement* (mimeo). (Research Report no. 13) Kasetsart University, Bangkok.

BOSERUP, E. (1965). *The conditions of agricultural growth.* Aldine, London.

BOUCHER, L. J. (1967). *Surface irrigation* (mimeo). Food and Agriculture Organization, Rome.

BOUDET, G. (1978). *Le role et les limites de la recherche dans l'amélioration de la gestion des parcours Sahéliens* (mimeo). Maitrise de l'espace agraire et développement en Afrique au sud du Sahara-logique paysanne et rationalité technique. Colloque ORSTOM–CVRS, Ouagadougou.

BOURKE, M. R. (1974). *A long term rotation trial in New Britain, Papua New Guinea* (mimeo). IITA, Ibadan.

BRADFIELD, R. (1968). *Farming systems in Africa. Conference on Agricultural Research Priorities for Economic Development in Africa, Abidjan* 1968 (mimeo).

BRANDT, H. (1971). Die Organisation bäuerlicher Betriebe unter dem Einfluss der Entwicklung einer Industriestadt: Der Fall Jinja, Uganda. *Z. ausl. Landw.* Materialsammlung, Heft **16**.

BRINKMANN, TH. (1922). Die Ökonomik des landwirtschaftlichen Betriebes. In *Grundriss der Sozialökonomik*, Chapter 7 (Mohr), pp. 27–124. Tübingen.

—— (1924). Bodennutzungssysteme. In *Handwörterbuch der Staatswissenschaften*, vol. 1 (Fischer), pp. 959–73. Jena.

BROWN, L. H. (1963). *The development of the semi-arid areas of Kenya* (mimeo). Ministry of Agriculture, Nairobi.

BUCK, J. L. (1930). *Chinese farm economy*. Chicago University Press, Chicago.

BURINGH, P., VAN HEEMST, H. D. J., and STARING, G. J. (1975). *Computation of the absolute maximum food production of the World*. Wageningen.

CABANILLA, L. S. and HERDT, R. W. (n.y.). *An economic analysis of large-scale rice farms under General Order. No.* 47. IRRI, The Philippines.

CAMPBELL, D. J. (1978). *Coping with drought in Kenya Maasailand: pastoralists and farmers of the Loitokitok area, Kajiado District*. Working Paper 337, IDS, University of Nairobi, Nairobi.

CAPOT-REY, R. (1962). The present state of nomadism in the Sahara. In *Problems of the arid zones*. UNESCO, Paris.

CARANGAL, V. R. (1977). Asian cropping systems network—a cooperative program in Asia: A progress report. IRRI (unpublished).

CARRUTHERS, I. D. (1968). *Irrigation development planning; aspects of Pakistan experience*. Agricultural Development Studies, Wye College Report 2, Chapter 6, London.

CARVALHO, L. C. C. and GRACE, L. R. (1976). *Productividade agricole de cana-de-acucar no Estado de Sao Paulo* (mimeo). SOBER, 05-08/09/76.

CHANG, J. H. (1968). The agricultural potential of the humid tropics. *Geog. Rev.* **58**, no. 3.

CHIA-NAN IRRIGATION ASSOCIATION (n.d.), *Standard diagram of rotation irrigation*.

CHARREAU, C. (1974). *Systems of cropping in the dry tropical zone of West Africa. Seminar on Farming systems* (mimeo). ICRISAT, Hyderabad.

CHEN, CH. S. (1963). *Taiwan, an economic and social geography*. Taipei.

CLEAVE, J. H. (1974). *African farmers. Labour use in the development of smallholder agriculture*. Praeger Publishers, New York.

COLLINSON, M. P. (1962–5). *Reports on survey at Bukumbi, Usmao and other locations of Sukumaland, Tanzania* (mimeo). Ukiriguru.

—— (1977). *Demonstrations of an interdisciplinary farming system approach to planning adaptive agricultural research programmes* (mimeo). Cimmyt, Nairobi, Kenya.

CONKLIN, H. C. (1957). *Hanunóo agriculture. A report on an integral system of shifting cultivation on the Philippines*. Food and Agriculture Organization, Rome.

COOK, O. F. (1921). Milpa agriculture, a primitive tropical system. In *Smithsonian Report for 1919*. Washington.

COOPER, J. P. (1970). Potential production and energy conversion in temperate and tropical grasses. *Herb. Abst.* **40** no. (1), 1–15.

CORDOVA, V. and BARKER, R. (1977). *The effect of modern technology on labour utilisation in rice production*. IRRI, paper, no. 77–7, The Philippines.

COURTENAY, P. P. (1965). *Plantation agriculture*. Bell and Sons, London.

CROPPING SYSTEMS WORKING GROUP (1975). *Second Cropping Systems Working Group Meeting, Indonesia.* Publ., IRRI, Los Baños, Philippines.

—— (1976). *Third Cropping Systems Working Group Meeting, Thailand.* Publ., IRRI, Los Baños, Philippines.

—— (1978). *Seventh Cropping Systems Working Group Meeting.* IRRI, Los Baños, Philippines.

CROWDER, L. V. (1971). *Forage and fodder crops in farming systems* (mimeo). IITA, Ibadan.

CU-KONU, E. Y. (1978). *Plantations paysannes et stratégies de l'espace dans le sud-ouest du Togo* (mimeo). Maitrise de l'espace agraire et développement en Afrique au sud du Sahara—logique paysanne et rationalité technique. Colloque ORSTOM–CVRS, Ouagadougou.

CUTIE, J. T. (1975). *Diffusion of hybrid corn technology: the case of El Salvador.* Abridged by CIMMYT. Centro Internacional de Mejoramiento de Maiz y Trigo, Mexico City.

DAHL, G. and HJORT, A. (1976). *Having herds. Pastoral herd growth and household economy.* University of Stockholm.

DALRYMPLE, D. (1971). *Survey of multiple cropping in less developed nations.* USDA, Washington.

DART, P. J. and DAY, J. M. (1975). Non-symbiotic nitrogen fixation in the field. In *Soil microbiology* (ed. N. Walker). Butterworth Scientific Publications, London.

DAVIES, W. and SKIDMORE, C. L. (1966). *Tropical pastures.* Faber and Faber, London.

DELVERT, J. (1972). Remarques sur les agricultures de l'Asie de la Mousson. In *Etudes de géographie tropicale offertes à Pierre Gourou.* Mouton, Paris.

DENT, J. B. and ANDERSON, J. R. (eds) (1971). *Systems analysis in agricultural management.* Wiley, Sydney, New York, London and Toronto.

DIAZ, R., PINSTRUP-ANDERSEN, P., and ESTRADA, R. D. (1974). *Costs and use of inputs in cassava production in Colombia.* CIAT, series. EE-no. 5, Cali.

DION, H. G. (1950). *Agriculture in the Altiplano of Bolivia.* Food and Agriculture Organization, Rome.

DITTMAR, K. (1954). *Allgemeine Völkerkunde.* Viehweg, Braunschweig.

DOPIERALLA, D. (1974). Pflanzenbauliche Möglichkeiten zur Steigerung der Produktivität kleinbäuerlicher Betriebe in der Zentralregion Malawis. *Z. ausl. Landw. Materialsammlung,* 25.

DUCKHAM, A. N. (1974). Biological efficiency in crop livestock production. *Chem. Ind.* Dec. 7.

—— (1976). Environmental constraints. In *Food production and consumption* (eds A. N. Duckham, J. G. W. Jones, and E. H. Roberts). North-Holland Publishing Company, Amsterdam and Oxford.

—— (1978). *The geography of agricultural food productivity* (mimeo). Conference of the International Geographical Union, Ile-Ife, Nigeria.

—— MASEFIELD, G. B. (1970). Farming systems of the World. Chatto & Windus, London.

—— JONES, J. G. W., and ROBERTS, E. H. (1976). The function and efficiency of human food chains. In *Food production and consumption* (eds A. N. Duckham, J. G. W. Jones, and E. H. Roberts), pp. 3–11. North-Holland Publishing Company, Amsterdam and Oxford.

DUMONT, R. (1966). *Types of rural economy.* Methuen, London.

DÜRR, G. (1977). *Production and marketing of potatoes in Kobirichia Location, Meru District, Kenya* (ms.). IPC, Nairobi.

Dyson, R. and Hudson, N. (1969). Subsistence herding in Uganda. *Scient. Am.* **220**, 76–89.

Edwards, D. (1961). *An economic study of small-scale farming in Jamaica.* University Press, Glasgow.

El-Hakim, A. H. (1973). Ökonomik der Bodenentsalzung. *Z. bewässerungsw. Sonderh.* 2.

Emery, F. E. and Frist, E. L. (1971). Socio-technical systems. In *Systems thinking* (ed. F. E. Emery). Penguin Books, Harmondsworth, Middlesex.

Estado de Sao Paulo (1977). *Prognostica 77–78, Regiao Centro-Sul.* Instituto de Economica Agricola. Secretaria da Agricultura, Sao Paulo.

Fals-Borda, O. (1955). *Peasant society in the Colombian Andes.* University of Florida Press, Gainsville.

Faucher, D. (1949). *Géographie agraire.* Génin, Paris.

Federacion Nacional de Cafeteros de Colombia. (1976). *Análisis de los cambios entre costos y precios de differentes regiones agropecuarios* (mimeo). Bogota.

Flach, M. (1973). The sago palm, a potential competitor to root crops. *Third International root crop Conference, Ibadan* (mimeo).

Flinn, J. C. (1974). *Resource use, income and expenditure patterns of Yoruba smallholders* (mimeo). IITA, Ibadan.

—— (1975). *Agricultural economics (notes) (mimeo).* IITA, Ibadan.

—— Jellema, B. M., and Robinson, K. L. (1974). *Barriers to increase food production in the lowland humid tropics of Africa* (mimeo). IITA, Ibadan.

—— Lagemann, J. (1976). *Experiences in growing maize using improved technology in South-Eastern Nigeria* (mimeo). IITA, Ibadan.

Food and Agriculture Organization/SIDA (1974). *Report on Regional Seminar on shifting cultivation and soil conservation in Africa.* Food and Agriculture Organization, Rome.

Fong, N. K., Lian, T. C., and Wikkramatileke, R. (1966). Three farmers of Singapore. *Pacif. Viewpoint* 7, 81–118.

Forbes, T. R. (1975). *A West African soil climosequence and some aspects of foodcrop potential* (ms.). IITA, Ibadan.

Forrester, J. W. (1972). *Grundzüge einer Systemtheorie.* Gabler, Wiesbaden.

Fotzo, P. T. (1977). *Resource productivity and returns in rice production under alternative farming systems. A comparative study in the North-West Province of Cameroun* (ms.). University of Ibadan, Nigeria.

Fox, J. W. and Cumberland, K. B. (1962). *Western Samoa.* Whitcombe and Tombs, Christchurch, New Zealand.

Fremond, Y. (1968). *Coconut-palms selection. Conference on Agricultural Research Priorities for Economic Development in Africa, Abidjan* (mimeo), Report 8/10.

French, M. H. (1967). Animal production and savanna areas. *Report on meeting on savanna development, Khartoum* 25, Oct.–6 Nov. 1966, pp. 153–6. Food and Agriculture Organization, Rome.

Frey, H. J. (1976). *Intensivierung kleinbäuerlicher Betriebe durch angepasste Agrartechnik. Arbeitszeitstudien im Bahati Settlement Scheme, Kenya.* (Bericht, no. 55), IFO-Institut. München.

Friedrich, K. H. (1968). Coffee-banana holdings at Bukoba. *Smallholder farming and smallholder development in Tanzania* (ed. H. Ruthenberg), pp. 175–212 (Afrika-Studien, no. 24). IFO Institut, München.

——Slangen, A. V. E. and Bellete, S. (1973). *Initial farm management survey,* 1972–73 (mimeo). Institute of Agricultural Research, Addis Abeba.

FROELICH, J. C., ALEXANDRE, P., and CORNEVIN, R. (1963). *Les populations du Nord-Togo*. Presse Universitaire de France, Paris.

FUCHS, H. (1973). *Systemtheorie und Organisation*. Gabler, Wiesbaden.

FUNEL, J. M. (1978). *Les interventions planifiées des techniques de développement* (mimeo). Colloque ORSTOM-CVRS, Maîtrise de l'espace et développement en Afrique au sud du Sahara—logique paysanne et rationalité technique. Ouagadougou.

GAILLARD, J. P. and LOSSOIS, P. (1974). *Eléments du prix de revient de la culture d'ananas au Cameroun pour l'exportation frais* (mimeo). Institut Français de Recherches Fruitières Outre-Mer, Paris.

GALLAIS, J. and SIDIKOU, A. H. (1978). Stratégies traditionnelles, prise de décision moderne et aménagement des ressources naturelles dans la zone sahélo-soudanienne. In *Aménagement des ressources naturelles en Afrique: stratégies traditionnelles et prise de décision moderne*. UNESCO, Paris.

GAUCHON, M. J. (1976). *Upper Solo watershed management and upland development* (mimeo). FAO, Solo, Indonesia.

GEERTZ, C. (1963). *Agricultural involution. The process of ecological change in Indonesia*. University of California Press, Los Angeles.

—— (1969). The impact of the concept of culture on the concept of man. In *Man in adaptation* (ed. Y. A. Cohen). Aldine, Chicago.

GIBBON, D., HARVEY, J., and HUBBARD, K. (1974). A minimum tillage system for Botswana. *Wld Crops* **26**, no. 5.

GIGLIOLI, E. G. (1965). Staff organization and tenant discipline on an irrigated land settlement. *E. Afr. agric. For. J.* **30**, 202–5.

GILG, J. P. (1970). Culture commerciale et discipline agraire Dobadéné (Tchad). *Etudes rurales*, no. 37, 38, 39. Jan.–Sept., pp. 173–97.

GOMEZ, A. A. (1976). Cropping systems approach to production program: the Philippine experience. In *Proceedings of the Symposium on Cropping Systems Research and Development for the Asian Rice Farmer*. IRRI, Los Baños, Philippines.

GOSS, C. (1973). *Costs of production and return accruing to rubber smallholders in Changwats Trang and Yala Peninsular, Thailand* (mimeo). Rubber Research Centre, Hat Yai.

GOSWAMI, P. C. and BORA, C. K. (1977). *Studies in the farm management in Nowgong District, Assam* (mimeo). Assam Agricultural University, Jorhat, India.

GOUROU, P. (1966). *The tropical world*. Longmans, London.

—— (1969). Les cacaoyers en pays Yoruba. *Recl. Art. Revue belge Géog.* **93**, 1–3.

GRANIER, P. (1975). Plant–animal interaction in the Sahelian zone. In *Evaluation and mapping of tropical African range lands*. ILCA Seminar, Bamako, Mali.

GREENLAND, D. J. and HERRERA, R. (1975). *Shifting cultivation and agricultural practices* (ms.). IITA, Ibadan.

GRIGG, D. (1970). *The harsh land*. Macmillan, London.

GRIMBLE, R. J. (1973). *The central highlands of Thailand. A study of farming systems*. Overseas Development Administration, Wye College, London.

GRIST, D. H. (1959). *Rice*. Longmans, London.

GROENEVELD, S. (1968). Traditional farming and coconut-cattle schemes in the Tanga region. In *Smallholder farming and smallholder development in Tanzania* (ed. H. Ruthenberg) (Afrika-Studien, no. 24), IFO Institut, München.

GROENEWOLD, H. H. (1974). Communication. In *The husbandry and health of the domestic buffalo* (ed. W. R. Cockrill). Food and Agriculture Organization, Rome.

GUILLARD, J. (1965). *Golonpoui, analyse des conditions de modernisation d'un village du Nord-Cameroun*. Mouton, Paris.

GUILLEMIN, R. (1956). Evolution de l'agriculture autochthone dans les savannes de l'Oubangui. *Agron. trop.*, *Nogent* 12, nos. 1, 2, 3.

GUILLOT, B. (1970a). Le village de Passia, essai sur le systeme agraire Nzabi. *Cah. ORSTOM* 7, no. 1.

—— (1970b). Structures agraires Koukouya (Congo-Brazaville). *Etudes rurales*, nos. 37, 38, 39. Jan.–Sept., 1970, pp. 312–35.

—— (1977). Problèmes de développement de la production cacaoyère dans les districts de Sembé à Souanke (Congo). *Cah. ORSTOM, ser.Sci.Hum.*, vol. XIV, no. 2.

GWYNNE, M. D. (1977). Some ecological criteria for land evaluation of different types of rangeland. In *Proceedings of a seminar on land evaluation for rangeland use*. Kenya Soil Survey Misc. Papers no. 11, Nairobi.

HALLAIRE, A. (1971). *Hodogway, un village de montagne en bordure de plaine (Cameroun)*. Mouton, Paris.

HANSEN, S. (1969). *An outline of a rubber programme in Ceylon* (mimeo). Rubber Research Institute of Ceylon, Dartonfield.

HARWOOD, R. R. (1973). *The concepts of multiple cropping* (mimeo). IRRI, The Philippines.

—— (1974). *The resource utilization approach to cropping systems development* (mimeo). IRRI, The Philippines.

—— (1975). Farmer-oriented research aimed at crop intensification. In *International Rice Research Institute, Proceedings of the Cropping Systems Workshop*. Los Baños, Philippines.

—— (1979). *Small farm development*. Westview Press, Boulder, Colorado, USA.

—— PRICE, E. C. (1975). *Multiple cropping in tropical Asia* (mimeo). IRRI, The Philippines.

—— PLUCKNETT, D., and ROMANOWSKI, R. (1977). *Vegetable cropping systems in the People's Republic of China* (mimeo). Rodale Press Inc., USA.

HASWELL, M. R. (1953). Economics of agriculture in a savanna village (Gambia). *Colon. Res. Stud.* no. 8.

—— (1963). *The changing pattern of economic activity in a Gambian village* (mimeo). Department of Technical Co-operation, Overseas Research Publication no. 2, London.

—— (1975). *The nature of poverty*. Macmillan Press, London.

HAYAMI, Y., BENNAGEN, E., and BARKER, R. (1977). Price incentives versus irrigation investments to achieve food self-sufficiency in the Philippines. *Am. J. Agricult. Econ.* 59 (4), 718.

HERDT, W. R. and BARKER, R. (1977). Multi-site test environments and the breeding strategies for new rice technology. *IRRI Research Papers*, No. 7.

—— LASCINA, T. (1975). *The domestic resource costs of increasing Philippine rice production*. Stanford Food Research Institute–Joint Commission for Rural Reconstruction. Workshop, Taipei.

—— WICKHAM, TH. H. (1974). *Major constraints to rice production with emphasis on yields in the Philippines* (mimeo). IRRI, The Philippines.

HERRERA, W. T. and HARWOOD, R. R. (1973). *Crop interrelationship in intensive cropping systems* (mimeo). IRRI, The Philippines.

—— BANTILAN, R. T., TINSLEY, R. L., HARWOOD, R. R., and ZANDSTRA, H. G. (1976). *An evaluation of alternative cropping patterns on a rainfed lowland rice area in Pangasinan*. IRRI, Los Baños, Philippines.

HEYER, J. U. (1966). *Agricultural development and peasant farming in Kenya*. Dissertation. University of London.

HICKLING, C. F. (1961). *Tropical inland fisheries.* Longmans, London.

HILL, P. (1963). *The migrant cocoa farmer of Southern Ghana.* Cambridge University Press.

Ho, R. (1964). Mixed farming and multiple cropping in Malaya. In *Land use and mineral deposits in Hong Kong* (ed. S. G. Davis), Hong Kong.

—— (1967). *Farmers of central Malaya.* Australian National University, Canberra.

HOBEN, A. (1976). *Social soundness of the Masai livestock and range management project* (mimeo). USAID, Nairobi.

HOLLIDAY, R. H. (1976). The efficiency of solar energy conversion by the whole crop. In *Food production and consumption* (eds A. N. Duckham, J. G. W. Jones, and E. H. Roberts), pp. 127–46. North-Holland Publishing Company, Amsterdam and Oxford.

HURAULT, J. (1970). L'organisation du terroir dans le groupements Bamiléké. *Etudes rurales,* no. 37, 38, 39. Jan.–Sept., pp. 232–56.

ICRISAT (1977). *Annual report of the farming system research program, 1976–77.* Hyderabad, India.

INDIAN SOCIETY OF AGRONOMY (1973). *Multiple cropping.* Indian Agricultural Research Institute, New Delhi.

IRAT (1972). *Nécessité agronomique et intérêt économique d'une intensification des systèmes agricoles au Sénégal* (mimeo). Bambey.

IRRI (1974). *An agro-climatic classification for evaluating cropping systems potentials in south-east Asia rice growing regions.* IRRI, The Philippines.

—— (1975). *Research highlights for 1974* IRRI., The Philippines.

—— (1976). *Annual report for 1975.* IRRI, Los Baños, Philippines.

—— (1976). *Report of the cropping systems group* (mimeo). IRRI, The Philippines.

—— (1978). *Rice research and production in China.* IRRI, The Philippines.

Iso, E. (1964). *Rice and crops in its rotation in subtropical zones.* Japan Food and Agriculture Organization Association, Tokyo.

IVORY COAST (1972). *Autorité pour l'aménagement de la région du Sud-Ouest. Normes Techniques* (mimeo). Abidjan.

JAHNKE, H. E. (1976). Tsetse flies and livestock development in East Africa. A study in environmental economics. Afrika Studien, no. 87, IFO Institut, München.

—— RUTHENBERG, H., and THIMM, H. (1974). *Range development in Kenya. A review of commercial, company, individual and group ranches.* Studies in Employment and Development, no. 4. IBRD, Washington.

JÄTZOLD, R. and BAUM, E. (1968). *The Kilombero Valley* (Afrika-Studien no. 28). IFO Institut, München.

JONES, W. O. (1968). Plantation. *Int. Encycl. soc. Sci.* **2**, 154–9.

JOOSTEN, J. H. L. (1962). *Wirtschaftliche und agrarpolitische Aspekte tropischer Landbausysteme* (mimeo). Institut für landwirtschaftliche Betriebslehre, Göttingen.

JORDAN, H. D. (1967). Rice production. In *Conference on Agriculture Research Priorities for Economic Development in Africa* (mimeo). Abidjan.

JURION, F. and HENRY, J. (1967). *De l'agriculture itinérante à l'agriculture intensifiée* INEAC hors série, Bruxelles.

KAHLON, A. S. (1975). *Energy requirements for intensive agricultural production. Co-ordinated Project Annual Report 1973–74* (mimeo). Ludhiana.

—— MIGLIANI, S. S. (1974). *Studies in the economics of farm management. Ferozepur District (Punjab). 1967–68 to 1969–70.* New Delhi.

KAMPEN, J. et al. (1975), *Soil and water conservation in management in farming system research for semi-arid zones* (mimeo). ICRISAT, Hyderabad.

KANWAR, J. S. (1972). Cropping patterns, scope and concept. In *Proceedings of the symposium on cropping patterns in India.* ICAR, New Delhi.

KASSAM, A. H. (1977). *Net biomass production and yield of crops in West Africa* (mimeo). FAO, Rome.

——KOWAL, J. M. (1973). Productivity of crops in the savanna and rain forest zones in Nigeria. *Savanna* 2, no. 1.

KELLER, J., PETERSON, D. F., and PETERSON, H. B. (1973). *A strategy for optimizing research on agricultural systems involving water management.* Agricultural and Irrigation Engineering/Utah Water Research Laboratory, Utah State University, Logan, Utah.

KENYA COFFEE GROWERS' ASSOCIATION (1976). *Coffee estate expenses* (mimeo). Nairobi.

KHAN, K. K. S. *et al.* (1974). *Integrated rural development programme* (mimeo). Tana Farm Survey, no. 9, Ministry of Agriculture, Dacca.

KIMBER, A. J. (1974). *Crop rotations. Legumes and more productive arable farming in the highlands of Papua, New Guinea* (mimeo). University of Papua, New Guinea.

KIRSCH, W. (1974). *Betriebswirtschaftslehre: Systeme, Entscheidungen, Methoden.* Gabler, Wiesbaden.

KNIGHT, P. T. (1976). Economics of cacao production in Brazil. In *Cacao production* (ed. J. Simmons). Praeger Publishers, New York.

KOBY, A. Th. (1978). *Projection des formations sociales sur l'espace: exemple du pays Odzukru en Côte d'Ivoire* (mimeo). Maîtrise de l'espace agraire et developpement en Afrique au sud du Sahara—logique paysanne et rationalité technique. Colloque ORSTOM-CVRS, Ouagadougou.

KOSTROWICKI, J. (1974). *The typology of world agriculture. Principles, methods and model types.* International Geographical Union, Warszawa.

KRANTZ, B. A. (1977). *The farming system research program.* ICRISAT, Hyderabad, India.

—— *et al.* (1974). Cropping pattern for increasing and stabilizing agricultural production in the semi-arid tropics. *Seminar on farming systems* (mimeo). ICRISAT, Hyderabad.

—— KAMPEN, J. (1974). *Farming systems research: current program and 1973 highlights.* ICRISAT, Hyderabad, India.

KRISHNAMOORTHY, CH. (1974). Present cropping systems including trends of changes and advances of approach for improvement. *Seminar on farming systems* (mimeo). ICRISAT, Hyderabad.

KUHONTA, P. C., WIJEKOON, L. D., and ARIYARATNAM, E. A. (1973). *Pineapple production in Sri Lanka* (mimeo). (Food and Agricultural Organization report no. 3). Paradeniya.

KULTHONGKHAM, S. and ONG, SHAO-ER (1964). *Rice economy of Thailand.* Bangkok.

KUNG, P. (1975). Farm crops of China. *Wld Crops,* 27, no. 3.

KYEYUNE-SENTONGO, L. L. (1973a). *An economic survey of Rzumbura and Kinkizi Counties of the Kigezi District of Uganda, Report II* (mimeo). Ministry of Agriculture, Entebbe.

—— (1973b). *An economic survey of high-altitude areas of Ankole* (mimeo). Uganda Ministry of Agriculture, Kampala.

LACOMBE-ORLAC, M. (1967). *Contribution à l'étude de l'emploi du temps du paysan dans la zone arachidière, Hanene—Sénégal* (mimeo). Institut de Science Economique Appliquée, Dakar.

LAGEMANN, J. (1975). *Case study from a smallholder with oil-palms in Eastern Nigeria* (mimeo). IITA, Ibadan.

LAGEMANN, J. (1977). *Traditional African farming systems in Eastern Nigeria.* (Afrika-Studie, no. 98). IFO-Institut, Weltforum Verlag, München.

—— FLINN, J. C., OKIGBO, B. N., and MOORMANN, F. R. (1975). *Root crop and oil-palm farming systems: a case study from Eastern Nigeria* (mimeo). ITTA, Ibadan.

LAHUEC, J. P. (1976). *Zaougho, étude géographique d'un village de l'est Mossi* (mimeo). ORSTOM, Ouagadougou.

LAL, R. (1975). *Role of mulching techniques in tropical soil and water management.* Technical Bulletin no. 1, IITA, Ibadan.

LANG, H. (1978). *The economics of rainfed rice cultivation in West Africa. The case of the Ivory Coast.* Diss. Hohenheim.

LASZLO, E. (1972). *Introduction to systems philosophy.* Gordon and Breach, New York, London, and Paris.

LAVINIA, G. S. (1974). *Studies in economics of farm management in Deoria, Uttar Pradesh 1966–69.* Ministry of Agriculture, New Delhi.

LEACH, G. (1976). Industrial energy in human food chains. In *Food production and consumption* (eds A. N. Duckham, J. G. W. Jones, and E. H. Roberts). North-Holland Publishing Company, Amsterdam and Oxford.

LEAKY, L. S. B. (1934). *Science and the African* (mimeo). Nairobi. (Re-edited by C. L. A. Leaky, 1970, Nairobi.)

LERICOLLAIS, A. (1972). *SOB, études géographiques d'un terroir Sérèr (Sénégal).* ORSTOM, Mouton, Paris.

LESHNIK, L. L. (1966). The system of dry-farming in the West Nimar District of Central India. In *Jahrbuch des Südasien Instituts der Universität, Heidelberg.* Harrassowitz, Wiesbaden.

LIBOON, S. P. and HARWOOD, R. R. (1976). *The effect of crop damage in corn–peanut intercropping* (mimeo). Paper presented at the 7th Annual Meeting of the Crop Science Society of the Philippines (1975). Davao City, Philippines.

LIEBIG, J. VON (1878). *Chemische Briefe* (6th edn). Winter'sche Verlagshandlung, Leipzig and Heidelberg.

LIH-YUH SHY TSAI and HERDT, R. W. (1976). *Economic changes in typical rice farms in Taiwan, 1895–1976* (mimeo). IRRI, Philippines.

LIM, Y. (1965). *Export industries and pattern of economic growth in Ceylon* (micro-film). Dissertation, Michigan.

LITSINGER, J. A. and MOODY, K. (1976). Integrated pest management in multiple cropping. In *Multiple cropping* (eds R. I. Papendic, P. A. Sanchez, and G. B. Triplet). Am. Soc. Agron. Spec. Publ. no. 27.

LITTLE, I. M. D. and TIPPING, D. G. (1972). *A social cost benefit analysis of the Kulai oil palm estate.* OECD Development Centre Studies (Series on Cost Benefit Analysis, no. 3), Paris.

LIZANO, C. L. and JEFE, P. (1977). *Costos de producción del café en Costa Rica (ms).* Oficina del Cafe, Costa Rica.

LUDWIG, D. H. (1967). *Ukara—Ein Sonderfall tropischer Bodennutzung im Raum des Victoria-Sees.* (Afrika-Studien, no. 22) IFO Institut, München.

—— (1968). Permanent farming on Ukara. In *Smallholder farming and smallholder development* (ed. H. Ruthenberg). Weltforum Verlag, München.

LUNING, H. A. (1961). *An agro-economic survey in Katsina Province (Nigeria).* Government Printer, Kaduna.

MACARTHUR, J. D. (1964). *Some economic studies of African farms in Rift Valley Province, Nandi District, Elgeyo, and West Pakot District* (mimeo). Nakuru, Kenya.

—— (1966). *Some economic aspects of agricultural development in Nyeri District 1963* (mimeo). Nakuru.

—— (1969). *The economics of perennial crops* (mimeo). Institut für ausländische Landwirtschaft, Berlin.

—— ENGLAND, W. T. (1963). *A report on an economic survey of farming in the Molo-Mau-Narok area, 1962/63* (mimeo). Nakuru, Kenya.

MCCONNEL, D. J. (1972). *The place of livestock in the farming systems of Sind (Mirwah) farms, Pakistan* (mimeo). Food and Agriculture Organization Report TA. 3069, Rome.

MCINTOSH, J. L. (1975). *Explanation and implementation of multiple cropping* (mimeo). IRRI/IPI, Central Research Institute for Agriculture, Bogor, Indonesia.

MAGBANUA, R. D., ROXAS, N. M., RAYMUNDO, M. E., and ZANDSTRA, H. G. (1977). *Testing of rainfed lowland rice cropping pattern in Iloilo, 1976–77* (mimeo). IRRI, Philippines.

MANN, K. S. (1973). *Lowland farm development project Hodeidah*, agricultural economics. Food and Agricultural Organization Tech. Report SF/YEM 11, Rome.

MANSHARD, W. (1961). *Die geographische Grundlage der Wirtschaft Ghanas.* Steiner, Wiesbaden.

—— (1968). *Einführung in die Agrargeographie der Tropen.* Bibliographisches Institut, Mannheim.

MARCHAL, J. Y. (1977). Système agraire et évolution de l'occupation de l'espace au Yatenga (Haute Volta). *Cah. ORSTOM, sér.Sci.Hum.,* vol. XIV, no. 2.

MARIN GAMBOA, M. P. (1977). *Análisis económico de la producción de café en Costa Rica. Cosecha 1976–1977.* Oficina del Cafe, Boletin tecnico no. 8, San José.

MAYMARD, J. (1974). Structures africaines de production et concept d'exploitation agricole. *Cah. ORSTOM, sér. Biol.* no. 24.

MEILLASSOUX, C. (1964). *Anthropologie économique des Gouro de Côte d'Ivoire.* Mouton, Paris.

MIAN, SH. M. (1977). *Costs and returns of aman cultivation, 1976* (mimeo). Comilla, Bangladesh.

MILLEVILLE, P. (1974). Enquête sur les facteurs de la production arachidière dans trois terroirs de moyenne Casamance. *Cah. ORSTOM, sér. Biol.* no. 24.

MIRACLE, M. P. (1964). *Traditional agricultural methods in the Congo Basin* (mimeo). Food Research Institute, Stanford University.

—— (1967). *Agriculture in the Congo Basin.* University of Wisconsin Press, Madison, Milwaukee, and London.

MOHR, B. (1969). *Reiskultur in Westafrika* (Afrika-Studien, no. 44). IFO Institut, München.

MONACO, L. C. (1977). Consequences of the introduction of coffee rost into Brazil. *Ann. N.Y. Acad. Sci.* **287**, 51–71.

MONNIER, J. (1976). *Première approche agro-socio-economique de l'exploitation agricole en pays Wolof Saloum-Saloum* (mimeo). ISRA, Dakar and Paris.

—— DIAGNE, A., SOW, D., and SOW, Y. (1974). *Le travail dans l'exploitation agricole Sénégalaise* (mimeo). Centre National de Recherches Agronomiques de Bambey.

MOODY, K. (1975). Weeds and shifting cultivation. *Pans* **21**, no. 2.

MORGAN, W. B. (1969). The zoning of land use around rural settlements in tropical Africa. In *Environment and land use in Africa* (eds M. F. Thomas and G. W. Whittington). Methuen, London.

MOROOKA, Y., HERDT, R. W., and HAWS, L. D. (1979). An analysis of the labor-intensive continuous rice production system at IRRI. *IRRI Research Paper Series*, no. 29, May 1979.

NAGBISWAS, S. C. et al. (1972). *Study on farm management and cost of production of crops, West Bengal*, Vol. 10, 1967–68. Directorate of Agriculture, Calcutta.

NAKAJUD, A., ARCHAVASMIT, P., and BOONYAKOM, S. (1971). *Multiple cropping system which included corn and sorghum in Amphoe Phayuha Khiri, Changwat Nakhon Sawan in 1968*. Technical Report no. 2 (mimeo). Ministry of Agriculture, Bangkok.

NARAYANA, D. L. (1974). *Studies in the economics of farm management in Cuddapah District (Andra Pradesh), 1969–70*. Ministry of Agriculture, New Delhi.

NAVARRO, L. A. (1977). *Estudio de caso en Costa Rica* (Informe preliminar). CATIE, Turrialba.

NELLIAT, E. V., BAVAPPA, K. V., and NAIR, P. K. R. (1974). Multi-storeyed cropping. *Wld Crops* **26**, no. 6.

NICOLAISEN, J. (1963). *Ecology and culture of the pastoral Tuareg*. National Museum of Copenhagen.

NORMAN, D. W. (1967). *An economic study of three villages in Zaria Province. Land and labour relationship* (Samaru Miscellaneous Papers no. 19). Ahmadu Bello University Zaria, Nigeria.

—— (1970). *Traditional agricultural systems and their improvement*. Paper presented at a seminar on Agronomic Research in West Africa, 1970. The University of Ibadan, Nigeria.

—— (1972). *An economic survey of three villages in Zaria Province. Input–output study*. (Samaru Miscellaneous Papers no. 37). Zaria, Nigeria.

—— (1973). Crop mixtures under indigenous conditions in the Northern part of Nigeria. In *Factors of agricultural growth in western Africa* (ed. I. M. Ofori). Legon University, Legon, Ghana.

NOWAK, E. (1954). Land und Volk der Konso. *Geogr. Abh.* no. 14.

NYE, P. H. and GREENLAND, D. J. (1961). *The soil under shifting cultivation*. Commonwealth Agricultural Bureau, Reading.

OBI, J. K. and TULEY, P. (1973). The bush fallow and ley farming in the oil palm belt of South East Nigeria. *ODA Miscellaneous* report no. 161.

OBOLER, R. S. (1977). *Work and leisure in modern Nandi* (mimeo). IDS Paper no. 324, Nairobi.

OKIGBO, B. N. (1974a). *Fitting research to farming systems*. Based on observations and preliminary studies of traditional agriculture in Eastern Nigeria (mimeo). IITA, Ibadan.

—— (1974b). *The IITA farming system program* (mimeo). IITA, Ibadan.

—— (1978). *Cropping systems and related research in Africa*. Association for the Advancement of Agricultural Sciences in Africa, Ibadan.

OWEN, R. VON (1969). *Produktionsstruktur und Entwicklungsmöglichkeiten der Rindfleischerzeugung in Südamerika*. Dissertation, Göttingen, University.

PALACPAC, A. C. (1977). *World rice statistics*. IRRI, The Philippines.

PALADA, M. C., TINSLEY, R. L., and HARWOOD, R. R. (1976). *Cropping systems agronomy program for rainfed lowland rice areas in Iloilo* (mimeo). IRRI, The Philippines.

PEAT, J. E. and BROWN, K. J. (1962). The yield response of raingrown cotton at Ukiriguru in the Lake Province of Tanganyika. *Emp. J. Exp. Agric.* **30**, 313.

PELISSIER, P. (1966). *Les paysans du Sénégal*. Imprimerie Fabregue, Saint-Yrieux.

—— DIARRA, S. (1978). Stratégies traditionnelles, prise de décision moderne et aménagement des ressources naturelles en Afrique soudanienne. In *Aménagement des ressources naturelles en Afrique: stratégies traditionnelles et prise de décision moderne*. UNESCO, Paris.

Piggot, C. J. (1964). *Coconut growing*. Oxford University Press, London.

Pinxten, K. (1954). *De Inlands Landbouwbedrijven in Belgisch-Kongo en Ruanda-Urundi*, vols 1 and 2. Ministerie van Kolonien, Bruxelles.

Porteres, R. (1952). *Aménagement de l'économie agricole et rurale au Sénégal.* IRAT, Bambey.

Pössinger, H. (1967). *Plantagenbetriebe oder Bauernbetriebe am Beispiel des Sisals in Ostafrika* (Afrika-Studien no. 13). IFO Institut, München.

—— (1968). *Landwirtschaftliche Entwicklung in Angola and Mozambique* (Afrika-Studien, no. 31). IFO Institut, München.

Pradeau, CH. (1970). Kokolibou, Haute Volta, ou le pays Dagari à travers un terroir. *Etudes rurales*, no. 37, 38, 39, Jan.–Sept. 1970, pp. 85–112.

Pratt, D. J. and Gwynne, M. D. (1977). *Rangeland management and ecology in East Africa*. Hodder and Stoughton, London, Sydney, Auckland, and Toronto.

Price, E. C. and Barker, R. (1977). *A preliminary evaluation of the time distribution of crop labor as a criteria for design and testing of new rice-based cropping patterns.* Paper presented at the Symposium on Household Economics, 1977. The Regent of Manila, Philippines.

Programa Nacional del Banano y Fruitas Tropicales (1977). *Costos, Sector Sabana-Calichana* (ms.). Guayaquil, Ecuador.

Ptsipef (1975). *Smallholders rubber development and replanting project, West Sumatra* (mimeo). Sipef and F.R.G./A.D.P. Medan and Bukittinggi, Sumatra.

Raison, J. P. (1970). Paysage rural et démographie Leimavo (Nord du Betsileo, Madagascar). *Etudes rurales*, no. 37, 38, 39, Jan.–Sept. 1970.

Rajagopalan, V. and Balasubramanian, M. (1976). *Studies in the economics of farm management in Coimbatore District* (Tamil Nadu). Ministry of Agriculture, New Delhi.

Randall, A. (1975). Growth, resources and environment: some conceptual issues. *Am. J. Agric. Econ.* **57**, no. 5, 803–9.

Randhawa, N. S., Cheema, S. S., and Dev, G. (1972). Soil fertility aspects. In *Proceedings of the symposium on cropping patterns in India*. ICAR, New Delhi.

Rastogi, B. K. (1974). *Income and employment raising potential of the New Dryland Agriculture Technology*. Fifth Annual Workshop. All India Coordinated Research Project for Dryland Agriculture. Hyderabad, India.

Refeno, G. (1975). *Etudes économiques de la culture de vanille à Madagascar* (ms.). Station de la Vanille d'Antalaha.

Richards, A. I. (1961). *Land, labour and diet in Northern Rhodesia*. Oxford University Press, London.

Ripailles, C. (1966). *Ananas.* (mimeo). Ministère d'Agriculture, Ivory Coast.

Roberts, E. H. (1976). The efficiency of photosynthesis. In *Food production and consumption* (eds A. N. Duckham, J. G. W. Jones, and E. H. Roberts). North-Holland Publishing Company, Amsterdam and Oxford.

Rothenhan, D. von (1966). *Bodennutzung und Viehhaltung im Sukumaland, Tansania* (Afrika-Studien, no. 11). IFO Institut, München.

—— (1968). Cotton farming in Sukumaland. In *Smallholder farming and smallholder development in Tanzania* (ed. H. Ruthenberg), pp. 51–86 (Afrika-Studien no, 24). IFO Institut, München.

Rouamba, P. T. (1970). Terroirs en pays Mossi, à propos de Yaoghin (Haute-Volta) *Etudes rurales*, no. 37, 38, 39, Jan.–Sep. 1970.

Rounce, N. V. (1949). *The agriculture of the cultivation steppe of the Lake, Western and Central Provinces of Tanganyika*. Department of Agriculture, Tanganyika. Longmans, Salisbury.

ROURKE, B. E. (1974). Profitability of cacao and alternative crops in Eastern Region, Ghana. In *Economics of cacao production and marketing* (eds R. A. Kotey, C. Okali, and B. E. Rourke). University of Ghana, Legon.

RUBEL, E. (1935). The replaceability of ecological factors and the law of the minimum. *Ecology* 16, 336–41.

RÜMCKER, A. von (1972). Die Organisation bäuerlicher Betriebe in der Zentralregion Malawis. *Z. ausl. Landw. Materialsammlung*, 22.

RUTHENBERG, H. (1966). *Agricultural development policy in Kenya* (Afrika-Studien, no. 10). IFO Institut, München.

—— (ed.) (1968). *Smallholder farming and smallholder development in Tanzania* (Afrika-Studien, no. 24). IFO Institut, München.

—— (1974). Artificial pastures and their utilization in the Southern Guinea savanna and the derived savanna of West Africa. *Z. ausl. Landw.* 13, no. 3, and 4.

—— (1977). The development of crop research in the humid and semi-humid tropics. *Plant Res. Develop.* 6, Tübingen.

RUTTAN, V. W., SOOTHIPAN, A., and VENEGAS, E. C. (1966). Changes in rice growing in the Philippines and Thailand. *Wld Crops* 18, 18–33.

RYAN, J. G., SARIN, R., and PEREIRA, M. (1979). *Assessment of prospective soil- and water- and crop-management technologies for the semi-arid tropics of peninsular India* (mimeo). ICRISAT, Hyderabad.

SALCI (1970). *Coût de plantation d'un hectare d'ananas* (mimeo). Abidjan.

SANCHEZ, P. A. (1977). Advances in the management of Oxisols and Utisols in tropical South America, in *Proceedings of the International Seminar on Soil Environment and Fertility Management in Intensive Agriculture*. Tokyo, Japan.

SANDERS, J. H. (1979). *New agricultural technology in the Brazilian Sertao* (mimeo). CIAT, Cali, Colombia.

SANTEN, C. E. VAN (1974). *Selected economic aspects of expanding rice production in Liberia* (mimeo). WS. Fl. 325. Food and Agriculture Organization.

SANTOIR, Ch. (1977). *Les sociétés pastorales du Sénégal face à la sécheresse 1972–73.* ORSTOM, Dakar.

SATMACI (1973). *Coût de plantation d'1 ha de caféiers moderne et d'1 ha de cacao moderne* (mimeo). Abidjan.

SAUTTER, G. and MONDJANNAGNI, A. (1978). Stratégies traditionnelles, prise de décision moderne et aménagement des ressources naturelles en zones forestières et préforestières d'Afrique. In *Amenagement des ressources naturelles en Afrique: stratégies traditionnelles et prise de décision moderne*. UNESCO, Paris.

SAVONNET, G. (1970). *Pina (Haute-Volta)*. ORSTOM, Mouton, Paris.

SCHICKELE, R. (1931). Die Weidewirtschaft in den Trockengebieten der Erde. *Probleme Weltwirt.* 53.

SCHINKEL, H. G. (1970). *Haltung, Zucht, und Pflege des Viehs bei den Nomaden Ost- und Nordostafrikas*. Akademie Verlag, Berlin.

SCHLIPPE, P. DE (1956). *Shifting cultivation in Africa*. Routledge and Kegan Paul, London.

SCHULTZ, J. (1974). *The basically traditional land use system of Zambia and their regions* (ms.). University Trier-Kaiserslautern, 1974.

SEDES (Société d'Etudes pour le Développement Economique et Social) (1972). *Etude socioéconomique de la zone de Lac Horo et des Mares de Niafunke et de Dire (République du Mali)*, (mimeo). Vol. 2, Paris.

SELVADURAI, S. (1968). *A preliminary report on the survey of coconut smallholdings in West Malaysia* (mimeo). Ministry of Agriculture, Kuala Lumpur.

SEMINAR FÜR LANDWIRTSCHAFTLICHE ENTWICKLUNG (SLE) (1977–8). *Etude Agro-*

socio-économique de base sur les conditions de développment de la sous-préfecture de Paoua, Ouham-Pende, Empire Centrafricain. Fachbereich Internationale Agrarentwicklung (FIA), Berlin.

SERÉ, C. (1978). *Entwicklungsmöglichkeiten milchviehhaltender Betriebe in den subtropischen indgemässigten Klimazonen Südamerikas* (in preparation). Hohenheim.

SHACK, W. A. (1969). *The Gurage. A people of the Ensete culture.* Oxford University Press, London, New York, and Nairobi.

SHANG, C. Y. (1973). *Economic aspects of eel farming in Taiwan.* Joint Commission for Rural Reconstruction Fisheries Series, no. 14. Taipei.

SICABAM (1977). *Eléments pour le calcul du prix de revient des bananes* (mimeo). Fort de France, Martinique.

SICK, W. D. (1969). *Wirtschaftsgeographie von Ecuador.* Geographisches Institut, Stuttgart.

SIMPSON, M. (1979). *Alternative strategies for agricultural development of the Central Rainlands of the Sudan—with special reference to the Damazine area.* University Press, Leeds.

SINGH, A. (1972). Conceptual and experimental basis of cropping patterns. In *Proceedings of the Symposium on cropping patterns in India.* Indian Council for Agricultural Research, New Delhi.

SINGH, R. P., NANDAL, D. S., and SINGH, L. (1973). Economic analysis on multiple cropping in Haryana. In *Symposium on Multiple cropping.* Indian Society of Agronomy (eds), New Delhi.

SMET, R. E. DE (1972). Une enquête budget temps dans les Ulélé (République du Zaire). In *Etudes De Géographie Tropicales offertes à Pierre Gourou.* Mouton, Paris.

SMITH, R. W. (1970). The Malayan Dwarf supersedes the Jamaica tall coconut. Changes in farming practices. *Oléagineux* 25e *Année,* no. 11, Nov. 1970.

SNAYDON, R. W. and ELSTON, J. (1976). Flows, cycles and yields in agricultural ecosystems. In *Food production and consumption* (eds A. N. Duckham, J. G. W. Jones, and E. H. Roberts), pp. 43–60. North-Holland Publishing Company, Amsterdam and Oxford.

SODEFEL (1971). *Opération ananas. Estimation du Prix du Revient d'un Hectare d'Ananas* (mimeo). Ono, Ivory Coast.

SODEPALM (1973). *Coût de plantation d'un hectare de palmiers et d'un hectare de cocotier* (mimeo). Abidjan.

SOOKSATHAN, I. and HARWOOD, R. R. (1976). *A comparative growth analysis of intercrop and monoculture planting of rice and corn.* Paper presented at a Saturday Seminar, 1976. IRRI, Los Baños, Philippines.

SPEDDING, C. R. W. (1975). *The biology of agricultural systems.* Academic Press, London.

SPENCER, D. S. C. (1975). *The economics of rice production in Sierra Leone* (mimeo). Njala University College.

SPENCER, J. E. (1966). *Shifting cultivation in south-eastern Asia.* University of California Press, Berkeley and Los Angeles.

SRI LANKA, Ministry of Agriculture (1974, 1975, 1976, 1977). *Costs of production of tea, rubber and coconuts.* Department of Census and Statistics, Colombo.

STOTZ, D. (1977a). *Smallholder dairy enterprise recording scheme* (mimeo). Livestock Recording Centre, Naivasha, Kenya.

—— (1977b). Zero-grazing. *Kenya farmer,* June.

—— (1979). *Smallholder dairy development in Kenya.* Diss. Univ. Hohenheim.

STROEBEL, H. (1975). *Entwicklungsmöglichkeiten landwirtschaftlicher Kleinbetriebe im Kericho Distrikt, Kenia* (ms.). University of Stuttgart-Hohenheim.

SUDAN GEZIRA BOARD (1975). *Economic Survey, Report of Gezira 1973/74*, Barakat. (In Arabic.)

SURYATNA, E. S., and HARWOOD, R. R. (1976). *Nutrient uptake of two traditional intercrop combinations and insect and disease incidence in three intercrop combinations* (mimeo). IRRI, The Philippines.

SYARIFUDDIN, A., SURYATNA, E. S., ISMAIL, I. C., and MCINTOSH, J. L. (1974). *Performance of corn, peanut, mungbean, and soybean in monoculture and intercrop combinations of corn and legumes in dry season 1973*. Contrib. Cent. Res. Inst. Agric., Bogor, Indonesia.

TA CHENG, CHONG BOO JOCK, SIER HOOI KOON, and MOULTER, T. (1973). *A report on paddy and paddy-field fish production in Krian, Perak*. Ministry of Agriculture, Kuala Lumpur.

TAIWAN DEPARTMENT OF AGRICULTURE (1966). *Record of farm record-keeping families in Taiwan, 1965*. Taipei.

TEMPANY, H. and GRIST, H. D. (1958). *Tropical agriculture*. Longmans, London.

TERRA, G. J. A. (1957). Landbouwstelsels en bedrijsstelsels in de tropen. *Landbouwk. Tijdschr.*, *'s-Grav*. 6, 430–8.

—— (1958). Farm systems in south-east Asia. *Neth. J. agric. Sci.* 6, 157–82.

THAILAND MINISTRY OF AGRICULTURE (1959). *Report on economic survey of rice farmers in Nakom Phatom Province, 1955/56*. Bangkok.

THORNTON, D. S. (1973). *Agriculture in South East Ghana*. Summary Report, Vol. 1; Development Study, no. 12. University of Reading. June 1973.

THORNTON, D. and ROUNCE, N. V. (1963). Ukara island and the agricultural practices of the Wakara. *Tanganyika Notes Rec.* no. 1, 25 *et seq*.

TONDEUR, M. G. (1956). L'agriculture nomade au Congo Belge. Food and Agriculture Organization. *Agric. Nomade* 1, 15–108.

TORRY, W. (1973). *Subsistence ecology among the Gabra: nomads of the Kenya Ethiopian frontier*. Dissertation, Columbia University, New York.

TOURTE, R., GAUDEFROY-DEMOMBYNES, P., and FAUCHÉ, J. (1954). Perfectionement des techniques culturales au Sénégal. *Anns Centre Réch. agron. Bambey Sénégal*.

—— *et al.* (1971). Thèmes légers-thèmes lourds. Systèmes intensifs. Voies différentes ouvertes au développement agricole du Sénégal. *Agron. Trop., Nogent* 25, no. 5.

—— (1974). Réflexions sur les voies et moyens d'intensification de l'agriculture en Afrique de l'Ouest. *Agron. Trop., Nogent* 29, no. 9.

TOUTAIN, B. (1978). *Situation de l'élevage dans le Sahel Voltaique face à l'extension de l'espace agraire* (mimeo). ORSTOM-CVRS-IEMVT. Colloque, Ouagadougou.

TREWARTHA, G. T. (1968). *An introduction to climate*. H. H. Mcgraw Ltd., New York.

TROLL, C. (1966). *Seasonal climates of the earth. World maps of climatology*. Springer, Berlin, Heidelberg, and New York.

UPTON, M. (1967a). *Agriculture in southwestern Nigeria*. University of Reading Development Study, no. 3.

—— (1967b). Socio-economic survey of some farm families in Nigeria. *Bull. Rur. Econ. Sociol.* 2, Sep. 1967; 3, no. 1, Apr. 1968; and no. 3, 2, Oct. 1968.

VARELA, E. (1977). *Costs and returns of coffee production in Restrepo Valle* (ms.). CIAT, Cali, Colombia.

WADDEL, E. (1972). *The mound builders*. University of Washington Press, Seattle and London.

WALKER, H. (1976). *Die Ökonomie des Ölpalmenbaues in Sumatra und Westafrika* (ms.). University of Stuttgart-Hohenheim.

WALLACE, B. J. (1970). *Hill and valley farmers. Socio-economic change among a*

Philippine people. Schenkman Publishing Company, Cambridge (Mass.) and London.

WALTER, H. (1973). *Die Vegetation der Erde, Bd. 1: Die tropischen und subtropischen Zonen in öko-physiologischer Betrachtung.* Fischer, Stuttgart.

WANG (1967). *Development of rice and wheat breeding* (ms.). Agricultural Research Station, Taipei.

WATTERS, R. F. (1971). *Shifting cultivation in Latin America.* FAO Forestry Development Paper, no. 17. Food and Agriculture Organization, Rome.

WEBSTER, C. C. and WILSON, P. N. (1967). *Agriculture in the tropics.* Longmans, London.

WEIDELT, H. J. (1968). *Der Brandhackbau in Brasilien und seine Auswirkungen auf die Waldvegetation.* Diss., Hann-Münden.

WESTPHAL, E. (1975). *Agricultural systems in Ethiopia.* Centre for Agricultural Publicity and Documentation, Wageningen.

WHITE, G. (1974). *Natural hazards.* Oxford University Press, London.

WHYTE, R. O. (1967). *Milk production in developing countries.* Faber and Faber, London.

WIENS, TH. B. (1977). *The economics of municipal vegetable supply in the People's Republic of China* (ms.). National Academy of Sciences, USA.

WILLIAMS, C. N. and JOSEPH, K. T. (1973). *Climate, soil and crop production in the humid tropics.* Oxford University Press, Singapore.

WILLIAMSON, G. and PAYNE, W. J. A. (1965). *An introduction to animal husbandry in the tropics.* Longmans, London.

WINCH, F. E. (1976). *Costs and returns of alternative rice production systems in Northern Ghana.* Diss., Michigan State University.

WIT, C. T. de (1965). *Photosynthesis of leaf canopies.* Agricultural Research Report, no. 663, Wageningen.

WITTFOGEL, K. (1931). *Wirtschaft und Gesellschaft Chinas.* Hirschfeld, Leipzig.

WOERMANN, E. (1959). Landwirtschaftliche Betriebssysteme. *Handw. Sozialwiss,* **6,** 477 ff.

WONG, C. T. (1968). Enlarging the size of farm business through intensive land use in Hong Kong. *Fm Mgmt Notes Asia Far East* **4,** no. 2.

WOOD, G. A. R. (1974). Some aspects of cacao production costs on plantations. In *Economics of cacao production and marketing* (eds R. A. Kotey, C. Okali, and B. E. Rourke). University of Ghana, Legon.

WÖRZ, J. G. G. (1966). *Genossenschaftliche und partnerschaftliche Produktions-förderung in der sudanesischen Landwirtschaft.* DLG, Frankfurt.

WRIGLEY, G. (1971). *Tropical agriculture.* Faber and Faber, London.

ZANDSTRA, H. G. (1976a). Cropping systems research for the Asian rice farmer. In *Proceedings of the Symposium on Cropping Systems Research and Development for the Asian Rice Farmer.* IRRI, Los Baños, Philippines.

—— (1976b). *Cropping systems research at IRRI.* Paper presented at the Cropping Systems Seminar for Administrators, 1976. IRRI, Los Baños, Philippines.

—— (1976c). A framework in which to relate environmental factors to cropping systems potentials. In *Workshop on Environmental Factors in Cropping Systems,* 1976. IRRI, Los Baños, Philippines.

—— (1976d). *Terminologies in cropping systems* (mimeo). IRRI, The Philippines.

—— CARANGAL, V. R. (1977). *Crop intensification for the Asian rice farmer* (ms.). IRRI, The Philippines.

—— PRICE, E. C. (1977). *Research topics critical for the intensification of rice based cropping systems* (mimeo). IRRI, The Philippines.

—— (1978). *Cropping system research for the Asian rice farmer* (mimeo). IRRI, The Philippines.

—— SWANBERG, K. G. and ZULBERTI, C. A. (1975). *Removing constraints to small farm production: The Cacqueza Project.* IDRC-0582, International Development Research Centre, Ottawa, Canada.

ZEVEN, A. C. (1967). *The semi-wild oil-palm and its industry in Africa.* (Agricultural Research Report 689). Wageningen.

ZUCKERMAN, P. S. (1973). *Yoruba smallholders farming systems* (ms.). International Institute of Tropical Agriculture, University of Reading.

Personal informants

AVILA, M. (1978). Catie, Turrialba, Costa Rica.

BEMELMANS, P. F. (1978). Secretaria da Agricultura. Instituto de Economia Agricola, Sao Paulo, Brazil.

CHAMPION, J. (1974). 6, rue du Général Clergerie, Paris, France.

CHANDRASIRI, C. R. (1978). Rubber Research Institute of Sri Lanka, Dartonfield, Sri Lanka.

CHUDLEIGH, P. (1979). Lincoln College, Canterbury, New Zealand.

DIETZ, D. (1975). Project Tinajone, Chiclayo, Peru.

DIETZ, M. (1975). Bonua, Ivory Coast.

DISSANAYAKE, A. B. (1975). Rubber Research Institute of Sri Lanka, Ratmalana, Sri Lanka.

FELDMAN, D. (1977). Ministry of Agriculture, P.O.B. 30028, Nairobi, Kenya.

GUILLOT, B. (1978). Rue aux Toiles, 22450 La Roche-Derrien, France.

HAAN, K. DE (1978). International Livestock Centre for Africa, P.O.B. 5689, Addis Ababa, Ethiopia.

HATZIUS, TH. (1977). Casilla 186, Chiclayo, Peru.

HÖRNER, J. (1977). P.O.B. 171, Tamale, Ghana.

LAGEMANN, J. (1978). Catie, Turrialba, Costa Rica.

LUGOGO, J. A. (1978). Egerton College, Njoro, Kenya.

MANSHOLT, D. J. (1978). Holambra. C.P. 13.100, Campinas, Sao Paulo, Brazil.

MUMIAS SUGAR CO., LTD. (1977). Private Bag, Mumias, Kenya.

NAGEL, F. (1975). B.P. 103, Bujambura, Burundi.

PERIES, D. S. (1978). Rubber Research Institute of Sri Lanka, Dartonfield, Sri Lanka.

PUDSEY, D. M. (1977). P.O.B. 30577, Nairobi, Kenya.

REEVES, M. E. (1977). Magarini, Malindi, Kenya.

ROCHETEAU, G. (1978). ORSTOM, B.P. 81, 97200 Fort de France, Martinique.

SCHMIDT, H. (1978). Ministry of Agriculture, P.O.B. 30028, Nairobi, Kenya.

SCHNEIDER, K. (1975). PT-Sipef Medan, Sumatra, Indonesia.

SCHOENMAKER, K. (1978). Facenda Ribeiro, C.P. 528, Campinas, Sao Paulo, Brazil.

SPITZNER, R. (1975). Farm Harribes, Namibia (South-West Africa).

STOTZ, D. (1979). Ministry of Agriculture, P.O.B. 30028, Nairobi, Kenya.

SWOBODA, R. (1978). Ministry of Agriculture, P.O.B. 30028, Nairobi, Kenya.

THORWART, H. (1977). P.O.B. 82770, Mombasa, Kenya.

WALES, M. L. (1978). International Livestock Centre for Africa, P.O.B. 46847, Nairobi, Kenya.

WILLING, T. (1979). 64 Duke Street, East Fremanth, Western Australia 6158.

WINKLER, C. (1977). Hof 3, 3119 Secklendorf, BR.

Index